W9-CBH-392

Mammalian Population Genetics

Mammalian Population Genetics

EDITED BY MICHAEL H. SMITH
AND JAMES JOULE

The University of Georgia Press
Athens

Athens, Georgia 30602

Printed in the United States of America

Library of Congress Cataloging in Publication Data

Main entry under title:

Mammalian population genetics.

Papers presented at a symposium held during the 1978 annual meeting of the American Society of Mammalogists, Athens, Georgia.
Bibliography: p.
Includes index.
1. Mammals--Genetics--Congresses. 2. Mammal populations--Congresses. 3. Population genetics--Congresses. I. Smith, Michael Howard, 1938-
II. Joule, James. III. American Society of Mammalogists.
QL738.5.M36 599.01'5 80-24667
ISBN 0-8203-0547-2

CONTENTS

CONTENTS

FOREWORD

The American Society of Mammalogists at its 1978 annual meeting in Athens, Georgia hosted a symposium on mammalian population genetics. We organized the symposium and invited the speakers. There were four major talks and a series of shorter contributed presentations. The first four chapters of this book are extensions of the four featured talks. The other chapters came from some of the contributed papers at the meeting and also papers invited by the editors from other authors to obtain a better balance among the topics included in the book.

We felt there was a need for the symposium and this book because of the virtual explosion in the amount of information available on the genetics of natural populations of mammals during the last ten years. The development and refinement of electrophoretic techniques have allowed large amounts of relevant data to be easily collected. These techniques can be used without the necessity of sacrificing the animals, thus long term population studies designed to collect detailed genetic information became possible. Short term changes in the genetic composition of natural mammalian populations were commonly observed and these were frequently correlated with other demographic factors. The important question seemed to be whether genetic changes were simply a result of short term shifts in population composition. A more appropriate view of the complex systems we seek to understand seems to be emerging in which the question has become to what extent environmental and genetic factors interact in producing population changes. Such a holistic view is overdue and is an indication of a certain degree of maturity in this area of science.

Large amounts of genetic variability have been documented by the use of electrophoretic techniques. It seems inconceivable to some that most of this variation can be adaptive and the result of natural selection. These scientists argue for neutrality as an explanation for most of this variation and rightly demand a demonstration of the adaptive significance of variants rather than an uncritical acceptance of the selective basis for this variation. Neutrality is the null hypothesis, and an attempt must be made to reject it before accepting the alternative view. Mammals do have relatively low levels of genetic variability as a class, but the dilemma between neutrality and selection is still a problem. The resolution of this dilemma will require some reorientation in the experimental approach to the study of field populations, but the future will probably result in a better appreciation of the relative importance of both views rather than a rejection of one in favor of the other.

The contents of this book should document the state of the intellectual development and controversies inherent in the field of mammalian population genetics. The rapid progress in this field has created new frustrations but also satisfaction from our enlarged perspective concerning the potential importance of genetics to population biology over both short and long time periods. The classical nature-nurture argument will surely be stirred anew by some of the contents of this book. We expect the book to stimulate the formulation of new questions in this exciting area of research and thus serve an important function within the scientific community.

Michael H. Smith
Savannah River Ecology Laboratory
and Department of Zoology
University of Georgia

James Joule
Division of Natural and Physical
Sciences
University of Colorado, Denver

ACKNOWLEDGMENTS

Grateful appreciation is given to those who helped us in completing the final manuscript for this book. Ms. Eleanor Cato and Ms. Joan Lowery served as coordinators, contacting authors and reviewers, and following up on the many details necessary to provide a solid base for a quality publication. Mr. Ronald Blessing and Ms. Jean Coleman helped with necessary revisions to the art and photography throughout the volume. Ms. Vicki Malley and Ms. Irma Smith helped facilitate the review process. Ms. Sarah Collie and Ms. Sylvia Greenwald assisted in the preliminary proofreading and Ms. Sheri Belew assisted in the preliminary typing. Ms. Kitty Brakensiek was immensely helpful in scheduling the retyping of the manuscripts to minimize the delays always inherent in such an undertaking. Our thanks to Ms. Tonya Willingham for the monumental task of completing the final typing of the volume. We appreciate the efforts and cooperation put forth by the authors. Financial support was provided by the Division of Natural and Physical Sciences of the University of Colorado in Denver and by contract No. DE-AC09-76SR00819 between the U. S. Department of Energy and the University of Georgia for the operation of the Savannah River Ecology Laboratory, Aiken, South Carolina.

We thank the following people for their assistance in reviewing manuscripts:

Anderson, P. K. - University of Calgary
Aquadro, C. - University of Georgia
Baccus, R. - University of California-Berkeley
Bekoff, M. - University of Colorado
Berry, R. J. - University of London
Bryant, E. H. - University of Houston
Cameron, G. N. - University of Houston
Canham, R. - Southern Methodist University
Chesser, Ron - Savannah River Ecology Laboratory, Aiken, S.C.
Duback, J. M. - New Mexico State University
Fallon, D. - University of Colorado
Foltz, D. - University of Michigan
Gaines, M. - University of Kansas
Garten, C. T. - Oak Ridge National Laboratory, Tennessee
Greenbaum, I. F. - Texas A & M University

ACKNOWLEDGMENTS

Hamrick, J. L. - University of Kansas
Hsu, T. C. - University of Texas
Koopman, K. F. - American Museum of Natural History, New York
Millar, J. S. - University of Western Ontario
Mitton, J. T. - University of Colorado
Myers, J. H. - University of British Columbia
Nevo, E. - University of Hiafa
Patton, James - University of California-Berkeley
Patton, John - University of Georgia
Peters, J. - Harwell, England
Ramsey, P. R. - Louisiana Technical University
Schmidly, D. J. - Texas A & M University
Shaughnessy, P. D. - South Africa
Smith, J. D. - California State University-Fullerton
Soule, M. - University of California-La Jolla
Tamarin, R. H. - Boston University
Thaeler, C. S., Jr. - New Mexico State University

CONTRIBUTORS

Aquadro, Charles F., Department of Genetics, University of Georgia, Athens, Georgia, 30602

Armitage, Kenneth B., Department of Biological Sciences, University of Kansas, Lawrence, Kansas, 66044

Avise, John C., Department of Genetics, University of Georgia, Athens, Georgia, 30602

Baker, Robert J., Department of Biological Sciences and The Museum, Texas Tech University, Lubbock, Texas, 79409

Berry, R. J., Royal Free Hospital, University of London, WCIN IBP, England

Calkins, D. G., Department of Fish and Game, Anchorage, Alaska, 99502

Gaines, Michael S., Department of Systematics and Ecology, University of Kansas, Lawrence, Kansas, 66044

Gayden, Neil A., Department of Biological Sciences and Human Genetics Center, North Texas State University, Denton, Texas, 76203

Joule, James, Department of Natural and Physical Science, University of Colorado, Denver, Colorado, 80202

Kilpatrick, C. William, Department of Zoology, University of Vermont, Burlington, Vermont, 05405

Leamy, Larry, Department of Biology, California State University, Long Beach, California, 90840

Lidicker, William Z., Jr., Museum of Vertebrate Zoology, University of California, Berkeley, California, 94720

Massey, Dean R., Division of Natural and Physical Sciences, University of Colorado, Denver, Colorado, 80202

McClenaghan, Leroy R., Jr., Department of Biology, San Diego State University, San Diego, California, 92110

CONTRIBUTORS

Patton, John C., Department of Biology, Baylor University, Waco, Texas, 76706

Peters, Josephine, MRC Radiobiology Unit, Harwell, Didcot Oxon, OX11 ORD, England

Rose, Robert K., Department of Biological Sciences, Old Dominion University, Norfolk, Virginia, 23508

Sage, R. D., Museum of Vertebrate Zoology, University of California, Berkeley, California, 94720

Schnell, Gary D., Department of Zoology, University of Oklahoma, Norman, Oklahoma, 73069

Schwartz, Orlando A., Department of Basic Sciences and Mathematics, Whitman College, Walla Walla, Washington, 99362

Selander, Robert K., Department of Biology, University of Rochester, Rochester, New York, 14627

Straney, Donald O., The Museum, Michigan State University, East Lansing, Michigan, 48823

Zimmerman, Earl G., Department of Biological Sciences and Human Genetics Center, North Texas State University, Denton, Texas, 76203

IMPORTANCE OF GENETICS TO POPULATION DYNAMICS[1]

Michael S. Gaines

Abstract--If genetic changes influence demographic parameters in small rodent populations two predictions can be made: (1) Changes in gene frequency should be associated with density changes, and (2) Genetic changes should be a cause and not an effect of density changes. A survey of multiannual and annual cycling species indicated that genetic changes at a large number of structural loci were associated with density changes. Gene frequency changes at some of these loci could be attributed to natural selection. In three species of small mammals, average genic heterozygosity was positively correlated with spatial variation in population density. This relationship has been interpreted in terms of changes in population breeding structure. Average genic heterozygosity was positively associated with body weight, but there was no consistent relationship between heterozygosity and other fitness components. Although perturbation experiments suggested that genetic changes at a single locus were effects of demographic changes rather than vice versa, they were not designed to refute the hypothesis that changes in average heterozygosity over many loci were causing density changes. Chitty's (1967) and Pimentel's (1964) models of population regulation are discussed with respect to temporal changes in allozymic variation.

INTRODUCTION

Until recently, demographic studies have generally ignored genetic variation. Populations were assumed to be assemblages of equivalent individuals each responding to the exigencies of the environment in an identical manner. The basic demographic approach taken by popula-

[1]This paper is dedicated to Professor Dennis Chitty upon his retirement from teaching at the Department of Zoology, University of British Columbia, June 1978.

tion ecologists is best exemplified by the two classical theories of population regulation, the biotic (Nicholson, 1954) and climatic (Andrewartha and Birch, 1954) schools, neither of which considers the genetic structure of the population. Although several early review papers (Birch, 1960; Milne, 1962; Orians, 1962; Pimental, 1964) emphasized that natural selection can affect population dynamics, the admonition went unheeded for the most part. After Chitty (1967) formulated a genetic-behavioral hypothesis to explain multiannual cycles in microtine rodents, ecologists began to consider seriously the influence of genetics on population dynamics. Chitty's hypothesis is based on an r- and α-selection argument (Stenseth, 1978a). During the phase of increasing numbers when intraspecific interference is minimal, nonaggressive genotypes with a high reproductive effort have a selective advantage; whereas, during the peak phase when mutual interference is maximal, there is selection for aggressive behavior. The aggressive genotypes have an advantage in intraspecific competition for food and space, but their fitness is reduced in other ways, which sets the stage for the decline phase. Chitty's hypothesis assumes that natural selection, causing changes in the genetic composition of the population, is the driving force behind population cycles.

With the advent of electrophoresis as a tool to study genic variation (Lewontin and Hubby, 1966), it became possible to monitor gene frequencies in fluctuating mammal populations and to assess their role in population dynamics. If genetic changes influence demographic parameters, two predictions can be made: (1) Changes in gene frequency will be associated with density changes, and (2) Genetic changes will be a cause and not an effect of density changes. In this paper, I review results of electrophoretic studies in fluctuating small mammal populations to determine whether the evidence supports these predictions.

Genetics and Population Dynamics

TOPICAL REVIEW

Correlations of Genetic Changes with Density Changes

There have been numerous studies monitoring changes in gene frequency at electrophoretic loci in fluctuating populations of microtine rodents. These long-term fluctuations (which consist of an increase, peak, and decline phase with a periodicity of 2-4 yr) have been commonly called population cycles. Microtine rodents have become a focal point for genetic studies because the first step in testing Chitty's hypothesis is to demonstrate that changes in gene frequency are related to population density.

Semeonoff and Robertson (1968) observed changes in gene frequency at a plasma esterase locus in a declining population of *Microtus agrestis*. Canham (1969), studying transferrin and albumin polymorphisms in fluctuating populations of *Clethrionomys rutilis* and *C. gapperi*, found that heterozygote excess was correlated with population density. Changes in gene frequency at a transferrin (TF) locus (Tamarin and Krebs, 1969; Gaines and Krebs, 1971) and a leucine aminopeptidase (LAP) locus (Gaines and Krebs, 1971) were correlated with population density in *M. ochrogaster*. LeDuc and Krebs (1975), monitoring gene frequency at a LAP locus in a population of *M. townsendii* found a positive association between changes in the Lap^F allele and changes in population density. Kohn and Tamarin (1978) reported that changes in gene frequency at a TF locus were correlated with changes in density in *M. breweri*.

There has been some reluctance to apply Chitty's hypothesis to annually fluctuating species based on blind faith that mechanisms of population regulation in annual fluctuations must be different from those operating in multiannual cycles. I maintain that Chitty's hypothesis is potentially applicable to any population in which individuals exhibit spacing behavior. Thus, cycling, fluctuating, and stable populations could all be self-regulating. Similar sentiments were expressed in a recent review by Krebs (1978a) who states: "Chitty's view that cyclic populations are convenient vehicles for testing his hypothesis must not be confounded with the erroneous belief that only cyclic populations can be self-regulating."

In this context it is interesting that genetic changes are also correlated with population density in rodents that undergo annual fluctuations in numbers. Canham (1969) observed shifts in heterozygosity at the TF locus that were related to density in *Peromyscus maniculatus* populations. Berry and Murphy (1970), who monitored six electrophoretic loci in island populations of *Mus musculus*, reported seasonal shifts for two loci (hemoglobin and serum esterase) in the observed number of heterozygotes compared to the number expected based on Hardy-Weinberg equilibrium. There were also seasonal changes in population density. Brown (1977) found that changes in gene frequency at a phosphoglucomutase locus in *Apodemus sylvaticus* were associated with population density.

The most striking feature of these studies (Table 1) monitoring genetic changes in protein polymorphisms in fluctuating rodent populations is that changes in allozyme frequency at a large number of loci are associated with density changes. However, in each study, only one or at most two loci were monitored. Gaines et al. (1978) expanded the scope of earlier studies by monitoring changes in gene frequency at five loci: transferrin (TF), leucine aminopeptidase (LAP), 6-phosphogluconate dehydrogenase (6-PGD) and two esterases (EST-1 and EST-4). Four live-trapped populations of *M. ochrogaster* were studied over a three-year period in eastern Kansas. Densities and gene frequencies over the population cycle on four grids are presented in Gaines et al. (1978; Figs. 1-6). Three conclusions were drawn from the Kansas data: (1) There were differences in gene frequencies at the same locus among populations during some trapping periods; (2) The Tf and 6-Pgd alleles were apparently fixed during some trapping periods in most populations. Polymorphisms at these loci were probably restored by immigration; and, (3) The largest temporal changes in gene frequencies occurred at the EST-1 and EST-4 loci, whereas gene frequencies at the 6-PGD locus were the most stable during the population cycle. These observations were quantified by calculating averages of absolute changes in gene frequency (Δp) for each locus in each population (Gaines et al., 1978; Table 3).

The relationship between changes in gene frequency at the five loci and density was investigated by calculating correlation coefficients between these two variables. In this analysis, I assumed that the relation-

Genetics and Population Dynamics

Table 1. Summary of small mammal species in which genetic changes in protein polymorphisms detected by electrophoresis have been found to be associated with population density.

Species	Protein Polymorphism	Locality
Multiannual Density Cycles		
Clethrionomys spp.[1]	Albumin, Transferrin	Northwest Territories
Microtus agrestis[2]	Esterase	Scotland
Microtus breweri[3]	Transferrin	Muskeget Island
Microtus ochrogaster[4 5 6]	Transferrin, Leucine Aminopeptidase	Indiana and Kansas
Microtus pennsylvanicus[4 5]	Transferrin, Leucine Aminopeptidase	Indiana
Microtus townsendii[7]	Leucine Aminopeptidase	British Columbia
Annual Density Cycles		
Apodemus sylvaticus[8]	Phosphoglucomutase	England
Peromyscus maniculatus[1]	Transferrin	Northwest Territories
Mus musculus[9]	Hemoglobin, Esterase	Skokholm Island

[1]Canham, 1969; [2]Semeonoff and Robertson, 1968; [3]Kohn and Tamarin, 1978; [4]Tamarin and Krebs, 1969; [5]Gaines and Krebs, 1971; [6]Gaines et al., 1978; [7]LeDuc and Krebs, 1975; [8]Brown, 1977; [9]Berry and Murphy, 1970.

ship between gene frequency and density was linear. However, if heterozygosity or homozygosity was selected for during part of the density cycle but selection was relaxed or reversed at another time during the cycle, the relationship between gene frequency and density could have been curvilinear. Only the gene frequencies at the TF and LAP loci were consistently correlated with population density (Table 2). Thus, three general patterns of changes in gene frequency emerge from the Kansas data: (1) One locus (6-PGD) was stable over the population cycle; (2) Two loci (EST-1 and EST-4) were variable but unrelated to density; and, (3) Two loci (TF and LAP) were correlated with density.

Since changes in gene frequency at the TF and LAP loci had been followed in fluctuating prairie vole populations in southern Indiana (Gaines and Krebs, 1971), I was able to assess the spatial replicability of these patterns of variation. In Indiana, Tf^E frequencies in both sexes were positively correlated with density over the entire population cycle. In the Kansas populations, the gene frequencies in both sexes were positively correlated with density only during the increase phase. Changes in Lap^F frequency in the Indiana populations were negatively correlated with population density in males over the entire cycle, whereas these two variables were positively correlated in males during the decline phase in Kansas populations. Therefore, changes in gene frequency with density are repeatable both over time and space only at the TF locus.

In summary, there are many examples in a wide variety of species undergoing multiannual and annual cycles where genetic changes are correlated with changes in population density. These results are consistent with, but not sufficient to support, the hypothesis that genetic changes are influencing demography; it is impossible to separate cause from effect.

Effects of Natural Selection on Allozymic Variation

The question of interest here is: To what extent are observed density related changes in gene frequencies the result of natural selection, as opposed to stochastic processes such as genetic drift? The answer is pertinent to the current controversy surrounding the maintenance of structural protein polymorphisms in

Table 2. Correlation coefficients for changes in gene frequency at the transferrin and leucine aminopeptidase loci vs population density for Microtus ochrogaster on four grids during different phases of the population cycle, 1970-1973 (Gaines et al., 1978).

Location Phase	Number of Trapping Periods	Tf^E Allele		Lap^F Allele	
		Male	Female	Male	Female
Grid A					
Increase	27	0.73**	0.56*	-0.37	0.17
Decline	9	-0.42	-0.56	0.89**	0.10
Grid B					
Increase	23	0.59**	0.67**	0.20	0.51*
Decline	14	-0.90**	-0.08	0.77**	0.71**
Grid C					
Increase	14	-0.14	0.06	-0.26	-0.88**
Decline	27	0.18	0.16	0.48*	0.16
Grid D					
Increase	13	0.61*	0.10	0.32	0.20
Decline	14	0.02	-0.04	0.72**	0.22

*$P < 0.05$; **$P < 0.01$.

natural populations. The neutralist theory (Ohta and Kimura, 1974) suggests that electrophoretic variants are selectively neutral. Recently, the theory has been modified to account for maximum limits of heterozygosity in large populations by assuming there is a balance between mutation to slightly deleterious alleles and selection against these mutants (Ohta and Kimura, 1975). The selectionist theory assumes that different electromorphs are maintained in populations by some form of balancing selection (Ayala, 1974), although the unit of selection, whether a single allele or a larger linkage group, remains unsolved (Lewontin, 1970; Franklin and Lewontin, 1970).

Several different methods have been employed to measure the direction and intensity of selection in fluctuating small mammal populations. Tamarin and Krebs (1969) plotted change in gene frequency, Δp, against gene frequency, p, (Li, 1955). In a balanced polymorphism where fitnesses are constant, the relationship between these two variables is defined as a curve with one nontrivial equilibrium point. The data confined to the portion of the curve around the equilibrium point approximates a straight line with a negative slope. Tamarin and Krebs (1969) calculated regressions of Δp on p at the TF locus for three *M. ochrogaster* and two *M. pennsylvanicus* populations in southern Indiana. Data for the sexes were analyzed separately. In all but one case, the regression line was significant with a negative slope. Furthermore, when data were analyzed from different phases of the population cycle, equilibrium points changed in a consistent manner. Kohn and Tamarin (1978) performed the same analysis for TF and LAP polymorphisms in *M. pennsylvanicus* populations from Massachusetts and *M. breweri* on Muskeget Island and again concluded the two loci had balanced polymorphisms. Results of such analyses should be viewed with caution, since a critical assumption in the Δp vs p method is that gene frequencies in different sampling periods are independent of one another. This is clearly violated in vole populations where the same individuals appear in subsequent trapping periods.

A second approach, used extensively by Krebs and his co-workers (Tamarin and Krebs, 1969; Gaines and Krebs, 1971; LeDuc and Krebs, 1975), is to measure physiological components of fitness (i.e., viability, fecundity, developmental rates) for different genotypes. Minimum survival rates, measured for TF and LAP genotypes during *M. ochrogaster* population cycles, are given in Table 3. Although there is large variation in survival rates for a given genotype among studies, some trends are apparent. At the TF locus, the Tf^F/Tf^F homozygote was extremely rare in the Indiana populations. The Tf^E/Tf^F heterozygotes generally had higher survival rates than Tf^E/Tf^E homozygotes during the decline phase in both Indiana studies, the only exception being the females in the Tamarin and Krebs (1969) study. During the decline phase in the Kansas populations, individuals

Table 3. Minimum survival rates per 14 days of TF and LAP genotypes in *Microtus ochrogaster* populations during phases of increasing and declining density reported in two studies in southern Indiana and a study in eastern Kansas.

Locus Phase of Cycle	Locality					
	Indiana (1965-1967)[1]		Indiana (1967-1969)[2]		Kansas (1970-1973)[3]	
	Male	Female	Male	Female	Male	Female
Transferrin[4]						
Increase						
Fast Homozygote	0.84	0.92	0.78	0.87	0.82	0.83
Heterozygote	0.76	0.76	0.91	0.85	0.72	0.85
Slow Homozygote					0.86	0.87
Decline						
Fast Homozygote	0.59	0.68	0.70	0.69	0.65	0.65
Heterozygote	0.75	0.33	0.77	0.74	0.66	0.69
Slow Homozygote					0.75	0.82
Leucine Aminopeptidase[4]						
Increase						
Fast Homozygote			0.60	0.84	0.80	0.85
Heterozygote			0.79	0.91	0.88	0.87
Slow Homozygote			0.85	0.86	0.80	0.86
Decline						
Fast Homozygote			0.70	0.62	0.51	0.58
Heterozygote			0.75	0.58	0.59	0.58
Slow Homozygote			0.75	0.77	0.63	0.63

[1]Tamarin and Krebs, 1969; [2]Gaines and Krebs, 1971; [3]Gaines et al., 1978; [4]Genotypes for TF locus were EE, EF, and FF, and for the LAP locus were FF, SF, SS, respectively.

with at least one Tf^F allele had higher survival rates than Tf^E/Tf^E homozygotes. Those individuals with two Tf^F alleles had the highest survivorship of any genotype. Data for the LAP genotypes during the decline phase are also consistent between the Indiana and Kansas studies. The Lap^S/Lap^S homozygotes had the highest survival rates for males and females in both studies.

In addition to survival rates, there were statistically significant differences among genotypes in other physiological components of fitness for Kansas but not Indiana prairie voles. Gaines et al. (1978) found differences in the frequency of breeding among TF and LAP genotypes in Kansas prairie vole populations. There were also differences in growth rates among genotypes (Table 4). The relative differences in components of fitness among genotypes change during different phases of the cycle. For example, Tf^E/Tf^E homozygotes have higher growth rates during the increase phase of the cycle, whereas Tf^E/Tf^F heterozygotes had the highest growth rate in the decline phase.

A third method used to detect selection is to compare observed genotypic frequencies with the expected frequencies based on Hardy-Weinberg equilibrium. This approach is not entirely satisfactory for two reasons: (1) Such evolutionary forces as genetic drift and inbreeding also could cause deviations from Hardy-Weinberg equilibrium, and (2) Selection must be completed in the interval between which gene frequencies and expected genotypic frequencies are estimated.

Berry and Murphy (1970) sampled *Mus musculus* populations on the island of Skokholm during the spring and autumn in two successive years (1968-1969) and found seasonal shifts in heterozygosity at a hemoglobin (HBB) and esterase (EST-2) locus. During the summer months when populations were increasing, there was a large excess of heterozygotes compared to expected numbers at the HBB locus at the end of the breeding season in autumn of both years. During the winter in both years when populations were declining, there was a decrease in this excess. The reverse trend occurred at the EST-2 locus for males with an excess of heterozygotes at the end of the winter which decreased during the summer breeding season. These results were interpreted as evidence for stabilizing selection during the summer and destabilizing selection during the winter at the HBB locus, and vice versa for the EST-2 locus.

Genetics and Population Dynamics

Table 4. Average growth rates (% per day ± SE) during phases of increasing and declining density in males adjusted by regression to a standard 35 g prairie vole (*Microtus ochrogaster*). Data were pooled over four populations, 1970-1973. Sample sizes are in parentheses (Gaines et al., 1978).

Locus and Phase of Cycle	Genotypes		
	Fast Homozygote	Heterozygote	Slow Homozygote
Transferrin			
Increase	0.84 ± 0.03 (749)	0.59 ± 0.06 (168)	0.52 ± 0.17* (221)
Decline	0.73 ± 0.05 (464)	0.90 ± 0.06 (228)	0.12 ± 0.31 (22)
Leucine Amino-peptidase			
Increase	0.53 ± 0.09 (102)	0.96 ± 0.05 (268)	0.82 ± 0.01* (569)
Decline	0.87 ± 0.12 (69)	0.93 ± 0.08 (198)	0.76 ± 0.05 (446)

*Comparison among genotypes that are statistically significant at $P < 0.05$.

Finally, Lewontin and Krakauer (1973) developed a statistical test to distinguish between the effects of natural selection and genetic drift on electrophoretic loci. Their test is based on the estimate of Wright's (1965) F_{st}, the effective coefficient of inbreeding. Assuming random mating and no migration, the estimate is:

$$\hat{F}_e = \frac{s^2_p}{\bar{p}(1-\bar{p})}$$

where

$\hat{F}_e$ = estimate of effective inbreeding,

s^2_p = variance in the frequency of one or two alternative alleles from population to population, and

$\bar{p}$ = mean frequency of the allele over the ensemble of populations.

If loci are selectively neutral, the $\hat{F}$'s calculated for many loci over an ensemble of populations will be homogeneous. Conversely, if selection is operating on some or all of the loci, the $\hat{F}_e$'s will be heterogeneous because of distortion by selection. However, Nei and Maruyama (1975) and Robertson (1975) criticized Lewontin and Krakauer's test, contending that it could lead to erroneous conclusions if the subpopulations are related historically. Lewontin and Krakauer (1973) responded to the criticism by distinguishing between universe and samples. They argue "that although populations in the universe may be correlated, populations in a random sample may not be."

I know of only one case where Lewontin and Krakauer's test has been applied to fluctuating mammal populations. Massey (1977) monitored changes in gene frequency at eight electrophoretic loci in *P. maniculatus* populations from five localities in Colorado. The sites, chosen to represent three habitat types (xeric, mesic and intermediate) were sampled four times over a one-year period. Chi-square contingency tests revealed significant heterogeneity in gene frequencies among populations during some sampling periods. The $\hat{F}_e$ values were calculated for each sampling period to determine whether the heterogeneity was due to selection or genetic drift. All $\hat{F}_e$ values were statistically homogeneous, suggesting that selection is not a major factor in the maintenance of this spatial heterogeneity.

Although there are considerable data on temporal gene frequency changes in mammal populations, no one has used the Lewontin and Krakauer (1973) test on this variation. The basic methodological problem in using the temporal variation test is the assumption that the same individuals must not appear in the different sampling periods. This assumption is clearly violated in small mammal populations that undergo multiannual cycles but may not be in annual cycling species if the sampling periods are in the fall and spring.

In summary, there is some indirect evidence that changes in gene frequency at electrophoretic loci may be due to natural selection. In some cases, such as certain physiological components of fitness, the effect of selection on specific alleles is repeatable both in time and space.

Multiple Locus Studies

Only a few attempts have been made to determine whether there is any association between average heterozygosity over many electrophoretic loci and the demography of small mammal populations. Smith et al. (1975) found a positive linear relationship between population density and average percent genic heterozygosity for 32 structural loci in *P. polionotus* from five areas along a north-south transect (Fig. 1). Since the populations were sampled from different localities under diverse ecological conditions, the apparent relationship may be confounded by other effects. Patton and Yang (1977) studying genic variation at 21 electrophoretic loci in *Thomomys bottae* populations from 23 localities, found a significant positive correlation between population density and average heterozygosity ($r = 0.42$; $P < 0.01$). The same result has been obtained for *T. umbrinus* (Patton, pers. comm.). Again, populations were sampled spatially but not temporally. Massey (1977) monitored average genic heterozygosity at eight electrophoretic loci in *P. maniculatus* populations. The study was not designed to estimate population densities, but he did observe seasonal variation in mean heterozygosity. The highest values attained in all populations were in June, concommitant with increased reproductive activity.

Smith et al. (1975) offered a biological explanation for the positive correlation between genic heterozygosity and density in *P. polionotus* populations. They suggest that at low densities inbreeding decreases heterozygosity, whereas at high densities there is increased dispersal and the population becomes relatively more outbred, leading to increased heterozygosity. This interpretation is supported by Patton and Yang's (1977) analysis of the relationship of breeding structure in *T. bottae* populations and genic heterozygosity. Pocket gopher populations were arbitrarily divided into

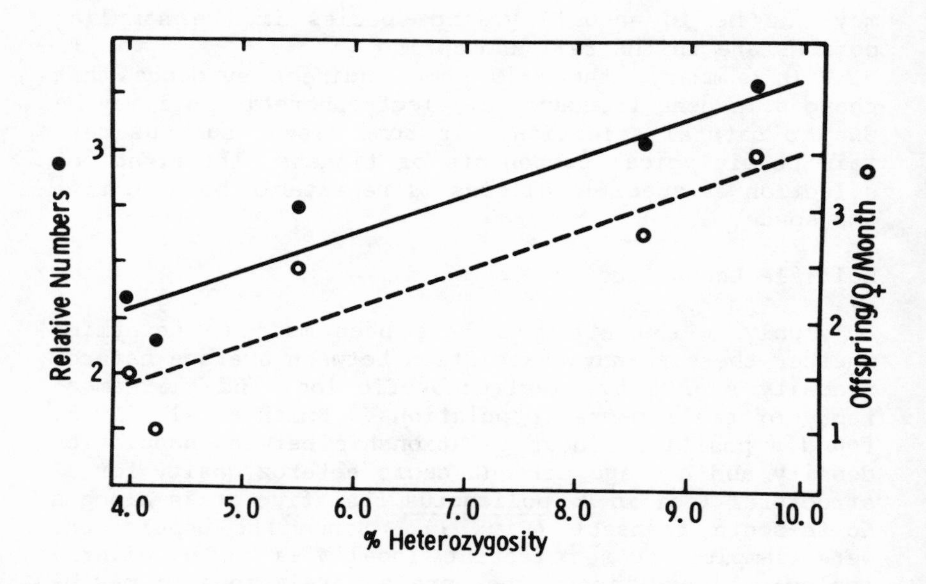

Fig. 1. Relative numbers of Peromyscus polionotus, expressed as numbers of mice captured per hour by digging burrows, and reproductive rates of adult female old field mice in relation to average percent heterozygosity from five localities in South Carolina, Georgia and Florida. Correlation coefficients for relative numbers and reproductive rates vs percent heterozygosity were $r = 0.96$ ($P < 0.01$) and $r = 0.91$ ($P < 0.05$), respectively (Smith et al., 1975).

three categories of increasing social structure: (1) large populations rather uniformly distributed; (2) populations exhibitiing dendritic geographic distributions (e.g., as along a river system); and, (3) populations reasonably isolated in all directions from adjacent ones (e.g., those localized on mountain tops or at desert springs). The measure of population structure was highly correlated with heterozygosity ($r = 0.69$; $P < 0.01$) in such a manner that large, uniformly distributed populations had higher levels of heterozygosity than small isolated ones.

Conversely, Gaines et al. (1978) found no correlation between average genic heterozygosity over five electrophoretic loci and density in four populations of *M. ochrogaster*. They also compared the number of observed heterozygotes (H_O) and expected heterozygotes (H_E)using the following relationship: $D = (H - H_E)/H_E$ (Selander, 1970a). A positive or negative value of D would indicate an excess or deficiency of heterozygotes, respectively. An average D value ($\bar{D}$) was calculated during each trapping period by summing D over polymorphic loci and dividing by the number of loci (5). D was negatively correlated with density for each population over the study (grid A---r = -0.26, $P < 0.05$; grid B---r = -0.33, $P < 0.05$; grid C---r = -0.33, $P < 0.05$; and, grid D---r = -0.41, $P < 0.01$). Thus, as population density increased, there was a greater deficiency of observed heterozygotes compared to the expected number based on the Hardy-Weinberg equilibrium.

There are two possible explanations for the contradictory results obtained for *M. ochrogaster* compared to *P. polionotus* and *T. bottae*. First, since no information is available on the social dynamics of any of these species, individuals of a given species may have unique social responses to changes in density. For example, at high population density, voles may have a more rigid social structure characterized by small demes with little gene flow between them, which would reduce D due to inbreeding or the Wahlund effect. This interpretation differs from Smith et al. (1975), who assume increased outbreeding due to dispersal at high densities in *P. polionotus* populations. New methods need to be developed to study the breeding structure of small mammals in the field. A second explanation for the inconsistencies is that populations of each species were sampled differently. *M. ochrogaster* populations were sampled temporally, whereas *P. polionotus* and *T. bottae* were sampled spatially. When Smith et al. (1975) examined the relationship between density and heterozygosity within local populations of *Peromyscus* (Ramsey, 1973) and Indiana *Microtus* (Gaines and Krebs, 1971), they found an inverse relationship between these two variables. This inconsistency in results between spatial and temporal studies might be related in part to the number of loci under consideration. Generally, in temporal studies, animals are live-trapped for demographic data which limits the investigator to alleles that can be assayed in blood samples. In spatial studies, animals are usually sacrificed and a greater number of structural proteins can be examined in a variety of tissues. Thus, estimates of heterozygosity may be more reliable for spatial studies.

It has been suggested that individuals heterozygous at electrophoretic loci have higher fitness than homozygous individuals (Berger, 1976). There is some evidence in support of this hypothesis from fluctuating small mammal populations. Smith et al. (1975) found a positive correlation between the mean reproductive rate (number of offspring produced per adult female per month) and average genic heterozygosity in P. polionotus populations (Fig. 1). Garten (1976) found a statistically significant positive relationship between average genic heterozygosity and body weight for males in this species. There was also a positive linear relationship between heterozygosity and aggressiveness (Fig. 2). In addition, Garten (1977) found a positive correlation between heterozygosity and exploratory behavior. Thus, old field mice with the highest heterozygosity tended to be larger, more aggressive and more exploratory in arena tests. Although Garten could not separate the confounding effects of genic heterozygosity and body size on behavioral parameters, his study is the first to link genetics and behavior of individuals in rodent populations.

The positive relationship between average genic heterozygosity and body weight may be a general phenomenon. Massey (1977) found a positive correlation between these two variables in males ($r = 0.26$; $P < 0.01$) from P. maniculatus populations. Gaines et al. (1978) observed an increase in mean body weight with percent heterozygosity during the increase phase of fluctuating M. ochrogaster populations. The mean body weights and standard errors over four populations for four intervals of percent heterozygous individuals for males during the increase phase were as follows: 0-20%, 35.7 g ± 0.3; 21-40%, 36.6 g ± 0.4; 41-60%, 37.8 g ± 0.7; and > 60%, 40.1 g ± 5.3.

In addition to body weights, Gaines et al. (1978) compared other components of fitness for individuals with different levels of heterozygosity. Heterozygous individuals did not consistently have the highest values for all fitness components. For example, minimum survival rates summed over the four populations for different intervals of heterozygosity for males during the decline phase were as follows: 0-20%, 0.65; 21-40%,

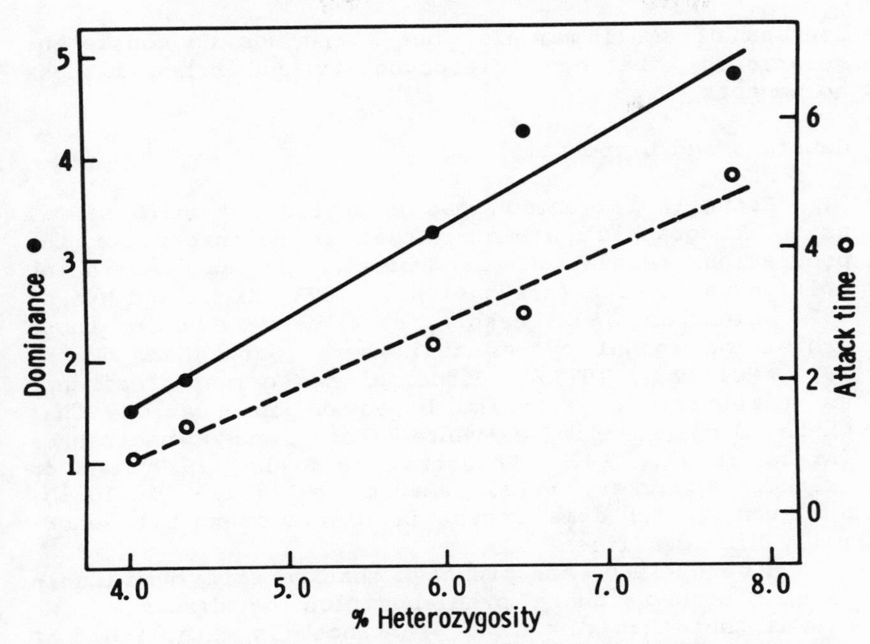

Fig. 2. Regressions of the number of matches resulting in a fight and mean social dominance scores in intralocality tests against mean percent genic heterozygosity of male Peromyscus polionotus from five sample locations. Regression equations for dominance and attack time vs percent heterozygosity were Y = -2.1 + 0.9X ($F_{1,3}$ = 95.4; P < 0.01) and Y = -3.1 + 1.0X ($F_{1,3}$ = 106.8; P < 0.01), respectively (Garten, 1976).

0.71; 41-60%, 0.51; and > 60%, 0.55. Males which were heterozygous at 0-40% of the five electrophoretic loci had higher survival rates than males that were heterozygous at greater than 40% of those loci.

In summary, average genic heterozygosity is positively correlated with population density as it varies spatially in P. polionotus, T. bottae and T. umbrinus populations. These correlations may be explained in part by changes in the population breeding structure with density. However, the two variables were not correlated in temporally fluctuating Kansas populations of M. ochrogaster. Average genic heterozygosity was positively associated with body weight in three different species of small mammals, but there was no consistent relationship between heterozygosity and other fitness components.

Genetics and Dispersal

Recently, attention has been focused on dispersal as a demographic parameter that is necessary for the population regulation of rodent species undergoing multiannual cycles (Krebs et al., 1973; Krebs and Myers, 1974; Lidicker, 1975; Bekoff, 1977) as well as for those exhibiting annual cycles in numbers (Garten and Smith, 1974; Sullivan, 1977). Dispersal has gained prominence as a regulatory mechanism based on observations that where dispersal is prevented, in fenced enclosures (Krebs et al., 1969; Boonstra and Krebs, 1977) and on islands (Lidicker, 1973; Tamarin, 1977a, 1977b, 1978b; Sullivan, 1977) some rodent populations exhibit abnormally high densities.

Howard (1960) has proposed that certain individuals have a strong innate predisposition to disperse. He claims that this dispersal "instinct" is established at birth and is genetically determined. If Howard is correct, a genetic influence on demography could be mediated through dispersal. Several electrophoretic studies of small mammal populations have demonstrated that dispersers are genetically different from residents, which is consistent with Howard's hypothesis. Myers and Krebs (1971a) studied the genetics of dispersal in *Microtus* by removing all animals caught on experimental grids in two days of trapping every second week. Voles that colonized the areas for the 12 days between trapping periods were defined as dispersers. The TF and LAP genotypes of the dispersers were compared to residents on neighboring control grids. At the TF locus there was a higher percentage of heterozygotes in dispersers than in residents during the increase phase of *M. pennsylvanicus*. Pickering et al. (1974) found that an esterase phenotype was correlated with dispersal in a *M. ochrogaster* population. Massey (1977), using a removal grid technique in a study of *P. maniculatus* populations, found a mean value of genic heterozygosity for residents of 0.14, compared to a value of 0.16 for dispersers but the difference was not statistically significant. Heterozygosity was high among the dispersers despite a precipitous drop in heterozygosity in the residents towards the end of the study. Although results from these electrophoretic studies are consistent with Howard's (1960) hypothesis for genetic control of dispersal, it cannot be accepted as prima facie evidence, since it is impossible to separate cause and effect.

Finally, Smith et al. (1978) developed a conceptual model which relates genetics to dispersal in fluctuating small mammal populations. They divide populations into primary and secondary types. Primary populations are stable in numbers because individuals have high survivorship and low reproductive effort. Conversely, secondary populations are unstable in numbers because individuals have low survivorship and high reproductive effort. All populations can be arranged somewhere along a continuum between these two extremes. Primary populations, which occur in optimal habitats continuously export individuals to secondary populations in marginal habitats. According to Smith et al. (1978), it is selectively advantageous for founders to be heterozygous since they would avoid the deleterious effects of inbreeding during the initial stages of expansion of the secondary population. However, their rationale does not follow for two reasons. First, maximum heterozygosity would be achieved in a secondary population, if primary populations exported individuals that are homozygous for alternative alleles. They are aware of this problem and assume, since populations are spatially heterogeneous, that dispersing heterozygotes are likely to be variable for different alleles at some of their loci. Second, their argument is based on group selection.

Sex Ratio and Population Regulation

Krebs and Myers (1974) suggest that shifts in the population sex ratio might regulate its density but their argument also implicates group selection. They propose that natural selection might favor a higher percentage of females during phases of increasing density which would maximize reproductive output, while at peak densities the sex ratio might equalize or favor males. The influence of sex ratios on density and recruitment has been assessed in two species of *Microtus* by experimentally altering the sex ratio in favor of males or females. Redfield et al. (1978a) manipulated the sex ratio towards 80% males and 20% males in different populations of *M. oregoni*. The only consistent effect was that raising the percentage of males increased survival of both males and females. In an analogous and simultaneous series of experiments on *M. townsendii* (Redfield et al., 1978b), the demography of the manipulated populations changed very little compared with the control population with 50% males. More studies, utilizing different experimental designs (Boonstra, 1978) on other species, need to be conducted before we can make any generalizations about how sex ratios affect demography.

If sex ratios affect demography, then some interesting implications arise with respect to the transferrin polymorphism in *M. ochrogaster* populations. Myers and Krebs (1971b) found a positive association between the number of Tf^F alleles in the parents in laboratory crosses and the percentage of male offspring. Myers and Krebs (1971b) concluded that variation in selection on the transferrin alleles could alter the mating frequencies of different genotypic pairs which in turn may have demographic consequences.

The Relationship Between Genetics and Population Dynamics: A Problem of Cause vs Effect

The major problem in assessing the importance of genetics to population dynamics is disentangling cause and effect. There are two opposing ways to view correlations between genetic and demographic changes: (1) Genetic changes cause demographic changes, and (2) Genetic changes are effects or correlates of demographic changes. I will now review some results from a theoretical model and empirical observations on field populations which suggest that the latter is the case.

Charlesworth and Giesel (1972), using computer simulations, demonstrated that a correlation between gene frequency and density could be produced by simply changing the rate of population growth. This result is dependent on the changing age structure of the population. A genotype which reproduces early relative to other genotypes will have an advantage during phases of increasing density when younger age classes predominate. Conversely, in declining populations where age structure is weighted towards older individuals, such a genotype is at a disadvantage. Thus, fluctuations in population age structure can alter relative contributions of early and late reproducing genotypes to the gene pool.

Genetics and Population Dynamics

Charlesworth and Giesel (1972) concluded "changes in gene frequency produced by changes in population growth rates may occur even if environmental agents which affect population size have no specific effects on genotypes (in terms of age-specific mortality and fecundity factors)."

There are two kinds of observations from field populations which support Charlesworth and Giesel's interpretation. First, it is unlikely that changes in gene frequency at a large number of loci for a variety of species are all causally related to population cycles. Gene frequencies at the TF and LAP loci were correlated with density in both Kansas and Indiana prairie voles, but the direction of changes in gene frequency at the LAP locus and some components of fitness among TF genotypes were different between localities. If the TF and LAP loci were causally related to population cycles, I would expect more consistency between geographic areas. There are two possible explanations for these inconsistencies: (1) The environmental regimes in eastern Kansas and southern Indiana are not the same, which gives different genotypes selective advantages at each locality; since the functional significance of different electromorphs is not known, this explanation is not appealing, and (2) If loci are associated with different linkage groups in the two localities, then I would expect differences in fitness parameters for the same genotype. Natural selection may operate on groups of spatially associated genes or on entire chromosomes rather than on specific allozyme loci. Berry (1978) attributed changes in genotypic frequencies at an HBB locus in *Mus musculus* populations on Skokholm to this "hitch-hiking effect."

Another source of evidence in support of Charlesworth and Giesel's hypothesis comes from perturbation experiments. Gaines et al. (1971) introduced three different transferrin genotypes of *M. ochrogaster* into three separate fenced enclosures. There was no significant effect of genotype on rate of population increase, percentage of lactating females, recruitment index, or survival rates of voles. In a similar experiment, LeDuc and Krebs (1975) maintained divergent allelic frequencies at an LAP locus in two field populations of *M. townsendii*. The altered allelic frequencies in the two populations did not produce any consistent effects on demography. Although the evidence indicates that genetic changes at a single locus are effects of density, it is still not possible to refute the hypothesis that changes in average heterozygosity over many loci are driving the density cycles.

Kohn and Tamarin's (1978) island study of a noncycling microtine population provides additional evidence that genetic changes are effects of density changes. There were similar changes in gene frequency at the TF locus in a cycling mainland population of *M. pennsylvanicus* and a noncycling island population of *M. breweri* over the same time period.

Now, at the risk of being accused of using a "bait and switch" technique, I will transpose the title of this paper and briefly consider the importance of population dynamics to genetics. While ecologists have ignored the genetics of populations, many geneticists have studied populations in ecological vacuums. A major concept which has considered the influence of demography on the genetic structure of the population is that of hard and soft selection (Wallace, 1968, 1975). Hard selection occurs when the fitness of a genotype is independent of other individuals in the population. For example, lethal genes generally kill individuals under all conditions. Soft selection occurs when the probability of survival and reproduction for one genotype depends upon population density and/or the frequency of other genotypes. Electrophoretic data from fluctuating vole populations suggest that fitnesses of genotypes are a function of population density. Both Tamarin and Krebs (1969) and Gaines and Krebs (1971) found that certain LAP and TF genotypes were favored during the increase phase of a *Microtus* cycle, while others were favored in the decline phase. These "increase" and "decline" genotypes at the TF locus were consistent over both studies. Again, there is a problem in distinguishing between cause and effect. If genotypic fitness is a function of density, then a change in density will change genotypic fitness, which in turn may affect density. This is the kind of genetic feedback mechanism hypothesized by Pimentel (1968) and illustrated in Fig. 3a.

Since the genetic feedback mechanism was originally applied to predator-prey interactions and under restrictive conditions results in oscillatory behavior (Lomnicki, 1974, 1977), it raises an interesting point with respect to the cause and effect dilemma. In predator-prey cycles, I view the question of whether prey density is a cause or an effect of predator density as uninteresting because the dynamics of the two populations are intimately related to one another. The interesting aspect is the set of simultaneous equations, one for changes in prey density and the other for changes in predator density, which describe the dynamics of the system. The same logic can be extended to the genetics and demography of fluctuating populations which also exhibit oscillatory behavior. We would like to know whether there is a set of simultaneous equations that describes the interactions between genetics and demography. The interesting task is to determine the form and relevant parameters of these equations.

The assumption in this predator-prey analogy is that the genetic structure and demography of fluctuating rodent populations are associated in a closed feedback loop without the interference of extrinsic factors, such as weather, predation, disease, and dispersal. Thus, the population described in Fig. 3a is completely self-regulating. Lidicker (1973) has taken a multifactorial approach to population regulation of small mammals. He has proposed that, depending on the species and geographical location, several factors may be involved. If this were the case, then the genetic feedback model would be a gross oversimplification of what happens in natural populations. The complexity of the model can be increased by incorporating extrinsic factors as illustrated in Fig. 3b. This model suggests that both genetic and demographic changes might be affected by some third causal agent. For example, dispersal may cause changes in population density and concommitantly cause changes in gene frequency. The problem of cause and effect is now compounded even further because not only must we consider the interaction of genetics and demography but also how demographic and genetic changes covary with each extrinsic factor or combination of factors. The same problem arises in a modified version of Chitty's hypothesis proposed by Stenseth (1977). He assumes that an interaction of extrinsic and intrinsic factors are needed for population cycles. According to Stenseth (1977) extrinsic factors such as weather create adverse environmental conditions which trigger the decline.

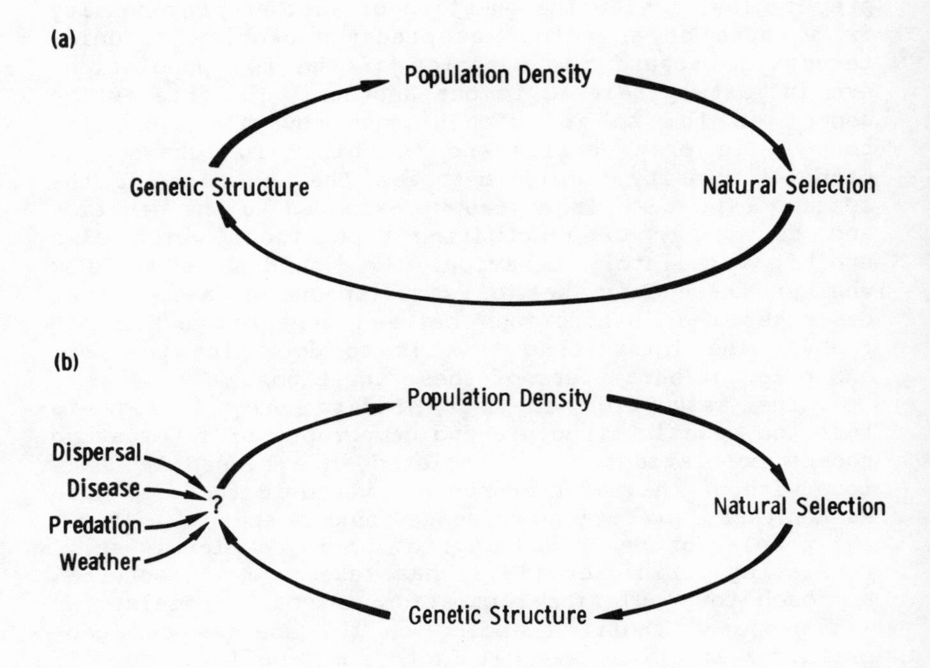

Fig. 3. (a) The genetic feedback model without the influence of extrinsic factors as illustrated by Krebs (1978b).
(b) The genetic feedback model with the incorporation of extrinsic factors.

CONCLUSION

During my review of the influence of genetics on population dynamics of small mammals, I have relied exclusively on allozymic variation. This approach has been unsatisfactory primarily because the functional significance of different electromorphs is not known. At this time, the only justification for examining the relationship of a specific protein polymorphism to the demographic performance of a population is that the locus may be in linkage disequilibrium with other ecologically relevant loci. Thus, protein polymorphisms have been relegated to the role of genetic markers. This also has been the case in studies relating pelage color polymorphisms to density changes in rodent populations (Smith et al., 1975; Gill, 1977).

The most promising results to date are from the few multiple locus studies which show that average genic heterozygosity is related to demographic parameters. The assumption here is that population heterozygosity measured by electrophoretic analysis of structural loci is a valid indicator of the total population genic heterozygosity. There is convincing evidence from three different rodent species (Garten, 1976; Massey, 1977; Gaines et al., 1978) for a positive association between average genic heterozygosity and individual body weight. Furthermore, Garten (1976, 1977) found an increase in aggressive and exploratory behavior with increasing heterozygosity. Taken together, these studies support Chitty's (1967) argument, that an increase in heterozygosity from outbreeding as the population increases results in increased aggression, which in turn is followed by increased mortality, decreased reproduction, and a population decline. Clearly the relationship between average genic heterozygosity and population parameters deserves to be investigated in other species.

If there is a consistent relationship between average genic heterozygosity and density for different species, the next step would be to do experimental manipulations. Populations can be maintained at different levels of average genic heterozygosity for electrophoretic loci, using similar methods developed for single locus perturbations (Gaines et al., 1971; LeDuc and Krebs, 1975) to assess the effect on demographic performance.

Future research must also be directed toward elucidating genetic systems that play a greater role than electrophoretic markers in the demographic process. A good example of such a system would be the t locus in _Mus musculus_ (Lewontin and Dunn, 1960) where changes in gene frequency have profound effects on deme size. However, this is an unusual case because the polymorphism is maintained by meiotic drive.

If we use Chitty's (1967) hypothesis relating genetics to population dynamics as our model, then a place to start is an examination of the inheritance of spacing behavior in fluctuating small mammal populations. Although Garten's (1976) study showing an association between aggressive behavior and nonadditive genetic variance (average genic heterozygosity) represents a promising beginning, this is a formidable endeavor because most behavioral traits are polygenic and difficult to measure. In addition, since the environment is continually changing during a population fluctuation, estimates of heritability of behavioral traits in different density phases may not be comparable. Nevertheless, at the very least we should be able to obtain ranges in heritability of a given trait over a population fluctuation. Only one attempt has been made to measure heritability of spacing behavior in fluctuating small mammal populations. Anderson (1975) found that heritability of spacing behavior in *M. townsendii* was uniformly low. Similar information needs to be obtained for other behavioral traits that may influence demography such as dispersal tendency.

Before any comprehensive statement can be made about the interactions between population genetics and demography, we will need information about the social dynamics of small mammal populations. More specifically, we want to know how the breeding structure of the population changes with density. Both Getz (1978) and Stenseth (1978b) have speculated about the social dynamics of microtine populations but they do not have field data to support their conclusions. All the genetic studies referred to in this review have treated populations as large panmictic units and by doing so the investigator may not discern "the trees from the forest." If field populations form demes, as has been reported for *Mus musculus* populations in barns (Selander, 1970a) and corn cribs (Anderson, 1970), it would certainly affect any interpretation of the influence of the genetic structure on population dynamics. It is interesting that in the only study of social structure in feral *Mus musculus* populations, Myers (1974) found no evidence for deme formation.

In concluding, I do not want to leave the impression that the application of electrophoresis to the study of population dynamics has been a futile exercise. Although electrophoresis has not provided conclusive evidence that genetic changes cause demographic changes, it has demonstrated that the genetic composition of populations changes in regular predictable ways as density fluctuates and that some of this change reflects the effect of natural selection on genetic and demographic processes.

Acknowledgments--I thank the following individuals for their critical comments on the manuscript: J. H. Brown, R. S. Hoffmann, M. L. Johnson, C. J. Krebs, M. N. Manlove, L. R. McClenaghan, Jr., R. K. Rose, M. L. Rosenzweig, W. M. Schaffer, O. A. Schwartz, M. H. Smith, R. H. Tamarin, T. S. Whittam, and two anonymous referees. It is a pleasure to acknowledge the National Science Foundation (GB 29135) for their continued support. This paper was written while I was a visiting scholar in the Department of Ecology and Evolutionary Biology at the University of Arizona. I am grateful to the graduate students, faculty, and staff for their warm hospitality. I owe special thanks to Michael and Carole Rosenzweig for making me part of their family. Finally, I am eternally indebted to Janalee P. Caldwell who has given me emotional and intellectual sustenance during my entire professional career.

GENETIC STRUCTURE OF INSULAR POPULATIONS

C. William Kilpatrick

Abstract--Genetic data based upon electrophoretic variation of proteins controlled by 17 or more loci are now available from 96 insular populations of mammals, excluding man. A total of 20 species are represented among these populations with eight species being endemic to islands and the remaining 12 species inhabiting both mainland and insular areas. The primary effect of isolation, which is apparent in nearly all non-chiropteran populations, is a reduction in genetic variability. This reduction in genetic variability primarily results from an increased proportion of monomorphic loci, absence of many rare or low frequency mainland alleles, and increased frequency of common mainland alleles. Analysis of levels of heterozygosity and biogeographic variables suggests that among insular populations this reduction in genetic variability is primarily the result of founder effect during colonization and subsequent immigration. Data on extent of genetic differentiation and reduction of genetic heterozygosity suggest that founder effect is the major force in structuring the insular gene pool. The occurrence of unique alleles and the increased frequencies of minor alleles in many island populations suggest the removal of certain selective constraints on islands. The alteration of selective pressures allows the occurrence of mutant alleles in detectable frequencies, the increased frequencies of minor alleles, probably through genetic drift, and the occasional fixation of different alleles by founder effect, genetic drift, and bottlenecking.

INTRODUCTION

Studies of the biology of island fauna have played a major role in the development of evolutionary theory (Darwin, 1859; Wallace, 1869). The importance of biogeography of islands to the understanding of evolutionary and ecological phenomena has been reviewed by Simberloff (1974). Development of two major theories of insular biogeography, the colonization model (MacArthur and Wilson, 1967) and the vicariance model (Croizat, 1958; Rosen, 1976, 1978), has stimulated additional research on the distribution of species on islands (Vuilleumier, 1970; Brown, 1971), experiments on colonization (Simberloff and Wilson, 1969; Crowell, 1973), and studies of the genetic structure of insular populations (Berry and Murphy, 1970; Wheeler and Selander, 1972; Soule and Yang, 1973; Webster et al., 1972; Yang et al., 1974; Berry and Jakobson, 1975b; Gorman et al., 1975; Patton et al., 1975; Gill, 1977; Berry and Peters, 1977; Browne, 1977; Berry et al., 1978a; Schmitt, 1978).

Islands provide natural laboratories in which many evolutionary and ecological hypotheses may be tested. Insular populations can be divided into three major groups, based upon the kinds of isolation involved: (1) oceanic islands; (2) continental shelf islands; and, (3) habitat islands. Oceanic islands are those islands separated from mainlands by great distances and generally have not been connected to a mainland area during recent geological periods with reduced sea levels. Continental shelf islands include coastal islands located on the continental shelf and islands on freshwater lakes and rivers. These islands characteristically have been connected to mainlands during periods of recent geological history and generally are less isolated than oceanic islands. Habitat islands for terrestrial species include areas of suitable habitat surrounded by suboptimal or unsuitable habitat. Habitat islands are characterized by various levels of isolation, and migrations by terrestrial species are assumed to occur at much higher rates across unsuitable parts of terrestrial environments than across the aquatic barriers surrounding other islands.

Recently, information has been obtained concerning the degree of genetic differentiation during the process of speciation in *Drosophila* (Ayala et al., 1974), fish of the family Centrarchidae (Avise and Smith, 1977), and rodents of the genus *Peromyscus* (Zimmerman et al., 1978) by comparing the genetic similarity and divergence of populations at various evolutionary stages. In this paper the effects of isolation were studied by comparing the genetic variabilities, genetic similarities, and genetic divergences between insular and mainland populations of the same or closely related taxa. Nei's (1972) index of identity, I, was utilized as a measure of genetic similarities, while genetic distance, D = $-\log_e I$, was utilized to estimate genetic divergence. Estimates of genetic distances are useful measures of the average number of electrophoretically detectable allelic substitutions that have occurred since the populations have been isolated (Nei, 1971). Since Sarich (1977) has presented evidence that values of D are strikingly different between loci which are highly variable and more rapidly evolving and those loci which are less variable and more slowly evolving, separate D values were calculated for these "fast" and "slow" evolving loci. The "fast" evolving loci included the esterase, general protein, prealbumin, and hemoglobin loci, in addition to single loci for albumin, transferrin, haptoglobin, α_2 macroglobulin, amylase, ceruloplasmin, 6-phosphogluconate dehydrogenase, leucine aminopeptidase, and acid phosphatase. All other loci were considered "slow" evolving. The genetic variability within populations was quantified by measurements of the proportion of loci polymorphic, P, and the proportion of heterozygous loci per individual, H (Lewontin, 1967).

Data obtained from a review of the literature on the genetics of insular populations of mammals were utilized to characterize the genetic structure of insular mammalian populations. Data collected from habitat island populations were excluded from these analyses due to my inability to estimate the degree of isolation of these populations. In addition, the relative contribution of island size and degree of isolation to the determination of the observed levels of genetic variability were analyzed by the statistical methods utilized in the studies of insular lizards by Soule and Yang (1973) and Gorman et al. (1975). Finally, I attempted to determine the relative importance of selection, founder effect, gene flow, genetic drift, and bottlenecking on the structure of the gene pools of insular populations of mammals.

GENETIC CHARACTERISTICS OF INSULAR POPULATIONS

Reduced Variability

An absence or severe reduction of genetic variability has characteristically been associated with isolated populations. Reduced levels of genic heterozygosity have been observed and reported in many populations inhabiting islands or caves, including insects (Prakash et al., 1969; Prakash, 1972; Saura et al., 1973; Laing et al., 1976), arachnids (Johnston and Carmody, 1980), fish (Avise and Selander, 1972), salamanders (Highton and Webster, 1976), and lizards (Soule and Yang, 1973; Webster et al., 1972; Gorman et al., 1975). In mammals this phenomenon has been observed in continental populations inhabiting isolated or patchy environments (Nevo and Shaw, 1972; Patton et al., 1972; Nevo et al., 1974; Selander et al., 1974; Penney and Zimmerman, 1976; Glover et al., 1977), as well as in insular populations surrounded by water (Selander et al., 1971; Avise et al., 1974a; Berry and Peters, 1977; Browne, 1977; Cothran et al., 1977; Schmitt, 1978). However, several insular populations have failed to demonstrate reduced levels of genetic variability (Berry and Murphy, 1970; Ayala et al., 1971; Johnson, 1971; Johnson et al., 1972; Wheeler and Selander, 1972; Berry and Peters, 1975; Greenbaum and Baker, 1976).

Reduction in genetic variability could result from a number of factors including founder effect, genetic drift, bottlenecking, changes in selective pressures, or some combination of these forces. Estimates of genetic variability from insular and mainland populations are available for 12 species of mammals (Table 1). Among the insular populations, only <u>Macrotus waterhousii</u> failed to demonstrate lower mean levels of heterozygosity than conspecific mainland populations. However, Greenbaum and Baker (1976) have suggested that the number of polymorphic systems and the low levels of heterozygosity observed at all but one locus indicate a reduction of genetic variation in this Jamaican population of chiroptera. Calculation of standardized percent reduction in mean genetic heterozygosity for each non-chiropteran species inhabiting islands (the difference between mean heterozygosity of mainland and insular populations divided by the mean heterozygosity of the mainland populations; Table 1) indicates that insular populations of non-chiropteran mammals demonstrate 42% less genetic heterozygosity than mainland populations. The lowest standardized reduction in heterozygosity, 4.6%, was observed among the populations of <u>Sigmodon hispidus</u>, while the greatest standardized reduction in heterozygosity was 90% in <u>Spermophilus spilosoma</u> from

Table 1. Heterozygosity ($\bar{H}$) and number (N) of insular and mainland populations of mammals.

Species	Mainland Populations			Insular Population			Reduction in $\bar{H}$%	Ref.*
		Heterozygosity			Heterozygosity			
	N	Mean	Range	N	Mean	Range		
Macrotus waterhousii	3	.021	.000-.043	1	.040			1
Macaca fuscata	13	.019	.000-.035	5	.013	.003-.018	31.6	2
Spermophilus spilosoma	12	.090	.049-.160	1	.009		90.0	3
Peromyscus eremicus	44	.040	.006-.079	2	.009	.000-.022	77.5	4
Peromyscus leucopus	3	.080	.076-.084	3	.071	.052-.078	11.3	5
Peromyscus maniculatus	22	.088	.054-.124	11	.068	.010-.131	22.7	6,7,8
Peromyscus polionotus	26	.063	.050-.086	4	.052	.018-.086	17.5	9
Sigmodon hispidus	4	.022	.017-.025	1	.021		4.6	10
Microtus pennsylvanicus	4	.142	.120-.171	9	.056	.023-.114	60.6	11
Mus musculus	16	.091	.032-.114	20	.041	.000-.079	55.0	12,13
Rattus fuscipes	3	.047	.020-.100	9	.011	.000-.040	76.6	14
Rattus rattus	1	.031		11	.026	.008-.056	16.1	15

* (1) Greenbaum and Baker, 1976; (2) Nozawa et al., 1975; (3) Cothran et al., 1977; (4) Avise et al., 1974a; (5) Browne, 1977; (6) Aquadro, 1978; (7) Avise et al., 1979a; (8) Gill, 1977; (9) Selander et al., 1971; (10) Johnson et al., 1972; (11) Kilpatrick and Crowell, 1980a; (12) Berry and Peters, 1977; (13) Selander and Yang, 1969; (14) Schmitt, 1978; (15) Patton et al., 1975.

Padre Island in the Gulf of Mexico. The greatest variation in the reduction of heterozygosity was observed among species with a small sample of either insular or mainland populations.

Comparison of electrophoretic data and estimates of genic heterozygosity of eight endemic insular species of rodents with closely related mainland taxa also revealed a general trend of reduced genic variation among insular forms (Table 2). Seven of eight species demonstrated lower levels of genic variation than their mainland relatives with an average reduction in genic heterozygosity of 77%. *Dipodomys compactus* was the only endemic insular species which did not demonstrate a lower value of heterozygosity than the mean of its mainland relatives (Johnson and Selander, 1971).

Genetic Similarity and Genetic Distance

MacArthur and Wilson (1967) predicted that isolation would increase the importance of immigration and emigration as factors affecting the genetic structure of a population and, potentially, permit rapid evolutionary change. Rates at which insular populations diverge from mainland populations depend upon several factors including founder effect, rate of gene flow, population size, and duration of isolation. I have analyzed the amount and patterns of genetic divergence of insular from mainland populations by comparing their genetic characteristics.

Comparison of Nei's (1972) mean genetic identity ($\bar{I}$) and genetic distance ($\bar{D}$) between insular populations of seven species of mammals is presented in Table 3. Among insular populations of a taxa, $\bar{I}$ was 0.934 with the highest $\bar{I}$ observed between insular populations of *Macaca fuscata* from the oceanic islands of southern Japan and the lowest $\bar{I}$, 0.841, observed among insular populations of *Rattus fuscipes* from continental shelf islands of southern Australia. Analysis of $\bar{D}$ indicates an average difference of eight amino acid substitutions per 100 loci between conspecific insular populations. Analysis of the genetic differentiation of "fast" and "slow" evolving loci (Sarich, 1977) indicates that 83% more genetic differentiation occurred at "fast" evolving loci as compared to "slow" evolving loci in insular populations.

Table 2. Heterozygosity and number (N) of endemic insular rodent species as compared to related mainland species.

Insular Species	N	Heterozygosity		Mainland Species			Ref.*
		Mean	Range	N	Heterozygosity Mean	Heterozygosity Range	
Dipodomys compactus	3	0.023	0.0-.069	41 (*Dipodomys*)	0.021	0.000-.071	1
Peromyscus eva	1	0.000		25 (*Haplomylomys*)	0.030	0.000-0.106	2
Peromyscus dickeyi	1	0.000		25 (*Haplomylomys*)	0.030	0.000-0.106	2
Peromyscus interparietalis	1	0.000		25 (*Haplomylomys*)	0.030	0.000-0.106	2
Peromyscus guardia	1	0.014		25 (*Haplomylomys*)	0.030	0.000-0.106	2
Peromyscus stephani	1	0.000		68 (*boylii* species group)	0.032	0.000-0.1176	2,3,4
Peromyscus sejugis	2	0.017		81 (*maniculatus* species group)	0.081	0.033-0.124	2,5,6,7
Microtus breweri	1	0.020		8 (*Microtus*)	0.098	0.012-0.186	8,9

* (1) Johnson and Selander, 1971; (2) Avise et al., 1974a; (3) Avise et al., 1974b; (4) Kilpatrick and Zimmerman, 1975; (5) Aquadro, 1978; (6) Avise et al., 1979a; (7) Selander et al., 1971; (8) Kilpatrick and Crowell, 1980a; (9) Kilpatrick and Crowell, 1980b.

Genetics of Insular Populations

Table 3. Mean genetic identity, $\bar{I}$ (± SE), and genetic distance, $\bar{D}$ (± SE), between insular populations (above) and between mainland and insular populations (below) of mammals. Genetic distances are also given for "fast" and "slow" evolving loci.

Species*	$\bar{I}$	D Fast	D Slow	$\bar{D}$
Macaca	.994 ± .002	.009	.002	.006 ± .002
fuscata	.996 ± .002	.005	.001	.004 ± .002
Peromyscus	.968 ± .017	.052	.017	.030 ± .020
leucopus	.976 ± .011	.039	.012	.025 ± .012
Peromyscus	.952 ± .019	.063	.036	.055 ± .022
maniculatus	.939 ± .022	.084	.044	.073 ± .029
Microtus	.930 ± .027	.094	.059	.124 ± .049
pennsylvanicus	.899 ± .032	.174	.066	.125 ± .041
Mus	.905 ± .028	.107	.094	.111 ± .034
musculus	.901 ± .034	.095	.108	.078 ± .067
Rattus	.841 ± .013	.286	.142	.179 ± .016
fuscipes	.855 ± .016	.332	.093	.172 ± .019
Rattus	.951 ± .004	.083	.031	.050 ± .005
rattus				
Means	.934 ± .017	.099	.054	.079 ± .021
	.928 ± .020	.122	.054	.080 ± .023

* *M. fuscata*, Nozawa et al., 1975; *P. leucopus*, Browne, 1977; *P. maniculatus*, Aquadro, 1978, and Gill, 1977; *M. pennsylvanicus*, Kilpatrick and Crowell, 1980a; *Mus musculus*, Berry and Peters, 1977; *R. fuscipes*, Schmitt, 1978; *R. rattus*, Patton et al., 1975.

A comparison of genetic differentiation between insular and mainland populations of six species of mammals revealed an $\bar{I}$ between insular and mainland populations of 0.928 (Table 3). Again, the greatest similarity in genetic structure was observed between insular and Japanese mainland populations of M. fuscata with an $\bar{I}$ of 0.996; the lowest level of genetic similarity was observed between Australian mainland and insular populations of R. fuscipes with an $\bar{I}$ of 0.855. The $\bar{D}$ indicated that an average of eight amino acid substitutions per 100 loci occurred during the divergence of insular and mainland mammalian populations. Analysis of the loci indicated that 126% more genetic differentiation occurred at "fast" than at "slow" evolving loci during the evolution of insular and mainland gene pools.

The $\bar{I}$ for mainland populations of eight species was 0.944 (Table 4). Analysis of $\bar{D}$ indicated that an average of eight amino acid substitutions per 100 loci occurred during the divergence of conspecific mainland populations. However, 360% more genetic divergence occurred at the "fast" than at the "slow" evolving loci.

The levels of genetic identity and total amounts of genetic differentiation observed among mainland populations, among insular populations, and between mainland and insular populations are not significantly different. However, there appears to be a major difference in the genetic divergence of "slow" evolving loci among insular populations and among mainland populations. Patterns of genetic divergence for "slow" evolving loci among insular populations and between mainland and insular populations indicated that about 83 to 126% more differentiation occurred at the "fast" evolving loci. Alternatively, approximately 360% more differentiation occurred among "fast" than "slow" evolving loci among mainland populations. This difference appears not to be the result of reduced genetic differentiation at "fast" evolving loci among insular populations (Tables 3 and 4), but rather increased levels of genetic differentiation at "slow" evolving loci among insular populations. Increased differentiation at "slow" evolving loci appears to result from increased isolation since additional decrease in the rate of gene flow would be expected to produce a greater effect on the genetic divergence of "slow" evolving loci. The initial genetic divergence between two populations would predominantly involve "fast" evolving loci. However, at some point in

Genetics of Insular Populations

Table 4. Mean genetic identity, $\bar{I}$ (± SE) and genetic distance, $\bar{D}$ (± SE), between mainland populations of mammals. Genetic distances are also given for "fast" and "slow" evolving loci.

Species*	$\bar{I}$	D Fast	D Slow	$\bar{D}$
Macaca fuscata	.999 ± .0001	.001	.001	.001 ± .0001
Peromyscus boylii	.991 ± .001	.028	.000	.009 ± .001
Peromyscus pectoralis	.964 ± .017	.078	.005	.037 ± .017
Peromyscus leucopus	.981 ± .011	.042	.001	.021 ± .013
Peromyscus maniculatus	.880 ± .047	.130	.068	.168 ± .067
Microtus pennsylvanicus	.920 ± .039	.149	.044	.177 ± .103
Mus musculus	.936 ± .023	.063	.014	.073 ± .027
Rattus fuscipes	.884 ± .019	.316	.046	.123 ± .021
Means	.944 ± .015	.101	.022	.076 ± .024

***M. fuscata*, Nozawa et al., 1975; *P. boylii* and *P. pectoralis*, Kilpatrick and Zimmerman, 1975; *P. leucopus*, Browne, 1977; *P. maniculatus*, Aquadro, 1978, and Avise et al., 1979a; *M. pennsylvanicus*, Kilpatrick and Crowell, 1980a; *M. musculus*, Berry and Peters, 1977; *R. fuscipes*, Schmitt, 1978.

time, with continuous isolation, further genetic divergence would be contributed mostly by "slow" evolving loci.

Pattern of Genetic Differentiation

The distribution of loci with various I values has demonstrated patterns of genetic differentiation during the process of speciation (Ayala et al., 1974; Avise and Smith, 1977; Zimmerman et al., 1978). Comparisons of distributions of loci with various I values between insular populations and mainland populations of cricetid rodents demonstrate a general increase in the homogeneity of genetic identity among loci of continental shelf insular populations (Fig. 1). Among these insular populations it was observed that 83 to 77% of the loci have I values above 0.95 while 76 to 73% of the loci of mainland populations have I values within this range. Among mainland populations, 4 to 8% of the loci were observed to have I values below 0.50 while no loci were observed with I values this low among the insular populations of cricetid rodents.

The distribution of the proportion of loci with various I values in oceanic insular populations of *Mus musculus* and continental shelf insular populations of *R. fuscipes* compared with mainland conspecific populations revealed a different pattern of genetic differentiation (Fig. 2). Only 55 to 63% of the loci of insular populations were found to have I values above 0.95 while 68 to 75% of the loci of mainland populations were within this range. A greater proportion of loci with I values below 0.70 was observed among the insular populations (0.18 to 0.24) than were found among the mainland populations (0.18 to 0.05).

In both *P. maniculatus* and *Microtus pennsylvanicus*, F_{max} revealed that the distribution of genetic identity among loci was significantly less heterogeneous among insular than among mainland populations. Among these insular populations of the continental shelf an increase was observed in the proportion of loci with high I values primarily resulting from an increase in the proportion of monomorphic loci (Fig. 1). However, the distributions of genetic identity among the loci of insular populations of *Mus musculus* and *R. fuscipes* were not significantly different from the mainland popula-

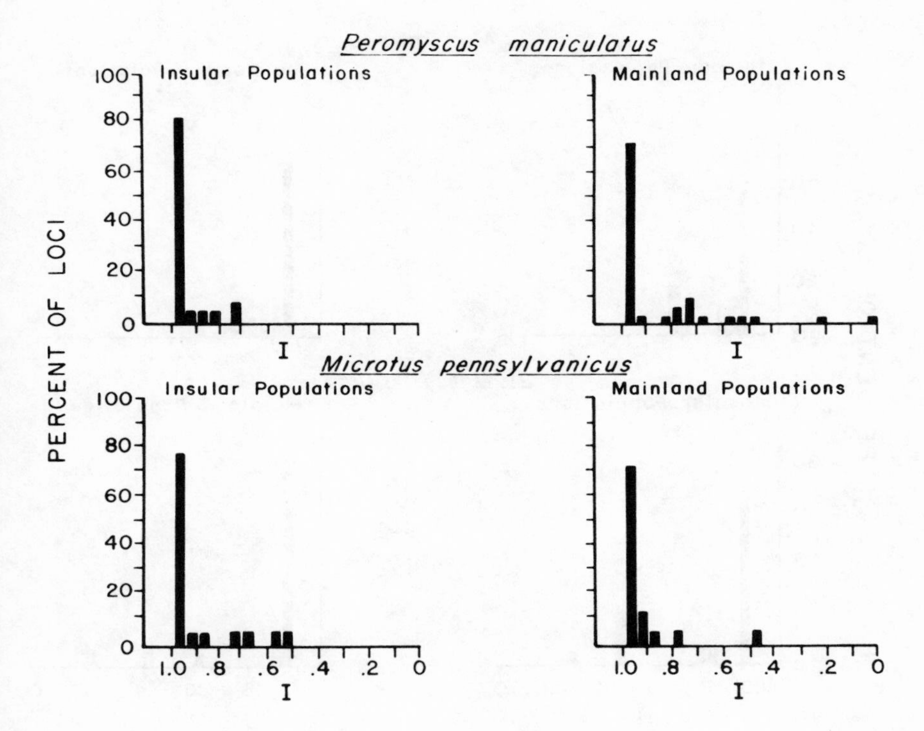

Fig. 1. Distribution of genetic identities, I, relative to loci for mainland and insular populations of cricetid rodents.

tions. A reduction in the proportion of loci with high I values was observed among these insular populations (Fig. 2). This reduction, along with an increase in the proportion of loci with I values below 0.70, appears to be the result of two conditions, the occurrence of unique alleles at moderate frequencies among certain insular populations and the predominance or fixation of minor mainland alleles in a few insular populations.

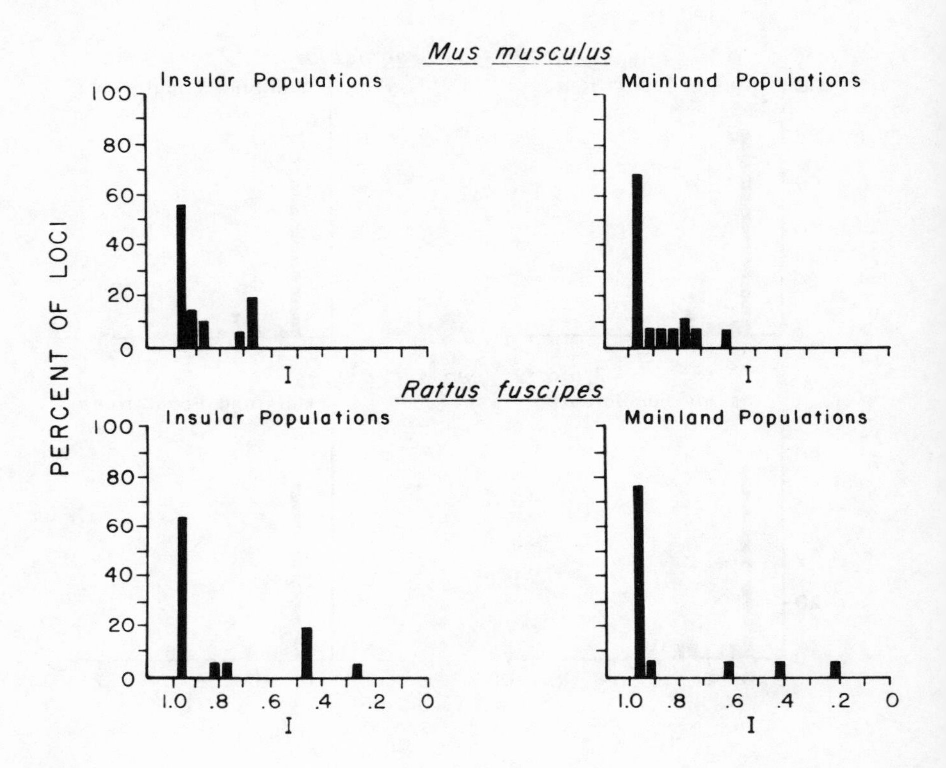

Fig. 2. Distribution of genetic identities, I, relative to loci for mainland and insular populations of murid rodents.

Unique Alleles

A total of 33 different alleles was observed among the insular populations of six species of mammals which were not observed in any mainland population (Table 5). Alleles that are unique to insular populations comprised an average of 8.6% of the total allelic composition of insular gene pools and ranged from 0 to 0.19%. A mean number of 1.22 unique alleles per insular population was observed for the six species of mammals in Table 5. These generally occurred at moderate frequencies (mean =

0.25) in those populations in which they were observed but were predominant or fixed in populations of half of the species (Table 5). Approximately 42% of the continental shelf island populations possessed unique alleles, while only 7.9% of the oceanic island populations were observed to have alleles not found among mainland populations. This difference observed between the continental shelf island and oceanic island populations may reflect the harsher environmental conditions of the more northern latitude of the oceanic islands analyzed by Berry and Peters (1977).

BIOGEOGRAPHICAL CONSIDERATIONS

A steady state model of genetic polymorphism in insular populations, developed by Jaenike (1973) suggests the number of polymorphic loci in an isolated population can be changed by four mechanisms. The number of polymorphic loci of a population may be increased by mutation occurring at a monomorphic locus, or by immigration of individuals carrying new alleles into the gene pool. A decrease in numbers of polymorphic loci may occur by fixation of one allele at a locus or by emigration of all individuals carrying a particular allele. Jaenike (1973) suggested that the number of polymorphic loci in an insular population at equilibrium is a function of immigration and fixation. Although estimates of rates of immigration and rates of fixation for mammalian populations are not well established, the rate of immigration is dependent upon the distance of island from colonizing source as well as island size. The rate of fixation is dependent upon population size which is, among other things, dependent upon the size of an island. Based upon the Jaenike (1973) model of insular polymorphism, the proportion of loci polymorphic would be predicted to be directly proportional to island size, since the smaller the island, the smaller the target for immigration and the greater the probability of fixation due to the smaller population size. This model also predicts that the proportion of loci polymorphic would be inversely proportional to the distance of the island from the colonizing source, the greater the distance, the smaller the probability of immigration.

In an attempt to test the predictions of this model, two ratios (area of island/distance to mainland

Table 5. Number of insular populations (N) and occurrence of unique alleles (A) in six species of mammals. The mean number of unique alleles per insular population was calculated by multipyling A times the number of islands on which the allele occurred and dividing by N.

			Unique Alleles				
				Mean Number per Insular	Frequency		% N with Unique
Species*	N	A	A/Total Alleles	Population	Mean	Range	Alleles
Macaca fuscata[1]	9	3	.08	0.33	.06	.003-0.150	44
Peromyscus polionotus[2]	4	2	.05	0.50	.18	.140-0.220	25
Sigmodon hispidus[3]	2	6	.15	3.00	.03	.010-0.100	50
Microtus pennsylvanicus[4]	11	10	.18	1.82	.13	.020-0.660	18
Mus musculus[5]	19	6	.13	0.48	.22	.012-1.00	8
Rattus fuscipes[6]	10	6	.19	1.20	.88	.120-1.00	70

*No unique alleles were reported for the insular populations of *Spermophilus spilosoma*, Cothran et al., 1977; *Peromyscus leucopus*, Browne, 1977; *Peromyscus maniculatus*, Aquadro, 1978.

[1]Nozawa et al., 1975; [2]Selander et al., 1971; [3]Johnson et al., 1972; [4]Kilpatrick and Crowell, 1980a; [5]Berry and Peters, 1977; [6]Schmitt, 1978.

and log of area of island/square of distance to mainland) were calculated for the insular populations of three species of rodents (P. maniculatus, M. pennsylvanicus, Mus musculus) from the data presented in Table 6 and their relationship with the proportion of loci polymorphic per insular population tested by regression analysis. Correlations observed ranged from 0.085 to 0.349 for the insular populations of these three species of rodents and indicated that either these populations have not yet reached equilibrium or that this model is either incomplete or incorrect.

The relative contribution of island size, habitat complexity, and various estimates of isolation to the determination of the observed levels of genetic variability have been analyzed by step-wise multiple regression (Soule and Yang, 1973; Gorman et al., 1975; Aquadro, 1978). The relationship between estimates of genetic variability and biogeographic variables of size and distance from the mainland (Table 6) were analyzed by step-wise multiple regression for four species of rodents. Browne (1977) found that among P. leucopus populations, average heterozygosity and land mass size were directly correlated while proportions of loci polymorphic and land mass size were not directly correlated. Analysis of Browne's (1977) data by step-wise multiple regression revealed slightly higher correlations for both estimates of genetic variability with measurement of distance from the mainland than with island size, but the number of islands sampled (n = 3) is too small to be meaningful. Seven insular populations of P. maniculatus from the coast of Maine were analyzed by step-wise multiple regression by Aquadro (1978). Distance to mainland or other colonizing source was the best single predictor of both the proportion of loci polymorphic and the mean heterozygosity. With the addition of an estimation of the time of separation of the island from the mainland and a measurement of the area of the island to the square of the distance from the mainland, a coefficient of determination of greater than 0.90 was observed by Aquadro (1978).

The log of the area was observed to be the best single predictor of the levels of genetic variability in the two additional species of rodents analyzed in this study. Step-wise multiple regression analysis indicated that log of the area was the single best predictor of genetic variability for the nine insular populations of

Table 6. Genetic variation of insular populations of rodents and size and distance from mainland for the islands they occupy.

Species Islands	Polymorphic Loci per Population: P	Heterozygous Loci per Individual: H	Island Area (km^2)	Distance from Mainland (km)
<u>Peromyscus</u> <u>leucopus</u>[1]				
North Bass	0.214	0.078	3.04	19.0
Middle Bass	0.179	0.052	3.08	10.4
South Bass	0.107	0.074	6.14	2.8
<u>Peromyscus</u> <u>maniculatus</u>[2]				
Matinicus	0.035	0.010	3.3	18.50
Vinal Haven	0.310	0.100	46.3	9.50
Calderwood Neck	0.207	0.086	8.1	0.05
Isle au Haut	0.172	0.066	27.9	7.60
Deer Isle	0.207	0.076	70.9	1.50
Swan's	0.069	0.024	28.1	5.00
Mt. Desert	0.138	0.062	275.9	0.50
<u>Microtus</u> <u>pennsylvanicus</u>[3]				
Prince Edward	0.125	0.023	5,656.600	13.70
Magdalen	0.250	0.106	264.200	148.20
Block	0.125	0.035	44.800	15.30
Newfoundland	0.188	0.066	404,519.600	20.00
Deer Isle	0.304	0.157	70.900	1.50
Fort	0.261	0.086	0.008	10.96
Second	0.217	0.109	0.015	11.35
Mark	0.500	0.121	0.023	12.12
Rock	0.401	0.095	0.008	10.76

Table 6 (continued)

Species Islands	Polymorphic Loci per Population: P	Heterozygous Loci per Individual: H	Island Area (km^2)	Distance from Mainland (km)
Mus musculus[4]				
Skokholm	0.169	0.060	1.0	3.5
Inchkeith	0	0	0.5	5.0
Isle of May	0	0	0.5	9.0
Orkney	0.221	0.041	298.0	28.0
S. Ronaldsay	0.263	0.090	36.0	14.0
Sanday	0.158	0.051	49.0	65.0
N. Ronaldsay	0.158	0.030	8.0	85.0
Shetland	0.211	0.054	979.0	169.0
Fair Isle	0.091	0.004	10.0	175.0
Bressay	0.154	0.040	28.0	205.0
Yell	0.250	0.057	215.0	235.0
Fetlar	0.125	0.053	36.0	245.0
Faroe Islands				
Hestur	0.045	0.014	5	398
Fugloy	0	0	8	416
Nolsoy	0.067	0.012	8	380
Sandoy	0.250	0.078	93	362
Streymoy	0.267	0.026	372	381
Mykines	0.045	0.010	8	411
Macquarie	0.316	0.065	119	1370

[1]Data from Browne, 1977; [2]Data from Aquadro, 1978; [3]Data from Kilpatrick and Crowell, 1980a; [4]Data from Berry and Peters, 1977.

M. pennsylvanicus (Table 6) with a coefficient of determination of 0.39. However, in a more local situation, such as Penobscot Bay where there is much less variation in insular size, the square of the distance from the mainland among five insular populations was observed to be the best single predictor with a coefficient of determination of 0.60 and a correlation coefficient of -0.777. Neither of these correlations between proportion of loci polymorphic and biogeographic variables among insular populations of *M. pennsylvanicus* were significant. Analysis of the estimates of genetic variability reported by Berry and Peters (1977) for 19 insular populations of *Mus musculus* (Table 6) again revealed that the best single predictor of genetic variation was log of the area of the island. A significant correlation was observed between log of the area of the island and heterozygosity with a correlation coefficient of 0.55, while a highly significant correlation was observed between log of the area and proportion of loci polymorphic with a correlation coefficient of 0.77, among the insular populations of *Mus*.

A nonsignificant correlation was observed for standardized estimates of genetic variability (insular variability/mean mainland variability) for pooled data from these four species of rodents and biogeographic variables of area and distance. Since significant correlations were observed between estimates of genetic variability of individual species and different biogeographic variables, different relationships between genetic variability and biogeographic variables may exist among the four species. However, two major sources of statistical bias are present and therefore one must be very careful in interpreting such data. An intercorrelation between distance from mainland and area of island exists among the islands from which populations of *Peromyscus* were sampled. Such intercorrelations detract from the reliability of the multiple correlation coefficient (Bryant, 1974b). The sample size of insular populations analyzed ranges from 3 to 19 for these four species and such small sample sizes may inflate the correlation coefficient (Bryant, 1974b). One possible explanation of this biogeographical data is that the relationship between genetic variability and biogeographic variables is dependent upon colonizing ability. The relative colonizing abilities of these three genera of rodents can be obtained, in part, from the work of Crowell (1973), which indicates that *Microtus* is a much better colonizer than *Peromyscus* of the continental shelf islands of northeastern United States. The colonizing ability of *Mus musculus* is well documented by the inadvertent introduction to most areas inhabited or visited by man (Holdgate and Wace, 1961; Newsome, 1969a,b).

The higher correlations observed between estimates of genetic variability and log of insular area than distance from the mainland for the two species, *M. pennsylvanicus* and *Mus musculus*, which are the better colonizers of the four species examined, suggest that the levels of genetic variability among insular populations of good colonizing species are more dependent upon size of populations and rates of fixation than upon immigration rates. However, both species of *Peromyscus* which were analyzed demonstrated higher correlations between genetic variability and square of the distance from the mainland. Among the insular populations of these species which are less efficient at colonization, the levels of genetic variability appear to be more dependent upon the rate of immigration which is an exponential function of distance from a colonizing source.

A second possible explanation of this biogeographical data which may be related also to differential colonizing ability is the range of biogeographic variables within each data set. Insular populations of *Peromyscus* are restricted to islands which are larger than 25 ha (Redfield, 1976) and which are not too distant from a colonizing source (less than 20 km). Since these populations are restricted to larger islands, the rates of fixation and population sizes are less variable and the probability of fixation is relatively small. Redfield (1976) suggested that small insular populations of *Peromyscus* were large enough to obliterate effects of random genetic drift. However, since these insular populations are all relatively close to a colonizing source the rate of immigration appears to be responsible for much of the variation in genetic variability among insular populations of *Peromyscus*.

Insular populations of both *M. pennsylvanicus* and *Mus musculus* are not restricted to islands of certain size (range: 0.8 ha to over 400,000 km^2) or restricted to a certain distance from the mainland (range: 0.82 to 1370 km). The rates of fixation and population sizes appear to be responsible for much of the variation in genetic variability among populations of small islands and populations of islands that are so distant from a colonizing source that the rate of immigration is very small. Positive and significant relationships between island area and genetic and morphological variation have been observed in several additional species (Soule, 1972; Soule and Yang, 1973; Gorman et al., 1975; Patton et al., 1975). Although the biological significance of this association remains to be fully understood, the time-divergence theory of variation (Soule and Yang, 1973; Gorman et al., 1975) has been hypothesized to explain this relationship. This hypothesis specifies that directional selection at rates inversely proportional to island size is responsible for the reduction in genetic variation (Gorman et al., 1975). However, if the reduction in genetic variability is dependent upon the rate of fixation, genetic drift as well as selection should be considered.

ROLES OF EVOLUTIONARY FORCES IN STRUCTURING OF INSULAR GENE POOLS

The origin of insular populations has been a subject of debate for a number of years and two major theories of insular biogeography have resulted, a colonization model (MacArthur and Wilson, 1967) and a relic or vicariance model (Croizat, 1958; Rosen, 1976, 1978). The evolutionary consequences of these two models are identical except during the initial stages of isolation. In the colonization model the genetic makeup of the colonizers produces a founder effect (Mayr, 1954), while in the vicariance model the relic population remaining may or may not be reduced to a small population size producing a bottleneck or founder effect.

Founder Effect

Handford and Pernetta (1974) have suggested that if a colonizing group contains only a fraction of the genetic variability of the parental population, then this fraction is likely to be nonrandom because of the highly selective nature of the colonizing process. Smith et al. (1978) indicated that the genetic content in a founding population is a truncated subset of the parental gene pool and that the genetic variability in this subset would be affected if animals with certain genotypes were the predominant founders. Recent studies have revealed differential dispersal with respect to genotype (Myers and Krebs, 1971a; Pickering et al., 1974), while exploratory behavior has been reported to be correlated with levels of genetic heterozygosity (Garten, 1974, 1976) but not with individual genotypes (Blackwell and Ramsey, 1972).

Founder effect has been identified to be of major importance in determining the characteristics of insular populations (Berry, 1964, 1969b, 1970; Berry and Peters, 1977; Berry et al., 1978a, 1978b). Berry and Murphy (1970) identified two major genetic consequences of a population founded by a small number of individuals, reduction in genetic variation and genetic differentiation brought about by large changes in gene frequency from the ancestral population. In addition to the direct effect that founder effect may have upon a gene pool by determining the initial allelic composition and allelic frequencies, this initial composition of the gene pool may also have indirect effects upon subsequent natural selection and random genetic drift. A severe reduction in the frequency of any allele that was not highly adaptive would increase the probability of that allele being lost from the insular population. In addition, the amount of genetic variation present in the founding gene pool is a major limiting factor in the possible adaptation of individuals to the insular environment.

Much, if not most, of the reduction of genic heterozygosity observed among insular populations is the result of founder effect (Berry, 1967; Prakash, 1972; Saura et al., 1973; Browne, 1977). In the insular gene pool, usually there is an absence of the rarer mainland alleles with a subsequent increase in the frequency or fixation of the more common mainland alleles. Analysis of the allelic composition of insular and mainland populations of nine species of mammals indicated that on an average 19% of the mainland allelic composition is absent from an insular population (Table 7). This reduction in allelic composition among insular populations is not only related to the degree of isolation of the islands but also appears dependent upon the colonizing abilities of the species. The greatest reduction in allelic composition occurs in the insular populations

Table 7. Reduction of genetic variability in insular populations of mammals.

Species	Mean Proportion						
		Loci Monomorphic				Standardized % Change in Mean Proportion of Monomorphic Loci	Data Source
	Mainland Alleles Absent from Insular Populations	Islands		Mainland			
		$\bar{X}$	SE	$\bar{X}$	SE		
Macaca fuscata	0.31	0.94	0.039	0.86	0.018	9.3	1
Spermophilus spilosoma	0.35	0.96		0.80	0.024	20.0	2
Peromyscus leucopus	0.06	0.83	0.034	0.80	0.050	3.8	3
Peromyscus maniculatus	0.15	0.79	0.036	0.77	0.035	2.6	4
Peromyscus polionotus	0.39	0.88	0.052	0.81	0.028	8.6	5
Sigmodon hispidus	0.15	0.78	0.092	0.90	0.020	-13.3	6
Microtus pennsylvanicus	0.13	0.80	0.024	0.61	0.020	31.1	7
Mus musculus	0.13	0.83	0.020	0.59	0.022	40.7	8
Rattus fuscipes	0.06	0.97	0.020	0.83	0.034	16.9	9

* (1) Nozawa et al., 1975; (2) Cothran et al., 1977; (3) Browne, 1977; (4) Aquadro, 1978; (5) Selander et al., 1971; (6) Johnson et al., 1972; (7) Kilpatrick and Crowell, 1980a; (8) Berry and Peters, 1977; (9) Schmitt, 1978.

of *S. spilosoma* and *P. polionotus*, both of which inhabit coastal barrier islands which are not too distant from the mainland. The observation that the oceanic island populations of *Mus musculus* do not demonstrate the greatest reduction in allelic composition (Table 7) may result in part from many of these populations by establishment and periodic reinforcement by commensalism with man.

The effects of the founding gene pool on the genetic composition of insular populations is also reflected in the increased proportion of monomorphic loci, although it is not possible to separate the founding effect from subsequent genetic drift or selection. A significant difference in the mean proportion of monomorphic loci in insular and mainland populations was found in 45% of the species analyzed (Table 7), and the mean proportion of monomorphic loci was greater in all insular populations, except for populations of *Sigmodon hispidus*, when compared with conspecific mainland populations. The mean proportion of monomorphic loci for mainland populations for the nine species was 77%, while continental shelf and oceanic insular populations had mean proportions of 87 and 83% monomorphic loci, respectively. A standardized percent change in the mean proportion of monomorphic loci (Table 7) was calculated by dividing the difference between the mean proportion of loci monomorphic in insular populations and mainland populations by the mean proportion of loci monomorphic in mainland populations. The significant correlation observed between the standardized percent reduction in mean heterozygosity (Table 1) and the standardized percent change in mean proportion of monomorphic loci (Table 7) suggests that the increase in mean proportion of these loci is primarily responsible for the reduction in mean heterozygosity.

Berry and Peters (1977) have indicated that founder effect alone can contribute substantially to the genetic differentiation of insular populations. The genetic distance between three samples of *Mus musculus* from Taunton (0.073, 0.084, 0.108) exemplify the extent of differences that can be produced by founder effect alone (Berry and Peters, 1977). The differences in skeletal traits observed between rick populations on a single farm (Berry, 1963; Berry and Berry, 1972) are also suggested to be the result of founder effect (Berry and Peters, 1977).

Lewontin (1974) has indicated that forces which decrease genetic variability within populations tend to increase variability between populations. A highly significant correlation was observed between standardized percent reduction in mean heterozygosity (Table 1) and the mean genetic distance between mainland and insular populations of six species of mammals (Table 3), thus suggesting that the forces which are responsible for the reduction in genetic variation are also responsible for much of the genetic differentiation between insular and mainland populations ($r^2 = 0.723$). Berry and Peters (1976) speculated that founder effect may produce enough difference from the parental population to result in instant subspeciation. However, founder effect does not appear to be of prime importance in determining the genetic structure of California populations of *Helix aspersa* (Selander and Kaufman, 1973b), Hawaiian populations of *Mus musculus* (Wheeler and Selander, 1972), and the Skokholm population of *Mus musculus* (Berry and Murphy, 1970). Berry and Peters (1977) concluded that founder effect is the main differentiating factor between insular populations of *Mus*, but this founder effect functions mainly as a mechanism of gene interchange and hence increases opportunities rather than decreases adaptation.

Selection

Data indicating that insular populations are exposed to different selective pressures than mainland populations have been reported for Bass Island water snakes (Camin and Ehrlich, 1958). Because species diversity decreases with decreasing island area and/or increasing distance from a colonizing source (MacArthur and Wilson, 1967), predation and interspecific competition might be expected to decline as the number of both predator and prey species declines. On very small islands predators frequently are not present and thus exert no selective pressure at all on the prey species of these islands (Foster, 1965; MacArthur and Wilson, 1967; Kurten, 1972).

Numerous traits and conditions have been observed to occur in higher frequencies in insular populations than they are known to occur elsewhere. Gill (1977) reported the occurrence of buffy colored voles on Brooks Island of the San Francisco Bay although this phenotype was not known from mainland populations of *M. californicus*. Additional analysis suggested that this polymorphism was maintained through heterozygote reproductive superiority and seasonal reduction of one of the homozygotes on the island, while selective predation might be an important factor in the absence of this phenotype from the mainland (Gill, 1977). Changes in selective pressures, such as reduced predation pressures in insular or isolated situations, may produce increases in genetic variation by stochastic processes.

Several skeletal abnormalities, reported by Berry and Jakobson (1975a), show higher frequencies in certain insular populations of *Mus* than in mainland populations. One example, dys-symphysis, an abnormality of the midline, has been found to occur at a frequency of 0.85 on the small Welsh island of Skokholm (Berry and Jakobson, 1975a). On this island, apparently with changes in selective pressures, the abnormality has become the norm. Another example, the absence of the third molar, has been observed in several insular populations of *Mus* (Berry and Jakobson, 1975a) and *P. maniculatus* (Aquadro, 1978), whereas this condition has not been reported from the mainland.

Change in selective pressure in insular populations also appears to be responsible for the fixation of rare alleles or directional shifts toward extremes of continuous traits. Berry and Peters (1976) concluded that unusual alleles are common in insular populations because selection acts on the available variation after founding. Blair (1951) suggested that the predominance of light colored *P. polionotus* on Santa Rosa Island was the result of directional selection for the lighter color form in this light beach sand habitat.

Numerous workers (Cameron, 1958; Corbet, 1961, 1964; Berry, 1964; Foster, 1965; Youngman, 1967; Berry, 1969a; Handford and Pernetta, 1974; Heaney, 1978) have reported that insular populations of small mammals tend to be larger and darker than their mainland relatives, while large mammals have generally smaller populations (Foster, 1965). Heaney (1978) has suggested that predation, competition, food limitation, and physiological efficiency are the major factors which may affect body size of insular mammals. Increased competition is associated with decreased mean body size (Schoener, 1974; Brown, 1975b), while McNab (1971) suggested that decreased competition may in part be responsible for increases in body size among squirrels and weasels of high latitudes. Grant (1965), however, has suggested that small mammals on islands may have evolved to a larger size in order to utilize a wider variety of food when their competitors have become extinct and food limitations become important.

Several electrophoretically detectable loci in murid rodents demonstrate variation which has apparently been affected by natural selection, acting on traits determined by the loci themselves or closely linked genes. The best evidence relates to the hemoglobin beta locus in Mus where genotypic frequencies have been observed to change with season and social status (Berry and Murphy, 1970; Myers, 1974; Berry and Jakobson, 1975a), and with age (Berry and Peters, 1975). In addition, this locus has been shown to respond to selection for size under laboratory conditions (Garnett and Falconer, 1975). Natural selection has been suggested to be the mechanism responsible for changes in genotypic frequencies at two additional loci (esterase-2 and dipeptidase-1) in Mus (Berry and Jakobson, 1975b) and for excess of heterozygotes at the isocitrate dehydrogenase-2 locus of Rattus (Schmitt, 1977).

Several workers (Soule and Yang, 1973; Gorman et al., 1975) have suggested that genetic variation within insular populations is reduced as a consequence of directional selection, while other workers (Berry and Murphy, 1970; Berry and Peters, 1977) have concluded that natural selection must have acted to increase the amount of genetic variation present in some insular populations. Relaxation of stabilizing selection alone is not capable of producing an increase in phenotypic variance among sexually reproducing organisms (Slatkin, 1970; Roughgarden, 1972). Endocyclic selection (opposing selective pressures at different stages of the life cycle), however, has been suggested to account for genetic differences between age classes of three insular populations of Mus (Bellamy et al., 1973; Berry and Jakobson, 1975a; Berry and Peters, 1975; Berry et al., 1978a).

Berry and Peters (1977) have suggested that the difference between the genetic distances (0.013, 0.029, 0.032) observed between samples collected in different years for house mice from Skokholm represent the magnitude of differentiation attributable to selection (Berry and Jakobson, 1975b; Berry and Peters, 1976). In addition, the high genetic similarity between the populations of *Mus* on the subantarctic Marion Island and of the North Atlantic Faroe Islands has been identified by Berry et al. (1978a) as illustrative of the importance of ancestry rather than contemporary selective pressures in determining the genetic composition of insular populations (Berry and Warwick, 1974; Berry, 1975).

Immigration and Gene Flow

Although founder effect may be the major force in shaping the initial gene pool of an insular population, many insular populations may receive additional genetic information from mainland populations and/or other insular populations in the genomes of immigrants which may further alter the insular gene pool. Alteration of the insular gene pool by this mechanism requires not only immigration of individuals onto the island but also incorporation of a portion of the immigrant genome into the insular gene pool. Small terrestrial mammals are capable of immigration onto continental shelf islands by swimming (Sheppe, 1965; Crowell, 1973), rafting (McCabe and Cowan, 1945; Crowell, 1973), and dispersal over ice bridges (Crowell, 1973). However, there is evidence (Anderson, 1965, 1970; Reimer and Petras, 1967, 1968) that the population structure of some rodents make it very difficult for immigrants to establish a territory, although sexually mature dominant male immigrants may be successful in establishment of territories even in high density populations (Adamczyk and Walkowa, 1971).

Little is known concerning the genetic aspects of dispersal. Differences in exploratory behavior or dispersal have been shown to be correlated with genomes (Myers and Krebs, 1971a; Pickering et al., 1974; Garten, 1974, 1976), but virtually nothing is known about differential establishment or survivorship of genomes.

Genetic Drift and Bottleneck

Genetic drift has been suggested as being a contributor to the reduction in genetic variability (Avise et al., 1974a; Nozawa et al., 1975), especially among the more distant, small insular populations. Supportive evidence for genetic drift is observed in fixation or predominance of alternate alleles on ecologically similar adjacent islands (Kilpatrick and Crowell, 1980a,

Berry and Peters, 1977; Aquadro, 1978; Schmitt, 1978). Wright's (1969) models of genetic drift predict that the frequency of fixation of an allele is dependent on the initial frequency distribution of that allele. Data for fixation of alternate alleles among insular populations of Rattus (Schmitt, 1978) and Mus (Berry and Peters, 1977) were compared with the predictions of Wright's (1969) models of drift assuming that the mainland allelic frequency is representative of the initial frequency distribution. Several loci could not be tested because the mainland populations were not polymorphic and hence no initial frequency distribution could be calculated. Comparison of observed with expected as predicted by Wright's (1969) models demonstrated little deviation for the LDH-2 and ME-1 loci of Rattus (Schmitt, 1978) and the IDH-1, MOR-1, HBB, and ES-2 loci of Mus (Berry and Peters, 1977). However, three loci (GPI-1, GOT-2, ES-3) of Mus (Berry and Peters, 1977) appear to demonstrate significant deviation although the number of observations is too small to test by Chi-square analysis.

Two insular populations of voles from Mark and Rock Islands in Penobscot Bay analyzed in 1974 demonstrated a 6-Pgd^{122} allele at frequencies of 0.143 and 0.133, respectively (Kilpatrick and Crowell, 1980a). In subsequent samples collected from these islands in 1975, the 6-Pgd^{122} allele was not observed in the Rock Island population. While no conclusive evidence is available to explain the loss of this allele from this population and genetic drift is certainly a viable explanation, selection is also viable especially since this allele has not beeen observed in any other vole populations.

Genetic drift resulting from a bottleneck has been used to explain the low genetic variability of elephant seals (Bonnell and Selander, 1974). Evidence of bottleneck within the past 15 years in three insular populations of M. pennsylvanicus (Mark Island, Rock Island, and Second Island) is apparent from the estimates of population densities published by Crowell (1973). In spite of population size of less than 23 individuals, these three insular populations do not demonstrate lower estimates of genetic variability (Table 6) compared to insular populations without such known reduction in population size (Kilpatrick and Crowell, 1980a). Second Island vole populations appear to have become extinct twice within the past 12 years (Crowell, 1973), but this population demonstrated similar levels of genetic variability (Table 6) observed in other insular vole populations. The level of genetic variability along with the observed reestablishment of a Microtus population on Second Island (Crowell, 1973) suggests that the rate of immigration among the insular populations of voles in Penobscot Bay is great enough to negate the consequences of genetic drift during bottlenecks.

While fixation of alternate alleles among insular populations in accordance with the predictions of Wright's (1969) models is supportive of genetic drift affecting insular gene pools, much, if not most, of this effect may have occurred when the insular population was founded. The similarity of genetic variability and lack of genetic differentiation between recently founded populations (Second Island) and older insular populations of voles (Kilpatrick and Crowell, 1980a) supports the conclusion by Berry et al. (1978b) that founder effect is responsible for most of the genetic differentiation of insular populations. Berry et al. (1978b) found that insular populations of Mus which had been in existence less than 40 years were as distinct as insular populations of Mus founded nearly 200 years ago and concluded that the divergence of the Faroe Island populations from British populations was apparently random.

CONCLUSIONS

Insular or isolated populations have been characterized by reduced levels of genetic variability. Significantly lower levels of genetic heterozygosity have been observed between insular and mainland populations of vertebrates but not between insular and mainland populations of invertebrates (Nevo, 1978). Most insular populations of mammals demonstrate lower levels of genetic heterozygosity than their mainland relatives. This reduction in genetic variability primarily results from an increased proportion of monomorphic loci, absence of many rare or low frequency mainland alleles, and increased frequency of common mainland alleles.

Founder effect seems to be the major evolutionary force responsible for structuring the gene pool of insular populations. This evolutionary force not only has a direct effect upon the gene pool but may also indirectly affect subsequent adaptation and genetic drift. Evidence for founder effect is observed in the increased proportion of monomorphic loci among insular populations, each usually being fixed for the more common mainland allele. Other loci demonstrate drastic differences in allelic frequencies between insular and mainland populations, usually with an increase in the frequency of the most common mainland allele among insular populations.

Significant correlation between standardized reduction in genetic heterozygosity and mean genetic distance between mainland and insular populations indicates that the forces responsible for the reduction in genetic variability are also responsible for the genetic differentiation between mainland and insular populations. The lack of greater genetic differentiation between older insular populations and recently founded insular populations indicates founder effect is the major force in structuring insular gene pools (Berry et al., 1978b). However, Thomas (1973) has demonstrated karyotypic differences between insular populations and suggested that some of these differences have resulted from chromosomal mutations within insular populations.

The effects of changes of selective constraints on insular gene pools are suggested by the increased variation at certain loci and the occurrence of unique alleles among certain insular populations. Changes in selective pressures may lead to increased variation at some loci by the predictions of the mutation equilibrium theory or by endocyclic selection (Berry et al., 1978a), while directional or stabilizing selection may decrease variation at other loci. Data collected by repeated sampling of the same population (Berry and Murphy, 1970; Berry and Jakobson, 1975a) indicate that, although selection is a factor in the structuring of isolated gene pools, changes produced are neither large nor continuous (Berry and Peters, 1976). The role of selection in structuring insular gene pools has been described by Berry (1978) as the expectant virgin syndrome, based upon the work of Elton (1927) who pointed out that animals are not always struggling for existence. Natural selection clearly plays a role in the structuring of insular gene pools; the major problem is determining the relative importance of selection in contrast to other evolutionary forces in determining the evolutionary outcome for an insular population.

Insular populations of mammals appear to be affected by gene flow in several ways. Increased differentiation of "slow" evolving loci among insular populations in comparison to conspecific mainland populations appears to be a result of reduced gene flow. In addition, a relationship between gene flow and genetic variability is suggested from the correlation between genetic variability and distance squared from the mainland. However, the major question is again the relative importance of gene flow to the evolution of the insular population. The lack of a significant correlation of biogeographic variables with standardized estimates of genetic variability from four species of rodents indicates that differences in the relative importance of gene flow and fixation exist among species.

Genetic drift appears to occur in many insular populations, although effects are especially obvious in populations of small islands. Evidence for genetic drift is strongly suggested by fixation or predominance of alternate alleles in adjacent insular populations. Most data which suggest genetic drift as an evolutionary force can also be explained by founder effect. Again, we are left with the question of the relative importance of these evolutionary forces in structuring an insular gene pool. Species which appear to be good colonizers demonstrate higher correlations between estimates of genetic variation and log of the area of the island than other biogeographic variables. This suggests that the gene pools of species which inhabit smaller and more distant islands, hence having reduced levels of immigration, are affected to a greater extent by loss of alleles through fixation than the gene pools of species which are restricted to larger islands relatively near a colonizing source.

All evolutionary forces play a role in structuring insular gene pools. While accumulated evidence suggests founder effect as the major evolutionary force responsible for the reduction of genetic variation and differentiation of insular populations, additional data are needed to substantiate the relative importance of founder effect and to determine the relative importance of other evolutionary forces. Manipulation of insular populations by simulating founding and immigration with individuals with partially known genomes and repeated sampling of these populations prior and subsequent to manipulations should allow determination of the relative importance of founder effect, genetic drift, gene flow, and selection in determining the genetic structure of insular populations.

ENVIRONMENTAL AND MORPHOLOGICAL CORRELATES OF GENETIC VARIATION IN MAMMALS

Gary D. Schnell and Robert K. Selander

Abstract--Intensive research on protein polymorphisms over the past decade has greatly enhanced our understanding of the genetic structure of populations of mammals and other organisms, but has yielded little understanding of the mechanisms by which the polymorphisms are maintained in natural populations. Much additional information will be required to evaluate competing hypotheses and to erect a better theoretical framework within which more precise and testable predictions can be made. Environmental correlates of variation at structural gene loci within species of mammals have been difficult to demonstrate and even harder to interpret. The "niche-width hypothesis," which postulates an adaptively-based association between the amount of genic variation in a population and the extent of temporal and spatial diversity in its environment, derives no support from mammalian studies. Rodents occupying a "monotonous subterranean niche" are not less heterozygous than other rodents, and purported examples of a positive relationship between level of genic variability in populations and extent of temporal variation in climate prove to be spurious. Apart from the singular case of sickle-cell hemoglobin in man, adaptive relationships between geographic variation in allele frequencies at structural gene loci and environmental factors have not been convincingly demonstrated in mammals. Studies of insular populations of rodents in which genic variability is severely reduced and of hybridizing populations of the house mouse suggest that there is no relationship between genic variability and morphological variability within populations. Mammalian studies provide considerable support for the theory that the processes underlying evolutionary divergence in structural genes, chromosomes, and morphology are essentially independent. Rates of chromosomal evolution and speciation may be related because both processes are accelerated when effective population size is small.

Environmental and Morphological Correlates

INTRODUCTION

The considerable difficulty evolutionary biologists have experienced in defining significant questions and generating testable hypotheses relative to molecular polymorphism and evolution has emphasized the inadequacies of our current theory of evolutionary genetics (Lewontin, 1974). It is apparent that the origin and dynamics of genetic variation in populations would not constitute a complete theory, even if the major deficiencies in our understanding of multilocus systems were corrected. For example, speciation, which may involve more than simple extensions of genetic processes occurring within species, and extinction are dealt with by present theory on only a most general and nonrigorous level (White, 1978).

The principles of population genetics obviously are expressed in terms of genotypic variables. But phenotypic variables also would be essential components of a sufficient and adequate overall theoretical framework because natural selection acts on the genotype through the phenotype; and for the most part we are able to work only with phenotypes. Although the interrelationships between genotype and phenotype at various levels of biological organization would be predictable from what Lewontin (1974) calls a dynamically sufficient theory (i.e., one containing all elements required to define the system under study), it would still remain to be demonstrated that these interactions can be measured with sufficient accuracy to evaluate the theory. In view of the unsatisfactory state of development of population genetics, it is not surprising to find little useful theory at the interface between populations genetics and other aspects of evolutionary biology.

Although we are attempting to analyze environmental and morphological correlates of genetic variation in mammals and other organisms in the context of an inadequate theoretical structure, our effort may be justified by the argument that the development of useful theory cannot proceed independently of considerations of the kinds of relationships that can be evaluated empirically with current or possible future techniques. When the critical tests of hypotheses involve variables that cannot be accurately measured, there is no progress. As P. B. Medawar reminds us, science is the art of the soluble.

A decade ago, when the Zoological Society of London sponsored a symposium on "Variation in Mammalian Populations" (Berry and Southern, 1970), population geneticists and mammalogists were just beginning to appreciate the magnitude of genic and karyotypic diversity in natural populations and the complexity of population structure being revealed by new techniques of studying variation. First efforts were being made to integrate genetics and ecology in an attempt to grasp the meaning of this extensive variation; and it was commonplace for evolutionary biologists to invoke adaptively based relationships between genetic variation and environmental heterogeneity. For example, Manwell and Baker (1970) took for granted a "general correlation between diversity of proteins and diversity of habitats," and Berry (1970) spoke of the "use" of this type of variation by populations of house mice (Mus musculus) to allow fine adjustment to changes in the environment. The first response was to interpret this newly found variation in conventional terms, applying existing evolutionary theory that had been generated largely from studies of morphology and physiology. Only later was the concept of selective neutrality of allozymic (electrophoretic) variation taken seriously by evolutionary biologists, who generally believed or hoped that the new variation would prove to be demonstrably of adaptive significance and maintained by various forms of balancing selection. Some still hold this view, but the current climate of opinion is different. We at least now realize the complexity of the problem of accounting for variation in allele frequencies and heterozygosity, to say nothing of the occurrence of the polymorphisms themselves. Having acquired a greater appreciation of stochastic processes, we have begun to question the dogma that all genetic variation and evolution in populations can be explained in terms of natural selection.

Studies of the genetics of natural populations of mammals other than man only recently have begun to make substantial contributions to evolutionary theory. Sumner's (1930) pioneering attempt to understand morphological variation in Peromyscus in genetic terms (continued by Dice, 1940; Blair, 1947, 1950; and other students) had only limited success largely because the inheritance of size and coat color proved to be complex; there were complicating environmental effects on the characters of laboratory populations. Much of this work recently has been critically evaluated by Wright (1978).

In this review we provide an indication of what is known about environmental and morphological correlates of genetic variation in natural populations of mammals and discuss reasons why associations are difficult to demonstrate and interpret. Rather than attempting to cover the whole field, we have emphasized a selected group of current problems for which studies of mammals may provide answers. Environmental correlates of genetic variation are considered first, followed by a discussion of genetic and morphological associations.

Since our knowledge of genic variation in natural populations of mammals is based almost entirely on studies of electrophoretically demonstrable variation in proteins, the term "genic variation," as used in this review, refers to that at structural gene loci as measured by frequencies of electromorphs (King and Ohta, 1975).

ENVIRONMENTAL CORRELATES OF GENETIC VARIATION

The Meaning of Protein Polymorphism

Despite a decade of intensive study of protein polymorphisms, the mechanisms of their maintenance in natural populations remain controversial (Lewontin, 1974; Nei, 1975); the full extent of allelic diversity has not been determined for any locus (Bonhomme and Selander, 1978). If most electrophoretically demonstrable polymorphisms are not selectively neutral (Kimura and Ohta, 1971; Kimura, 1977), they may be maintained by a balance between recurrent mutation and adverse selection of the order of the mutation rate (Ohta, 1976) or by weak balancing selection (frequency dependent or other, but probably not heterosis; Lewontin et al., 1978). That some pairs of alleles at particular loci may be maintained by strong selection pressures does not invalidate this generalization (Wright, 1978).

Apart from the sickle-cell hemoglobin polymorphism in humans, a selective basis for maintenance of polymorphic variants at structural gene loci has not been conclusively demonstrated in any mammal, notwithstanding claims to the contrary. For example, Berry and Peters (1977) maintain that "there are a number of electrophoretically detected loci in [house] mice whose variation is unequivocally affected by natural selection, acting on traits determined by the locus itself or closely linked genes." Elsewhere, Berry (1977b) claims that "a myth has grown up about such data [i.e., allozymes], that natural selection does not affect house mouse populations. However, we have seen that under certain conditions, selection may act very strongly on Hbb, and the assumption must therefore be changed. The proper conclusion from the existing data on HBB (and some other loci) is that selective pressures are inconstant." Selection coefficients obviously may be expected to vary spatially and temporally (Gillespie, 1975; Karlin and Levikson, 1974), but the critical question is whether the available data prove that selection maintains the polymorphisms at the HBB and other structural gene loci. As evidence that the HBB locus in the house mouse has "been shown to respond to selection for size under laboratory conditions," Berry and Peters (1977) cite the experiment of Garnett and Falconer (1975), in which the Hbb^s allele became fixed in six "large" lines of a laboratory population divergently selected for body weight. By their nature, experiments of this type cannot provide unequivocal evidence of the action of natural selection on particular loci. Examining possible causes of the fixation of the Hbb^s allele, Garnett (1976) excluded pleiotropy since there was a nonsignificant effect of hemoglobin genotype on body weight. Results of his work suggest linkage as the most plausible explanation for the fixation of the Hbb^s allele in the six "large" lines.

Genic Variability and Niche Width

Theoretical population geneticists have generated many models predicting stable polymorphism due to environmental heterogeneity (Hedrick et al., 1976), but few polymorphisms have been studied in the laboratory or other situations where there is any possibility of determining the type and strength of selection pressures involved. Consequently, the extent to which polymorphic loci are maintained by environmental heterogeneity is unknown.

A pervasive theme of ecological genetics is that there is an important, adaptively based correlation between the genetic variability of populations and the temporal or spatial diversity of the environment (see critical discussion in Soule, 1976). This intuitively appealing notion has been expressed in the niche-width variation hypothesis or the environmental amplitude theory. A related concept is the homeostasis theory (after Levins, 1968), which attempts to account for variation in level of heterozygosity in terms of environmental grain, i.e., the degree to which the environment is effectively variable with respect to individuals of a population, depending on their body size, mobility, and physiological homeostasis (Selander and Kaufman, 1973a). Finally, there is the persistent notion that variation is beneficial for the individual, the population, and the species, if not proximally then ultimately in terms of potential for evolutionary adaptation to changing conditions of the environment.

The environmental amplitude-genetic variation theme is variously expressed in mammalian population genetics. For example, Nevo and Shaw (1972) argue that fossorial mammals have relatively low levels of genic heterozygosity as a consequence of selectively mediated responses to a monotonous subterranean niche. It has also been argued that reduced variation in insular populations reflects the circumstance that they "face ecologically less variable, but stringent environments" (Nevo et al., 1974). This theory derived from Nevo and Shaw's (1972) study of allozyme variation in the mole rat *Spalax*, which yielded a relatively low estimate of individual heterozygosity (H), but they did not assay esterases which contribute, on the average, 43% of the overall heterozygosity in rodents (Selander, 1976). Although an analysis of available data by Selander et al. (1974) showed no statistically significant difference in mean level of heterozygosity between fossorial and nonfossorial rodents, this notion has persisted, again being advanced for the pocket gopher *Thomomys talpoides* (Nevo et al., 1974). Although a second statistical analysis again failed to support the idea (Selander, 1976), it continues to be invoked (Penny and Zimmerman, 1976; Nevo, 1978). Perhaps the recent demonstration of a relatively high level of heterozygosity in the pocket gopher *Thomomys bottae* (Patton and Yang, 1977) will be effective in laying the hypothesis to rest.

Genic Variability and Temporal Variation in Environments

For a variety of organisms, including three rodents, Bryant (1974a) claimed to have demonstrated that approximately 70% of the geographic variation in heterozygosity at structural gene loci can be accounted for by four measures of within-year environmental variability (SDT, standard deviation of mean monthly temperature; CVP, coefficient of variation of mean monthly precipitation; RT, mean daily temperature range; and, CVRT, coefficient of variation of mean daily temperature range). Referring to Levins' (1968) theory of fitness in heterogeneous environments, he concluded that geographic variation in heterozygosity reflects adaptive responses to corresponding changes in variability of specific components of the environment. Here obstensibly was verification of the niche-width variation hypothesis. However, as Bryant (1974b) acknowledged, the conclusions are undermined by inadequacies in the statistical analysis, involving the inherent unreliability of estimates of regression coefficients when predictor variables are highly intercorrelated and the chance inflation of multiple correlation coefficients by small sample sizes. Bryant's (1974a) analysis has been widely cited as supporting Levins' (1968) hypothesis. Since the hypothesis being tested explicitly predicts a positive association between genic and environmental variability, it is important to determine how closely the findings correspond with the expected results.

The data for heterozygosity in *M. musculus* analyzed by Bryant (1974a) involved only four loci in samples from nine localities in the United States. In this and other analyses, Bryant conducted a modified principal component analysis to extract heterozygosity factors, which in turn were regressed in a step-wise procedure on the four measures of environmental variability. For *M. musculus*, 81.6% of the variation in one heterozygosity factor (indexing H for three of the loci) was explained by CVP. This association contrasted the more northerly localities, having lower heterozygosities and lower CVPs with those in southern areas. However, as pointed out by Selander et al. (1969b), these differences are most likely a consequence of the fact that northern and southern populations represent introductions of different stocks (subspecies) from Europe. Also, the CVP measure does not take into account north-south differences in seasonal variation in the physical form of precipitation. Because of these problems and the fact that the species was only relatively recently introduced to North America, we believe that Bryant's (1974a) analysis and interpretation are unconvincing.

For the old-field mouse (*Peromyscus polionotus*) heterozygosity decreases clinally northward from the Florida peninsula (Selander et al., 1971). Bryant (1974a), using data for 12 loci in samples from 12 unspecified localities, found that almost all variation in his heterozygosity Factor I could be explained by SDT (94.2%) and CVRT (3.4%). But inspection of this factor in Bryant's Table 2 indicates that, while four loci have strong positive loadings, two are negatively associated. Hence, as variability in temperature increases, heterozygosity increases at four loci, but it decreases at two loci, hardly support for Levins' (1968) hypothesis.

In the case of the cotton rat *Sigmodon hispidus* (data from Johnson et al., 1972), one automatically becomes suspicious when 99.7% of the total variance in a factor is explained by two environmental parameters (CVRT and CVP). These data involve only five localities, and it clearly is statistically inappropriate to incorporate four dependent variables in the analysis. Furthermore, five of the loci with high positive loadings on the heterozygosity factor were polymorphic at only one locality (a Florida island), while a sixth was variable there and at only one other place. A seventh locus, with a negative loading, was not polymorphic at the Florida locality but was at the other four locations. Even if the analysis were statistically sound, the finding that six of the loci are more heterozygous in Florida (the environmentally most uniform region based on the parameters analyzed by Bryant), while one locus shows the opposite trend, would not support the hypothesis.

In sum, Bryant's (1974a,b) findings for mammals are unreliable and cannot be used to support hypotheses concerning associations between genetic and environmental variability. Furthermore, his results for invertebrates should also be reevaluated since the same types of analyses are involved.

Geographic Variation in Genic Heterozygosity

The most striking case of geographic variation in genic heterozygosity in mammals is provided by *P. polionotus* in the southeastern United States (Selander et al., 1971). On the mainland, mean heterozygosity increases clinally from north to south, but whether this pattern represents an adaptive response to environmental variation is problematical because information on niche-width and habitat distribution is not sufficient to analyze the situation. Populations occupying small barrier islands and peninsulas on the Gulf coast of western Florida are only one-fifth as variable as those on the adjacent mainland. Because the beach environment is relatively uniform and seasonally stable, ecological explanations for this reduced variation might be invoked (Nevo et al., 1974), but the ecological amplitude hypothesis was rejected by Selander et al. (1971) for several reasons: (1) Beach-inhabiting populations on the Atlantic coast of Florida, although similarly constrained in habitat distribution, do not manifest reduced levels of genetic variation in comparison with the neighboring mainland populations. (2) On the western beaches, the more strongly isolated Santa Rosa Island population is less heterozygous than are the nearby peninsular populations occupying essentially identical sand-dune habitats but in firmer genetic contact with populations on the mainland. (3) The beach populations are strongly heterogeneous in allele frequencies, despite the ecological similarity of the various beaches. The most probable cause of reduced variation in the western beach populations is sampling drift. Reasons for the retention of variability in the eastern beach populations should be sought in terms of differences in degree of isolation, population size and structure, and other aspects of historical demography.

Severely reduced levels of genic variation have been reported in populations of other rodents occupying small islands, including *M. musculus* (Hunt and Selander, 1973), *Peromyscus* spp. (Avise et al., 1974a), *Rattus fuscipes* (Schmitt, 1978), and *Spermophilus spilosoma* (Cothran et al., 1977). Reductions in all cases may reasonably be attributed to sampling drift, including founder effects.

Smith et al. (1975; see also Garten, 1976) have noted various geographic correlations between reproductive effort, behavior, morphological characters, and genic heterozygosity in P. polionotus, which "point to the importance of genic heterozygosity in population regulation of this species." However, causal relationships remain to be demonstrated.

Convincing evidence is provided by recent studies of parthenogenetic and strongly selfing organisms that extensive genetic variability is not requisite for a broad niche and wide geographic distribution. For example, some clones of the ubiquitous parthenogenetic earthworm Octolasion tyrtaeum are as broad-niched (with reference to soil pH and composition) and as widely distributed as sexual species of earthworms (Jaenike et al., 1979). Similarly, a monogenic strain of the selfing land snail Rumina decollata has a "general purpose genotype" conferring sufficient phenotypic flexibility to permit it to colonize much of North America (Selander and Hudson, 1976). There also is some supporting evidence from mammals. For example, the exceptional colonizing ability of the black rat (Rattus rattus) apparently is not associated with unusually high levels of genic or chromosomal variation (Patton and Myers, 1974; Patton et al., 1975).

In sum, a relationship between genic diversity (indexed electrophoretically) within populations and ecological amplitude or niche-width has not been established empirically. If current theory is valid, our inability to demonstrate correlations between diversity at structural gene loci and environmental heterogeneity may mean that we are not measuring genetic variation of the right kind and/or that existing relationships are too weak to measure, given the magnitude of "noise" generated by historical factors, genetic drift, and sampling error. Alternatively, it may be that most geographic and interspecific variation in heterozygosity at structural gene loci can be attributed to factors relating to the historical demography of populations.

Geographic Patterning of Allele Frequencies in Relation to Environmental Variation

For geographic variation in allele frequencies at individual structural gene loci, adaptive relationships to environment have not been established in mammals other than man. Actually, few studies have provided data sufficiently extensive to permit statistical analysis of relationships; but there are other difficulties relating to interpretation. The demonstration of a correlation between the frequency of an allele and an environmental factor or complex of factors does not constitute conclusive evidence of an adaptive relationship. Even if historical demographic influences could be eliminated as a cause of the correlation, an adaptive relationship involving the locus in question could not be inferred because genes occur in linked groups. Additionally, the highly subdivided population structure of many mammals introduces strong stochastic elements into geographic patterns of variation that may obscure adaptive trends in relation to the environment. Another complicating factor is that interpopulation variation in genetic characters may be determined not by direct response of allele or zygotic frequencies to external environmental factors but, rather, by interaction with different multifactorial genetic backgrounds, the development of which may be caused by sampling drift affecting many loci simultaneously. Wright (1978) has long maintained that this is the most important evolutionary consequence of sampling drift.

Finally, we note that frequencies of alleles at selectively affected loci in natural populations of mammals and other organisms may in general not be at equilibrium because of secular variation in habitat structure or other aspects of the environment, or because the populations are not old enough for equilibrium conditions to have been attained. Fincham (1972) has suggested that populations of most terrestrial organisms never reach genetic equilibrium.

The difficulties involved in interpreting correlations are illustrated by the work of Selander et al. (1969) on geographic variation at four loci in *M. musculus* in Texas. These workers had little success in attempting to relate geographic patterns of variation in allele frequencies to patterns of variation in particular climatic factors, independent of geographic effects. Employing data for 19 regions, they performed unweighted multiple regression analyses with the mean regional frequency of a particular allele as the dependent variable and altitude, latitude, longitude, annual precipitation, mean January temperature, mean July temperature, number of days in January with minimum temperature lower than 0 C, and number of days in July with maximum temperature higher than 32 C as independent variables. For the ES-2 locus, with the frequency of the $Es\text{-}2^b$ allele as the dependent variable, an analysis of variance yielded a significant F value of 4.70 and an R^2 value of 0.79. In step-wise regression analysis, only latitude was significant, and the correlation between the frequency of $Es\text{-}2^b$ and latitude was 0.50. Hence, the fact that the frequency of $Es\text{-}2^b$ was also correlated with mean temperature in January ($r = -0.47$) and with number of days in January with freezing temperature ($r = 0.37$) may reflect merely the strong correlations of these climatic variables with latitude.

In a similar analysis of the ES-5 locus, a significant regression was obtained only for annual precipitation. The correlation between the frequency of the $Es\text{-}5^a$ allele and annual precipitation was -0.54. However, annual precipitation is so closely correlated with longitude in Texas ($r = -0.95$) that Selander et al. (1969) could not conclude that annual precipitation per se influences the frequency of $Es\text{-}5^a$. For $Es\text{-}3^b$ and Hbb^s, no significant regressions were obtained.

Clines in Zones of Hybridization

Hybrid zones provide special opportunities to study relationships between genetic and environmental variation. For a zone involving *M. m. musculus* and *M. m. domesticus* on the Jutland peninsula of Denmark, Hunt and Selander's (1973) analysis of allele frequencies at seven polymorphic enzyme loci showed that the geographic position was the same as that defined earlier by Ursin (1952) on the basis of variation in morphological characters. The zone is asymmetrical, a marked increase in width in western Jutland (Fig. 1B) being associated with a more gradual geographic gradient in precipitation (Fig. 1A). Geographic variation in a hybrid index based on six loci is significantly associated with yearly precipitation (Hunt and Selander, 1973; Schnell et al., 1980), but there is considerable interlocus variation in geographic pattern, some alleles introgressing farther than others into the parental populations. ES-3, while showing some regional as well as local patterning with respect to interlocality distance, has an overall geographic pattern unrelated to that of the hybrid index or of any environmental factors tested (altitude, mean monthly temperature, and yearly precipitation). Variation at the MDH-2 locus is correlated with that of the

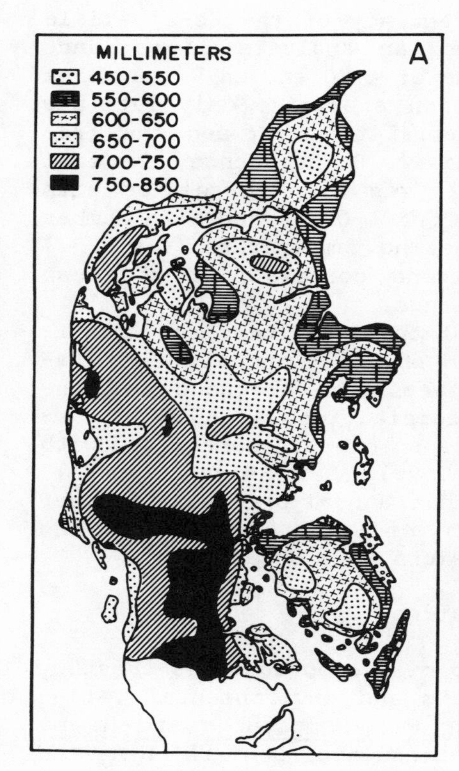

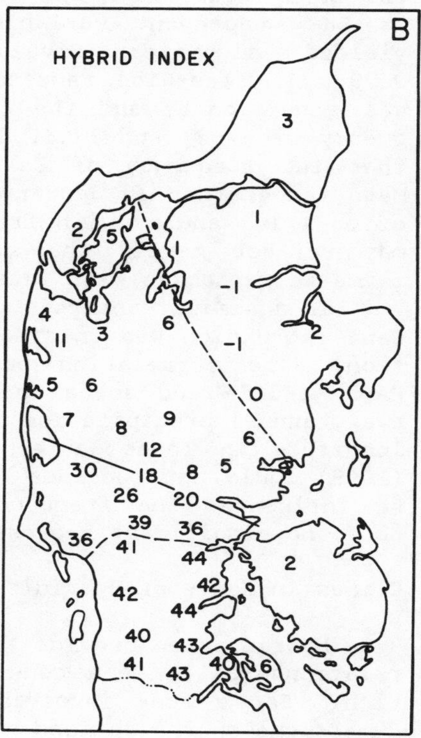

Fig. 1. (A) Average total yearly precipitation on Jutland peninsula, Denmark (Danske Meteorol. Inst., 1933; Davies, 1944). (B) Geographic variation in hybrid index score (x 10), based on allele frequencies at six polymorphic loci, in populations of *Mus musculus musculus* (northern form) and *M. m. domesticus* (southern form) studied by Hunt and Selander (1973). Dashed lines in B indicate limits of introgression; and solid line locates center of hybrid zone.

hybrid index, but its geographic pattern is markedly different from those of other loci in that it shows some relationship to altitudinal variation in Jutland (Fig. 2), even though the range in altitude for the 41 localities sampled is less than 100 m. Because multiple testing was involved in this analysis, the demonstrated

Environmental and Morphological Correlates

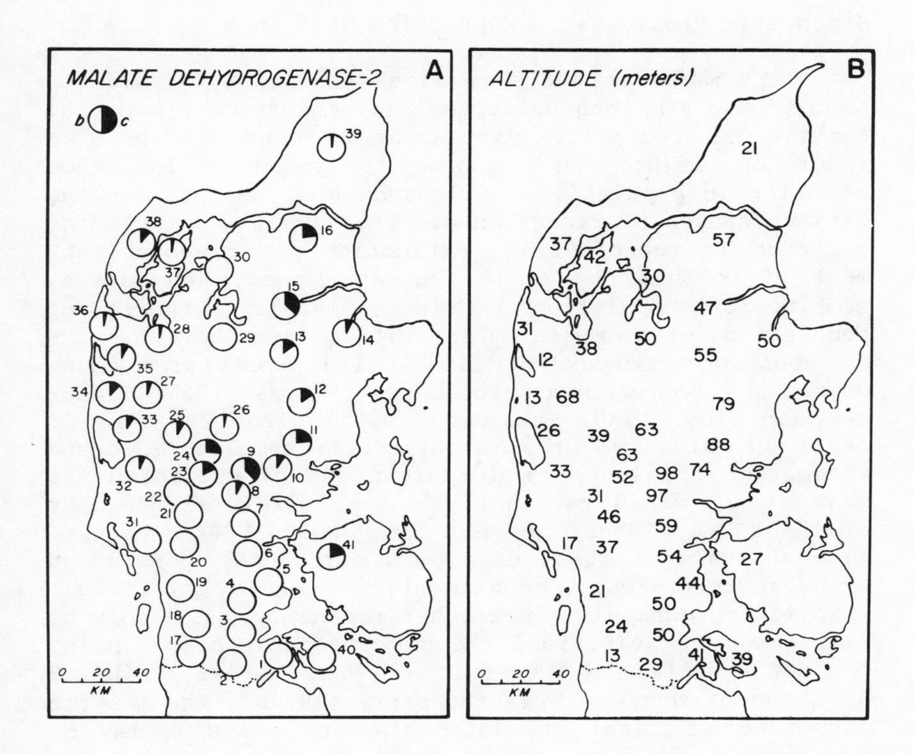

Fig. 2. (A) Variation in frequencies of alleles (a and c) at MDH-2 locus in zone of hybridization between populations of Mus musculus on Jutland peninsula. Numbers refer to sample localities. (B) Altitudes of localities at which populations were sampled.

association could be a statistical artifact, but, alternatively, MDH-2 may represent a marker indirectly reflecting selectively mediated differences between upland and lowland Mus populations. In sum, while there is a general correspondence between genetic and environmental variation in Jutland populations of house mice, patterns of geographic variation for individual loci are not identical. In this case, where on historical grounds alone one might expect close genetic-environmental correspondence, the interrelations are complex.

Stochastic Processes in Population Structure

Many mammals have deme sizes sufficiently small to permit wide sampling drift and the generation of spatial genetic heterogeneity over areas of considerable size (Anderson, 1970). In assessing the potential influence of drift in populations, allowance must be made for the various causes of reduction of effective size, including variance in reproduction, extinction and re-establishment of colonies by small founder groups, and bottlenecking of populations (Wright, 1978). Studies of populations of *M. musculus* inhabiting farms have provided abundant evidence of fine-scaled subdivision among barns, and even within large barns (Petras, 1967; Selander and Yang, 1969; Selander, 1970a). For 22 farms in Texas on which two or more barns were sampled (73 barns altogether), Selander and Kaufman (1975) estimated the mean standardized variance (F_{ST}) of allele frequencies at the EST-2, EST-3, and HBB loci as 0.025 (see comparable analysis of these data by Wright, 1978). As noted by Selander (1970a), Lewontin (1974), and Wright (1978), this microgeographic genetic heterogeneity can hardly be attributed to selection, the most probable causes being founder effects and inbreeding in territorial "tribes." As shown by Myers (1974) and Berry (1977b), the genetic structure of feral populations of *M. musculus* may be drastically different from that of commensal populations.

Evidence of microgeographic subdivision and differentiation of house mouse populations was cited by Ehrlich and Raven (1969) in support of their argument that gene flow within species is not extensive and therefore cannot have the cohesive influence often assumed for it (Mayr, 1963). But their reasoning is fallacious. Recent experimental work by Lidicker (1976) demonstrated cohesive social grouping in house mice and suggests a relatively rapid rate of gene flow, thus supporting the earlier interpretations of Anderson (1970) and Selander (1970b). As noted by Selander (1970b: 88), "mouse populations may experience repeated cycles involving the finding of a food supply by a few mice, establishment of one or more founding tribes, rapid proliferation of tribes from the founding stock, inbreeding and drift within tribes, and the eventual depletion of the food supply, leading to dispersal of tribe members and the initiation of a new cycle. Also to be considered is the very real possibility that an active, long-distance dispersal of young may be occurring while the parental populations remain statically subdivided. For this reason, the demonstration of subdivision cannot be interpreted as evidence of an absence or low level of gene flow. It is possible that gene flow within and between populations is only slightly impeded by subdivision."

The effects of social behavior on genetic population structure also are illustrated by long-term studies of the Japanese macaque (*Macaca fuscata*; Nozawa et al., 1975), in which inbreeding clearly plays a major part in generating heterogeneity in allele frequencies among local troops, which individually may be almost monogenic. Mean heterozygosity was only 0.017 in the 18 troops studied. Population census data indicated that the average troop size is approximately 66 individuals; and the average effective size was estimated by Nozawa (1972) as about 20. On the assumption of neutrality of alleles segregating at structural gene loci, Nozawa and his associates employed a two-dimensional stepping-stone model of population structure to estimate the genetic migration rate between troops as less than 5% per generation. Troops separated by more than 100 km were essentially independent in their genetic character.

In a study of population structure of the Yanomama Indian tribe, of South America (Neel et al., 1977), blood from members of 47 villages was analyzed for variation in 15 biochemical genetic systems. Mean F_{ST} over villages was 0.073 which indicates considerable differentiation apparently resulting primarily from the manner in which new villages are formed and augmented subsequently by sampling drift. Because new villages arise by fissioning of preexisting villages, usually to some extent along lineal lines, new demes arise through a succession of small, nonrandom samplings of the "gene pool." Migration eventually would be expected to reduce the original differences, but village alliances (and so migration patterns) are subject to sudden and frequent change. The net result of the Yanomama population structure is that the approach to genetic equilibrium at the deme level is constantly disturbed, being only fleetingly, if ever, attained at any locus in a village. This model of population structure has obvious implications for the origin and rapid evolution of tribal and racial differences in humans.

Ward and Neel (1970) have demonstrated clines in allele frequencies at several blood-group loci in the Yanomama Indian population. These are believed to be primarily related to population fissioning and a centrifugal population expansion that has occurred within the past century.

When adequate historical and demographic information is available, the testing of random models can yield powerful insights into the causal bases of population structure. For example, genetic drift alone is sufficient to account for differentiation at eight polymorphic loci between Icelandic and Norwegian cattle (Kidd and Cavalli-Sforza, 1974). Results from the application of migration-matrix models to a variety of human populations have been interpreted by Bodmer and Cavalli-Sforza (1974) as evidence that random drift is the major cause of variance in allele frequencies in relatively restricted geographic areas.

Karyotypic Variation Within Species

We know of no serious attempt to relate chromosomal variation in populations of mammals to environmental amplitude (niche-width variation hypothesis), although professions of faith that geographic patterns of chromosomal variation are adaptively related to those of environmental factors are common enough (Lawlor, 1974; Wahrman et al., 1969; Berry and Baker, 1971; Pizzimenti, 1976; Yoshida et al., 1971). Considering the overwhelming evidence that inversion polymorphisms in *Drosophila* have adaptive function, this attitude may be understandable, but it should be realized that the Robertsonian rearrangements and the inversions studied by mammalogists are different from those occurring in *Drosophila*. Duffy (1972) has questioned the prevalent view that pericentric inversions are a primary source of the observed karyotypic variation in *Peromyscus*, and, by inference, in other mammals (see also Pathak et al., 1973; White, 1978). His studies of the length of the chromosome arms and the localization of constitutive heterochromatin suggest that the changes are primarily the result of the addition or deletion of heterochromatic portions of chromosomes. This mechanism allows for marked chromosomal variation with relatively few rearrangements within euchromatic portions of the genome.

Environmental and Morphological Correlates

The interpretation of patterns of karyotypic variation within and between species with respect to those of environmental factors presents problems similar to those discussed previously in connection with genic variation. Correlation may be mistaken for causal relationship. For example, four chromosomal forms (2N = 52, 54, 58, and 60) of mole rats (*Spalax ehrenbergi*) inhabit regions of Israel along a north-south ecological gradient, their distributions being correlated with the humidity index and climatic regime (Wahrman et al., 1969); and the positions of narrow parapatric zones of hybridization between adjacent pairs are associated with changes in the humidity index (Nevo and Bar-El, 1976). However, there is no evidence justifying the conclusion that the chromosomal variants themselves have anything directly to do with climatic adaptation (Lawlor, 1974). Whether a karyotype of 52 chromosomes (vs 60 chromosomes) per se is advantageous in a relatively humid region remains to be determined.

In an extensive and particularly thorough analysis of chromosomal variation in the pocket gopher *Thomomys bottae*, Patton and Yang (1977) found no obvious correlations between karyotype and single environmental parameters such as elevation, soil texture, soil depth, or habitat type, except in localized situations (Patton, 1970), where interpretation is difficult. Historical biogeographic events are believed to have had an overriding influence on the development of chromosomal diversity in the species. The high degree of concordance between geographic patterns of chromosomal and allozyme variation (Patton, 1972; Patton and Yang, 1977) also more probably reflects events of historical demography rather than adaptive relationship.

MORPHOLOGICAL CORRELATES OF GENETIC VARIATION

Morphological Variability and Genic Heterozygosity

Relationships between morphological and genetic variability have rarely been studied in natural populations of mammals or other organisms. From comparisons of coefficients of variation in meristic characters (scale counts) and average genic heterozygosity in insular populations of lizards, Soule et al. (1973) developed the concept of a "genetic-phenetic variation correlation," implying that measures of morphological variability provide useful estimates of the amount of underlying genetic variation over the whole genome in populations. Indeed, Soule and Yang (1973) suggested that the mean coefficient of variation in meristic characters is superior to genic heterozygosity as an estimator because a much larger fraction of the genome may be involved in the development of the morphological characters. Similar notions have been expressed by Berry and Jakobson (1975b) in relation to nonmetric (quasi-continuous) morphological characters of the mammalian skeleton, and by Thorpe (1976) regarding the use of multivariate analyses of quantitative characters in assessing population affinities.

The "genetic-phenetic variation correlation" was based on the observation that variability in the number of subdigital scales on the second toe of the hind foot was strongly correlated with mean heterozygosity at 21 or 22 loci in eight species of *Anolis* in the West Indies, and by the demonstration of a weak correlation between heterozygosity at 19 loci and the mean coefficient of variation for five meristic characters in 13 populations of *Uta stansburiana* in California and Mexico.

Variation in meristic characters may indeed reflect the level of genic heterozygosity. Parker (1978) found that scale-count characters are less variable in uniclonal populations of the parthenogenetic lizard *Cnemidophorus tesselatus* than in multiclonal populations and in those of sexual species of the genus. However, variability in morphometric traits (continuously varying characters) was not reduced in uniclonal populations of *Cnemidophorus*. It is common knowledge that morphological traits often are more variable in inbred strains of rodents (mice, guinea pigs) than in F_1 crosses or even randombred strains (Wright, 1977). Genotype-environment interactions are different in inbreds and randombreds, and canalization (the location of developmental thresholds) varies with inbreeding.

The recent discovery that the pocket gopher *Geomys tropicalis* is invariable, both genically (34 loci assayed electrophoretically in 30 individuals by Selander et al., 1974) and karyotypically (Davis et al., 1971), provides an opportunity to test the "genetic-phenetic variation correlation" hypothesis for morphometric characters. Williams and Genoways (1977) analyzed 3 external and 13 cranial characters in 94 specimens (separated by age and sex). In Table 1, coefficients of variation for five of these characters in *G. tropicalis* are compared with those for three other species of pocket gophers and a variety of mammals. With the possible exception of the interorbital constriction, *G. tropicalis* does not exhibit reduced variation in morphological characters.

Further evidence of a lack of relationship between genic heterozygosity and variability in morphological characters is provided by Wright's (1978) analysis of Sumner's (1930) data for *P. polionotus*, contrasting the Santa Rosa Island form *P. p. leucocephalus* with the adjacent mainland populations of the subspecies *P. p. albifrons* and *P. p. polionotus* (Table 2). Average genic heterozygosity over 32 loci was 0.052 on the mainland versus 0.019 on Santa Rosa Island (Selander et al., 1971). In the insular population, tail length and ear length may be slightly less variable, but body length is not; and three characters of coat color (Ab, Red, and Hue) are more variable. The fact that the mainland samples were taken over larger geographic areas than that from Santa Rosa Island may account for the greater variability in tail length and ear length.

As a final example of the apparent independence of morphological and genic variability, we have compared variation in *M. musculus* on the Jutland peninsula of Denmark (Table 3), contrasting unintrogressed populations of *M. m. musculus* and *M. m. domesticus* with those in the center of the zone of hybridization, where genic heterozygosity is greatly increased (Hunt and Selander, 1973). The five morphometric characters analyzed were chosen because they exhibit marked mean differences between *M. m. musculus* and *M. m. domesticus*; and, in addition, geographic patterns of variation in their mean values closely correspond to that of the hybrid index score based on allozymic data (Fig. 1B). Clearly, there is no increase in variation in morphometric characters in the hybrid zone.

In view of this abundance of evidence against the hypothesis of a "genetic-phenetic variation correlation," the recent report by Patton et al. (1975) of a significant correlation ($r = 0.77$; $P < 0.05$) between morphological variability (mean C.V. for 14 characters of skin and skull) and genic heterozygosity (37 loci) in

Table 1. Coefficients of variation for morphological characters in *Geomys tropicalis*, other pocket gophers, and mammals in general.[a]

Character	*G. tropicalis*		*G. personatus*		*G. arenarius*		*Thomomys talpoides*	Mean for Mammals
	M	F	M	F	M	F		
Total Length	5.0	4.3	6.1	5.2	11.1	5.4	4.5	5.31
Greatest Length of Skull	3.9	2.8	3.9	2.9	5.3	2.6	3.1	3.21
Zygomatic Breadth	5.3	3.8	4.5	3.0	6.2	4.9	5.9	3.95
Mastoid Breadth	3.8	3.4	4.0	3.9	5.6	3.2		3.05
Interorbital Constriction	2.6	2.9	9.5	3.8	5.3	5.3	5.8	4.36

[a] Data as summarized by Williams and Genoways (1977), who obtained some values from Long (1968, 1969). Males (M) and females (F).

Environmental and Morphological Correlates

Table 2. Mean ($\bar{X}$), standard deviation(s), and coefficient of variation (C.V.) in subspecies of Peromyscus polionotus. Data from Sumner (1930), analyzed by Wright (1978).

Character Statistic	Peromyscus polionotus leucocephalus (N = 72)	albifrons (N = 41)	polionotus (N = 46)
Body Length			
$\bar{X}$	80.0	81.0	80.9
s	2.72	2.40	2.77
C.V.	3.40	2.96	3.42
Tail Length			
$\bar{X}$	54.2	53.5	51.3
s	2.49	3.20	3.19
C.V.	4.59	5.98	6.22
Ear Length			
$\bar{X}$	14.5	14.7	15.0
s	0.48	0.54	0.68
C.V.	3.31	3.67	4.53
Colored Area of Hair:Base			
$\bar{X}$	45.5	66.5	100.0
s	3.81	3.05	
C.V.	8.37	4.59	
Red			
$\bar{X}$	25.4	17.2	9.6
s	3.53	1.37	1.06
C.V.	13.90	7.98	11.10
Hue			
$\bar{X}$	27.7	42.0	35.2
s	4.82	5.09	3.88
C.V.	17.43	12.11	11.03

Table 3. Genic heterozygosity and coefficients of variation for morphological characters in *Mus musculus* from Jutland.

Character[a]	*Mus m. musculus*	Hybrid Zone	*Mus m. domesticus*
Genic Heterozygosity			
Four Loci[b]	0.090	0.343	0.023
Seven Loci[c]	0.191	0.336	0.132
Morphological Variability			
Frontal Length	6.1	6.2	5.6
Ulna Length	5.0	4.5	4.9
Pelvic Length[d]	8.6	7.3	7.2
Tail Length	7.9	7.8	7.6
Hind Foot Length	3.8	4.2	4.3
Mean	6.2	6.0	5.9

[a]All measurements of morphological characters corrected to average age by regression on tooth-index. Values of genic heterozygosity and coefficients of variation are averages based on samples from the following localities (Fig. 2): *Mus m. musculus*, 12-15; hybrid zone, 8, 9, 22-25, 32, 33; *Mus m. domesticus*, 2, 3, 17, 18. Average sample size is 68 per locality.

[b]ES-1, ES-2, IDH-1, and MDH-2; loci showing marked transitions in allele frequencies in the zone of hybridization. Locus abbreviations follow Selander et al. (1971).

[c]Loci listed in footnote b and ES-3, ES-5, HBB.

[d]Distal to acetabulum.

populations of R. rattus on the Galapagos Islands warrants careful evaluation. The log of island area was correlated with both genic heterozygosity ($r = 0.73$; $P < 0.05$) and morphological variability (apparently calculated on a per-island basis; $r = 0.95$; $P < 0.001$). These associations led Patton et al. (1975) to conclude that "no two operationally independent estimates [$\bar{H}$ and $\overline{C.V.}$] accurately reflect the same basic parameter, i.e., overall genetic variability." But if the estimates of $\overline{C.V.}$ pertain to islands rather than local populations, the analysis may be invalid as it is meant to determine whether there is a relationship between genic heterozygosity and variation in morphometric characters. With greater island size, populations are larger and there is greater opportunity for microdifferentiation through isolation by distance and social structuring into demes. Thus if samples from several localities on the larger islands were pooled, but only single localities on smaller islands were sampled, one would expect to find greater morphological variability on the larger islands. Coefficients of variation might also be inflated by pooling collections made at different times. Because Patton et al. (1975) did not provide an unambiguous description of their sampling procedures, we cannot judge whether they were appropriate for estimating intrapopulation variability in morphology. The observed correlation of genic heterozygosity with morphological variability may be statistically significant but biologically spurious.

In sum, we find no evidence supporting the hypothesis of a "genetic-phenetic variation correlation." Because morphological characteristics are in general determined through complex interactions of substantial numbers of loci and environmental factors, it is unrealistic to expect that morphological variability will directly and consistently reflect the underlying variability at structural gene level. Moreover, if allozymic variation is selectively neutral, correlations would be expected only if heterozygosity at the structural genes is itself correlated with that at other types of loci influencing morphological variation. At present, genetic variability can only be estimated by analyzing specific structural gene loci, and only a very small portion of the genome can be sampled.

Schnell and Selander

Independence of Genic, Karyotypic, and Morphological Variation and Evolution

The most significant generalization emerging from recent studies of variation in natural populations of mammals and other organisms is that the processes underlying divergence in structural genes, chromosome form and structure, and morphology are essentially independent. This concept is not entirely new, for almost 20 years ago, Matthey and van Brink (1960) suggested that chromosomal and morphological evolution are unrelated processes, an hypothesis being supported by recent work. There now is a substantial body of evidence indicating that, while mammals in general have been evolving faster than lower vertebrates in karyotype and in morphological and physiological phenotype, the rate at which point mutations have been accumulating in structural genes and unique DNA sequences is not greater in mammals (Wilson et al., 1974b, 1977a; Bush et al., 1977).

Numerous mammalian examples of the independence of genic and morphological evolution, both within and among species, are available. For example, the demonstration that protein and blood group loci in three races of man (Caucasian, Negro, and Japanese) apparently are remarkably similar, although differences in morphological characters such as pigmentation and facial structure are conspicuous, led Nei and Roychoudhury (1972, 1974) to conclude that the genes controlling these morphological characters were subjected to stronger natural selection than were "average genes" in the process of racial differentiation. A similar conclusion regarding the extent of genetic diversity between human races was reached earlier by Lewontin (1972), using a different method of analysis. Recently, Mitton (1977) has argued that these analyses are misleading because the "information" index employed by Lewontin (1972) is relatively insenstitive to the range of allele frequency differences normally found between populations of man, and because Nei's (1975) genetic distance averaged over loci may disguise substantial differentiation. Mitton's, (1977) argument has been refuted by Lewontin and others in a series of papers appearing in the American Naturalist (Vol. 112, No. 988, 1978). But even if accepted, Mitton's (1977) interpretation and reanalysis of Lewontin's data, using a modification of Hedrick's (1971) measure of genotypic identity, would not seem to invalidate the generalization that structural gene differences between human races are relatively minor. The recent hierarchical analysis by Wright (1978), using F statistics, of human racial differentiation in five blood groups and the haptoglobin locus also supports this contention. Diversification is much greater, on the average, among than within races, but largely because of the contribution of one strongly differentiated locus, the Duffy.

The deer mouse (*P. maniculatus*), the most widely distributed and abundant native North American mammal, is highly variable geographically in morphological characters, continental populations having been assigned to some 30 to 40 subspecies (Hall and Kelson, 1959; Hooper, 1968). There also is marked geographic variation in karyotype, with the number of biarmed autosomes ranging from 16 to 42 in various populations (Hsu and Arrighi, 1968; Kreizinger and Shaw, 1970). Yet an analysis of allozymic variation in samples from 71 sites over the vast continental range and remarkable variety of habitat types occupied by the species demonstrated that populations almost invariably share alleles at both polymorphic and monomorphic loci (Avise et al., 1979). The same electromorph predominates at 75% of the loci in all populations sampled; and allele frequencies at each of six strongly polymorphic loci are rather uniform throughout the range. The mean value for Rogers' (1972) coefficient of genetic similarity (S) was 0.934 (range: 0.874 to 0.974), based on 21 genetic loci; and there was no geographic pattern in average individual heterozygosity, which averaged 9.1% (range: 5.4% to 12.4%).

A notable finding of this work was that patterns of allozyme variation do not reflect the division of *P. maniculatus* into the two major morphological and ecological types long recognized by mammalogists: (1) a long-tailed, large-eared, large-footed forest type ranging in the Appalachians, across Canada, and into the mountains of the western United States; and (2) a short-tailed, small-eared, small-footed form occupying the interior grasslands of the continent. Where the two forms meet in northern Michigan, ecological differences are sufficient to isolate them where they occur together (Dice, 1968); and they show a mild degree of intersterility (Harris, 1954), thus behaving as a semispecies. In the western United States, where the situation is more complex because of intergradation and interdigitation of montane and grassland habitat types, the two forms intergrade in some areas and remain distinct in others (Blair, 1950). Even subspecies of the grassland form may occur sympatrically, as in north-central Texas, where *P. m. ozarkiarum* and *P. m. pallescens*, differing morphologically and karyotypically (11 subtelocentrics, 6 metacentrics, and 6 acrocentrics vs 10, 8, and 5, respectively) occur in the same habitat at two localities, without hybridizing (Caire and Zimmerman, 1975). Laboratory experiments have demonstrated partial reproductive sterility of hybrids between those forms, as a result of nondisjunction of chromosomes and the consequent production of unbalanced gametes.

Compared with *P. maniculatus*, the closely related *P. polionotus* has a relatively small geographic range, being confined to South Carolina, Georgia, Alabama, and Florida. Eleven subspecies are generally recognized (Hall and Kelson, 1959), but variation in morphological characters is conservative in relation to that in *P. maniculatus*, with the conspicuous exception of coat color, which lightens dramatically in mice occupying the white sands of coastal beaches (Sumner, 1932; Bowen, 1968; Bowen and Dawson, 1977). Karyotypic variation in *P. polionotus* also is relatively minor, with only one major polymorphism in a pair of the small autosomes, which may be acrocentric or metacentric, and some minor variations ("rabbit ear condition") in some of the other autosomes (Te and Dawson, 1971). But genic variation apparently is greater in *P. polionotus* than in the continental populations of *P. maniculatus* (Table 4), involving near-fixation of alternative alleles at several loci (Selander et al., 1971).

How can we account for the relative genic uniformity of populations of the widely distributed *P. maniculatus*? Gene flow would seem to be ruled out by the physical distances involved; and the fact that regional populations have distinctive karyotypes (in some cases there is no overlap in number of biarmed autosomes, even when populations are chromosomally polymorphic) additionally suggests considerable geographic restriction of gene flow. For this reason, an explanation has been sought in terms of historical demography. Presumably, it was at the time of retreat of Pleistocene glaciers (18,000 - 20,000 B.P.) that *P. maniculatus* began to occupy the northern part of its present range, extending its distribution by more than 1,600 km. Before the east-

Environmental and Morphological Correlates

Table 4. Standardized variance (F_{ST}) of leading allele at structural gene loci in Peromyscus maniculatus from 18 regions and P. polionotus from 9 regions.

Locus[a]	Peromyscus maniculatus[b]	Peromyscus polionotus[c]
6-PGD	0.214	0.645
αGPD	0.037	0.085
LDH-1		0.414
ADH	0.139	
PGI		0.054
GOT-1	0.046	
PGM-3		0.197
EST-1		0.067
EST-2		0.046
EST-3		0.186
ALB		0.753
TRF	0.165	0.030
HB	0.380	0.498
Mean	0.164	0.271

[a]Standard abbreviations taken from Selander et al. (1971).
[b]Avise et al. (1979).
[c]Data from Selander et al. (1971), analyzed by Wright (1978).

ern forests were cleared, the eastern grassland form was restricted to tall-grass prairies, only recently moving into Michigan, Pennsylvania, and other eastern states (Baker, 1968). If the invading populations remained large, the observed genic uniformity over major portions of the range of P. maniculatus could merely reflect these range expansions with little time for the development of greater genetic divergence (see discussion of population size effects, Nei et al., 1975; Chakraborty and Nei, 1977). A consideration of how markedly climate and habitat type vary over the range of this species also brings to mind the possible role of genetic environment in maintaining the relative uniformity of allele frequencies, a concept frequently invoked (although, admittedly, never with sufficient factual support) to account for geographic uniformity in many organisms (Clarke, 1968; Ayala and Anderson, 1973; Ayala and Tracey, 1974; Hunt and Selander, 1973; Wright, 1978).

To account for the relatively marked geographic variation in allele frequencies at enzyme loci in _P. polionotus_, the interpretation developed in the previous paragraph requires that we postulate a history of long-term geographic isolation of relatively small populations (most probably in the Pleistocene), with random drift and perhaps adaptive changes leading to differentiation. A situation apparently similar in some respects to that in _P. maniculatus_, but differing in scale, has been reported for _P. boylii_ (Lee et al., 1972; Avise et al., 1974b; Schmidley and Schroeter, 1974; Kilpatrick and Zimmerman, 1975; see also Carleton, 1977).

Earlier we examined variation in a zone of hybridization between races of semispecies of _M. musculus_ in Jutland for evidence of environmental-genetic correlations. This zone also provides an opportunity to study concordance between variation at structural gene loci and that in both nonmetric ("quasi-continuous") and metric morphological characteristics (Schnell et al., 1980).

Nonmetric characters have been studied extensively in several mammalian species, including mice and men (see Finnegan and Faust, 1974). The patterns of geographic variation in 23 nonmetric characters, selected from the 35 described in _Mus_ by Berry (1963), were evaluated for degree of concordance with the general pattern for polymorphic structural gene loci defined by the hybrid index (Fig. 1B). As in the case of the genetic-environmental associations, the interrelationships between morphological and genetic variation prove to be complex. Patterns for nine nonmetric characters are associated with that of the hybrid index, but there is substantial heterogeneity in pattern even in this character subset. For example, the pattern for the foramen hypoglossi (Fig. 3A) shows a sharp break in the hybrid zone, with little evidence of introgression, while that of the maxillary foramen II (Fig. 3B) manifests a broad transition zone, with considerable vari-

Environmental and Morphological Correlates

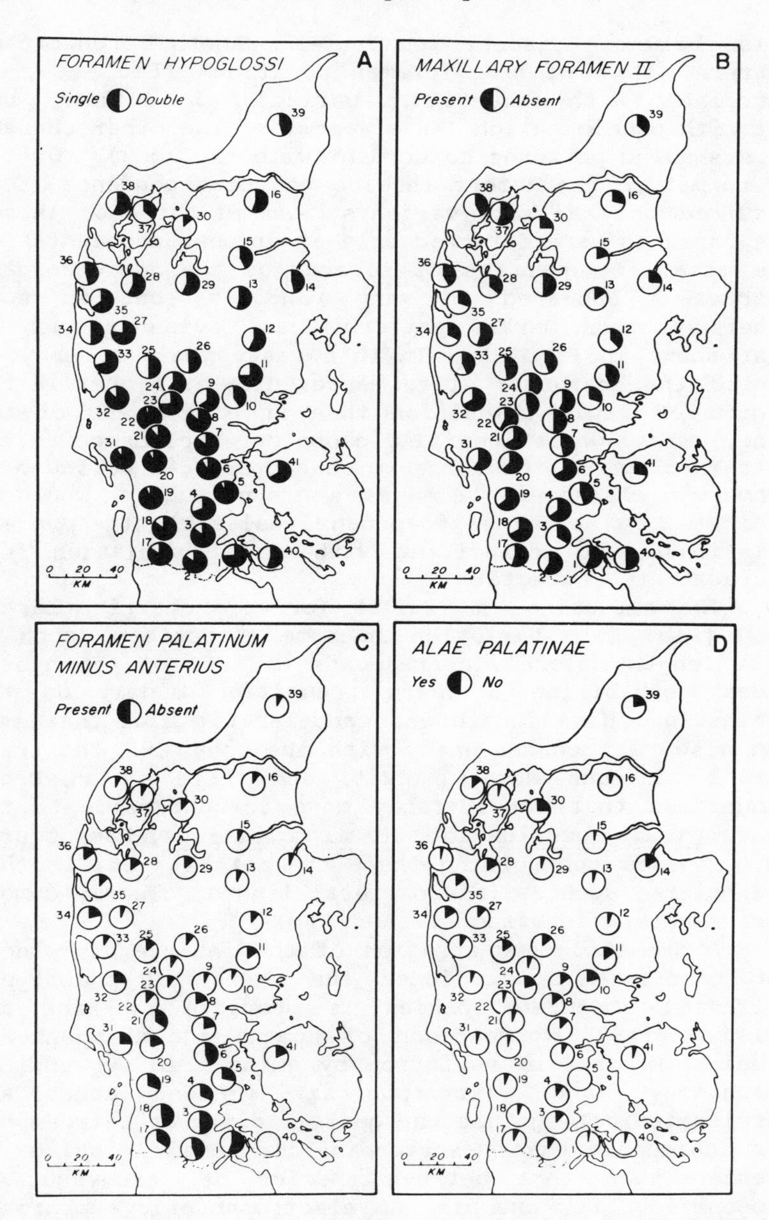

Fig. 3. Variation in frequencies of alternative states in four nonmetric cranial characters in zone of hybridization between Mus musculus musculus and M. m. domesticus on Jutland peninsula.

ation between adjacent localities. Another concordant pattern, that of the foramen palatinum (Fig. 3C), is correlated with temperature variation in Jutland, but not with precipitation (as are most of the other characters showing patterns concordant with the zone). Of the 14 nonmetric characters showing no correspondence, four occurred only as rare variants; and in three of these, the variants were detected only at or near the center of the zone. Frequencies of several of the nonconcordant characters appeared to vary randomly; but in some others, rather marked patterns were evident, such as that shown in Fig. 3D. In this case, the frequency of one of the character states apparently is higher in the region of rapid transition than at either end of the zone. Thus, while generally close correspondence to the pattern of variation in genic characters can be found in nonmetric characters taken as a group (such as would be indicated by principal component analysis), the overall relationship is coincident with marked variation from character to character.

The situation is similar for morphometric characters (Fig. 4). Variation in some characters, such as tail length, closely corresponds to that of the hybrid index, exhibiting a sharp transition midway on the peninsula. Ulna length and condylar width of the humerus also vary concordantly with the zone, but the transition is much more gradual, and there is a strong suggestion that considerably more introgression to the south (*M. m. musculus* into *M. m. domesticus*) has occurred in genes controlling these characters. Still other characters, such as interparietal length, clearly demonstrate heterotic effects.

A dramatic demonstration of the failure of molecular genetic data to index the degree of phenotypic difference between species is provided by King and Wilson's (1975) comparison of humans and chimpanzees. Studies of protein variation by amino acid sequencing, immunology, and electrophoresis yielded concordant results indicating that the genetic distance between man and chimpanzee is extraordinarily small, being no greater than that between species of *Peromyscus* or *Drosophila*. For example, an electrophoretic comparison of 44 proteins suggested that there is an average of only 0.62 detectable codon difference per locus. It seems unlikely that this relatively small divergence in the structural genes can account for the major anatomi-

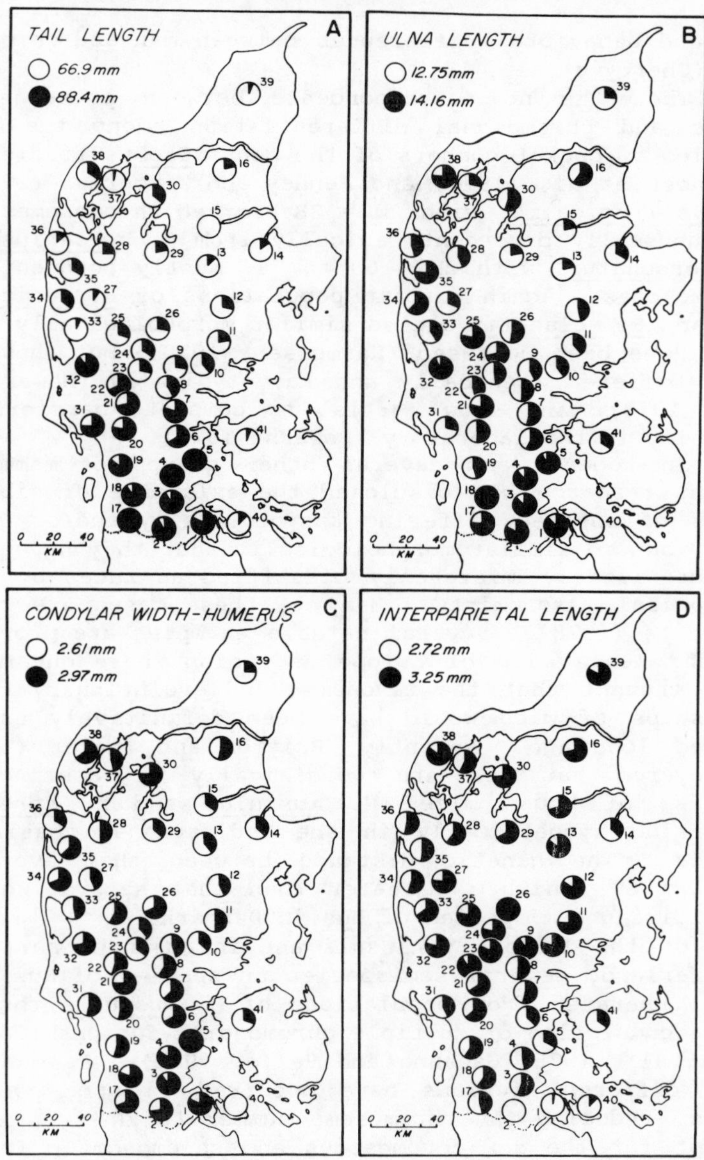

Fig. 4. Variation in mean values of four morphometric characters in zone of hybridization between Mus musculus musculus and M. m. domesticus on Jutland peninsula. (Measurements of individuals were adjusted for age, as estimated by degree of toothwear.)

cal and behavioral differences between man and chimpanzee (Cherry et al., 1978).

There is no correspondence between patterns of genic and chromosomal differentiation among the five species of pocket gophers of the genus *Geomys* studied by Selander et al. (1974) and Penney and Zimmerman (1976). *Geomys tropicalis*, with 2N = 38 biarmed chromosomes, is not unusually divergent genically from *G. bursarius* and *G. personatus*, with 2N = 68 to 74, mostly acrocentric, chromosomes. Again certain populations of *P. maniculatus* and *P. melanotis* are so similar morphologically that they have been confused (Rasmussen, 1970); but they are very different genically and karyotypically (Bowers et al., 1973), and show partial to complete barriers to crossing in the laboratory (Bowers, 1974).

In rodents and several other groups of mammals, recent research has disclosed the existence of sibling pairs of species differing karyotypically and/or genically but so similar morphologically that they have been unrecognized or mistakenly classified as races by morphological taxonomists. Many of these cases are cited by White (1978). Several notable examples are provided by the house mice of Europe, a region where one might have thought that the taxonomic and evolutionary relationships of mice would have been definitively established long ago. Recently, Britton and Thaler (1978) discovered that there are two genically very distinctive species of house mice (*M. musculus* and *M. spretus*) occurring sympatrically in the Mediterranean basin of France. The genetic distance between them averages about 0.60, which is greater than that separating the hybridizing semispecies of *Mus* in Denmark.

On the Italian peninsula and in adjacent parts of Switzerland, several semispecies or species of *Mus* have been generated from populations of *M. musculus* through the accumulation of multiple chromosomal fusions (Capanna et al., 1975; Capanna and Valle, 1977). As many as nine different fusions have occurred in some populations, reducing the chromosome number to 2N = 22; and, except for the sex chromosomes and chromosome pair 19, all the acrocentrics of the ancestral 2N = 40 chromosome karyotype have been involved in one or more fusions. The main populations in which multiple fusions occur are homozygous for them, heterozygotes occurring only in narrow zones where the forms meet and hybridize. There are no known ecological correlates of these reductions in chromosome number; and if morphological differences exist, they were not detected by earlier taxonomists and have not yet been described. According to V. M. Chapman (pers. comm.), the Alpine populations are genically very similar to, if not identical with, "normal" populations of M. musculus. Thus, the available evidence suggests a relatively recent history of extensive chromosomal rearrangement, unaccompanied by genic change or morphological differentiation.

A particularly good example of the nonconcordance of patterns of different types of variation among species is provided by the difficulties encountered by mammalogists attempting to classify the species of kangaroo rats of the genus Dipodomys into species groups on the basis of genic, karyotypic, and morphological data. In a recent study, Schnell et al. (1978) found no association at the level of interspecific comparison between phenetic groups based on data for 41 morphometric characters and those defined in Johnson and Selander's (1971) study of allozyme variation at 17 loci. When comparisons were made among various classifications, including the morphological taxonomic schemes of Grinnell (1921), Setzer (1949), Lidicker (1960), and Best and Schnell (1974) and one based on karyotypic variation (Stock, 1974), arrangements derived from genic characters were the most divergent. But associations between the morphometrically defined groups and those of other classifications also were relatively weak. For these reasons, Schnell et al. (1978) concluded that no one classification can accurately represent phenetic, cladistic, and phylogenetic affinities within the genus.

It should hardly be necessary to point out that the concept of casual independence of the processes of genic, karyotypic, morphological, and other phenotypic evolution is not in any way invalidated by observations that genic data sometimes yield dendrograms or other classifications that are closely similar to those based on morphology or karyology, or that average genetic distance increases monotonically from local populations, through subspecies and semispecies, to species in particular groups, such as Peromyscus (Selander and Johnson, 1973; Ayala, 1975; Avise, 1976; Zimmerman et al., 1978). Statistical associations are expected simply because time is a common factor in all processes (see comment by Wilson, 1975).

Genetic Basis of Morphological Evolution

Mounting evidence of differences in rates of evolution at the molecular and anatomical levels in mammals and other organisms has led many evolutionary biologists to conclude that variation in the structural genes is not the primary basis for organismal evolution. Attention recently has turned to "regulatory genes," those influencing rate and timing of structural gene activity. The hypothesis that important evolutionary changes require the addition of novel patterns of gene regulation or the reorganization of existing regulatory patterns was proposed by Britten and Davidson (1969, 1971), and recently has been invoked by Wilson and his colleagues (Wilson et al., 1974a, b; Wilson, 1975; King and Wilson, 1975; Wilson et al., 1977a, b) in relation to the evolution of mammals and other vertebrates. Because our present understanding of the nature of gene regulatory mechanisms (to say nothing of the structure and evolution of specific regulator loci) in higher organisms is meager, the principal argument for the regulatory gene theory is merely the apparent inadequacy of evolutionary changes in the structural genes themselves to account for evolution at the organismal level.

Evidence of genetic polymorphisms affecting quantitative variation in levels of enzyme activity or amounts of nonenzymatic proteins in mammalian tissues and organs is accumulating, although the actual mechanisms of control of this type of variation are poorly understood. In some cases, regulatory loci or functions may not be involved, the variation apparently being caused instead by polymorphism in number of closely linked structural loci present, as in the case of salivary amylase in bank voles (_Clethrionomys glareola_; Nielsen, 1977). In _P. maniculatus_, there are complex electrophoretic hemoglobin phenotypes generated by at least four, and probably five, globin structural loci (Snyder, 1978a); but according to Snyder (1978b), the continuous nature of individual variation in "partitioning" of the total hemoglobin among the various components excludes a simple explanation in terms of polymorphism for a variable number of duplicated structural loci. His extensive genetic data were interpreted as evidence that the range of phenotypes in the subspecies _P. m. sonoriensis_ is controlled by 10 or more alleles differentially regulating globin production at the HBD locus, the phenotypic differences between alleles being on the order of 1% of total hemoglobin. Although direct evidence eliminating the possibility that control is affected by structural gene alleles having inherent and characteristic rates of globin synthesis was not obtained, Snyder's (1978b) argument for variation at separate regulatory loci is strong. Whether variation in the hemoglobin components is functionally significant and, hence, expressed at the organismal level and potentially under selective control remains to be determined.

The genetics of β-glactosidase and β-glucouronidase have been intensively studied in inbred strains of the house mouse as models for organization of eukaryotic genomes (Breen et al., 1977; Swank et al., 1973). In the case of β-galactosidase, a site (Bgs) closely linked with the structural gene (Bge) on chromosome 9 regulates the level of the enzyme in all tissues. Another closely linked site (Bgt) determines the developmental expression of β-galactosidase in the liver; and excretion of the enzyme is controlled by several other unlinked loci.

There is good evidence that a marked qualitative difference in renin activity in the submaxillary gland between the SWr/J and C57BL/10J strains of the house mouse is controlled by a regulatory gene rather than by a difference in the structural gene encoding the enzyme (C. M. Wilson et al., 1977).

Variation in activity of 36 liver and erythrocyte enzymes, all concerned with glycolysis and gluconeogenesis, in seven inbred strains of the house mouse, recently was analyzed by Bulfield et al. (1978) in research motivated by the supposition that the observed enzyme differences, which are generally small, will be found to correspond to interstrain metabolic differences, thus perhaps providing a molecular basis for variation in morphometric characters. Such a connection between genes and morphometric characters is much to be desired, but, unfortunately for our understanding of developmental regulation, variation in none of the enzymes could be assigned to a single locus; no unequivocal evidence of segregation was obtained.

Significance of Karyotypic Changes in Speciation and Evolution

Wilson et al. (1974a, b; 1977a; see also Wilson, 1975) have proposed that gene rearrangement is a major cause of morphological evolution. The theory was generated in an attempt to solve the problem posed by apparent differences between rates of evolution at the molecular and other levels of biological organization. If point mutations have been accumulating in structural genes (and unique DNA sequences) at more or less constant rates (see Fitch, 1976; Sarich, 1977), how can variation in rates of morphological evolution and speciation among groups be explained? Noting a correlation between rate of speciation and rate of chromosomal evolution, Wilson et al. (1974b) hypothesized that the chromosomal rearrangements in themselves cause regulatory changes through gene position effects and the creation of new close-linkage groups (supergenes). The underlying supposition is that gene rearrangements act at the molecular level as regulatory mutations, producing altered patterns of gene expression at the phenotypic level. Thus, Wilson (1975) and Bush et al. (1977), who refer to "karyotypic facilitation of speciation and adaptive evolution," believe that there are important, causal relationships between karyotypic evolution and speciation beyond the likelihood that both processes are accelerated by population subdivision (White, 1968, 1978; Grant, 1971). In short, it is suggested that organismal evolution is driven by karyotypic changes through the production of regulatory mutations.

Extending an earlier study by Wilson et al. (1975), Bush et al. (1977) recently have estimated rates of speciation and chromosomal evolution in 225 genera of vertebrates, including 131 genera of mammals, and have demonstrated that they are strongly correlated. Among mammals, primates and horses apparently have been evolving very rapidly at the chromosomal level and have experienced especially high rates of speciation; average rates are intermediate for lagomorphs, rodents, and artiodactyls, and low for insectivores, carnivores, bats, and whales. Bush et al. (1977) are able to make a reasonably good case for the interpretation that the observed correlation arises in significant part because of variation among taxonomic groups in average effective population size (although few reliable estimates are available), both processes being accelerated when deme size is small, for reasons recently discussed at length by White (1978). While it might be argued that the "especially low" average level of genic heterozygosity in mammals is consistent with their hypothesis, this would leave unexplained the similar or even lower levels of genic heterozygosity in reptiles (Selander, 1976), which supposedly must have larger effective population sizes.

Deme Size and Rate of Fixation of Chromosomal Rearrangements

A singularly important contribution to our understanding of chromosomal evolution has been made by Lande's (1979) theoretical work on the population dynamics of negatively heterotic rearrangements (chromosomal translocations and inversions that are deleterious when heterozygous but have normal fitness as homozygotes). Because these chromosomal variants are selected against when in the minority (although favored when in the majority), they can become established in a local area only through genetic drift, a process normally requiring small population size. According to his analysis, the correlation of rate of speciation (indexing morphological evolution) with rate of karyotypic change found by Wilson et al. (1974b) and Bush et al. (1977) in comparing mammals and other vertebrates indeed can be explained by differences in population structure and phylogeny (Wright, 1977), although other factors, such as differences in spontaneous rearrangement rates and, because of parental care, the greater ability of mammals to compensate for gametic wastage, may also be important. But citing evidence that position effects are rare, and concluding, with Muller (1956), that the evolutionary potential of position-effect mutations is severely limited, Lande (1979) questions Wilson's (1975) hypothesis that gene rearrangement is the major cause of morphological evolution.

For most mammals, fixation rates (R) in phylogeny are 10^{-6} to 10^{-7} per species per generation (White, 1973; Van Valen, 1973; Bush et al, 1977). With spontaneous rates (M) of Robertsonian changes and inversions between 10^{-3} and 10^{-4}, the ratio of R/M is between 10^{-2} and 10^{-4}. For this range of parameters, long-term effective deme sizes of 30 to 200 individuals are indicated, which seems reasonable from what is known of deme sizes in contemporary mammals. Among lower vertebrates, fixation rates are roughly 10 times slower than in mammals (Bush et al., 1977), so that $R \simeq 10^{-7}$ to 10^{-8}

per species per generation. And assuming similar spontaneous occurrence rates and selection coefficients, effective deme sizes are only 30 to 300.

In sum, the long-term effective deme sizes estimated by Lande (1979) indicate: (1) the observed fixation rates in mammals and in other organisms can be explained by known genetic properties of spontaneous rearrangements and the common ecological processes of local extinction and colonization; and (2) there have been substantial opportunities for the action of random genetic drift during animal evolution.

CONCLUDING COMMENTS

Our examination of variation in natural populations of mammals has demonstrated that it is extremely difficult to define relationships between ecological factors and genic, karyotypic, and morphological variables other than those that might be reasonably expected on the basis of historical demographic principles. We have found no convincing evidence of adaptive relationships between genetic and morphological variability and temporal or spatial environmental heterogeneity; and structural gene variability and morphological variability apparently are uncorrelated.

Much of the theoretical framework within which recent studies of the genetics of mammalian populations have been undertaken is questionable if not demonstrably invalid, being based to a large extent on the unwarranted identification of correlation as cause. This is the major weakness of mammalian evolutionary genetics, which relies heavily on observational rather than experimental approaches.

Our interpretation of available evidence is that evolution proceeds more or less independently at the genic, karyotypic, and morphological levels in mammals, although there is a broad correlation between rates of chromosomal evolution and speciation, presumably because both processes are accelerated when effective population size is small. Even if structural gene polymorphism is not the type of variation on which evolution at the organismal level is based, electrophoretic studies of proteins may still be useful in providing evidence relating to population structure. But they will not tell us much that is new about the genetics of adaptive evolution in anatomy, physiology, behavior, and ecology, or the extent of genetic change involved in the process of speciation.

Acknowledgments--We thank Elizabeth Grobe for assistance in preparing the manuscript. The authors' research on variation in mammalian populations has been supported by grants from the National Science Foundation and the National Institutes of Health. This review was written while G. D. S. was a Visiting Research Associate at the University of Rochester, which provided partial support for his work.

THE STREAM OF HEREDITY: GENETICS IN THE STUDY OF PHYLOGENY

Donald O. Straney

Abstract--Systematists of the great synthesis of the 1930's and 40's saw in genetics the explanation of phylogeny: genes transmitted from generation to generation produce phylogeny. However, analyzing genetic data for the purpose of identifying genealogical information is difficult. Two general techniques are used today: distance and cladistic. There is no reason to believe that distance analysis of genetic data gives a reasonable phylogenetic estimate. The usual cladistic techniques generally fail, for they cannot handle characters that display high levels of population variation. Thus, there is a continued search for a general method of phylogenetic inference. A conservative form of phylogenetic inference from genetic data is possible: those relationships indicated by data sets different enough to be considered independent (e.g., genetic, chromosomal, and morphological) can be confidently identified as phylogenetic. Current techniques of allelic assay must be improved to obtain precise homologies of genetic units, but because the potential evolutionary behavior of alleles is fairly well understood, genetic information can make major contributions to the study of phylogeny.

INTRODUCTION

If there is a common thread in biology today, it must surely be a genetic one. Systematic biology has felt the impact of genetics longer than most biological disciplines. The "new systematics," formed from a pleasant mutualism between genetics and evolutionary biology, has fostered almost 40 yr of activity. In the full confidence of that early synthesis, Simpson (1945:6) observed: "The stream of heredity makes phylogeny; in a sense, it is phylogeny. Complete genetic analysis would provide the most priceless data for mapping that stream." Not all systematists today share this view. Rosen (1974), for example, questioned whether a genetic perspective is warranted above the species level. Only recently have techniques been available to assess how priceless genetic information might be to the systematic biologist. Biochemical methods that permit quick assay of genotypes and karyotypes of free-living organisms do not provide the complete analysis Simpson envisioned, but come close enough for an initial evaluation of the usefulness of genetic information in addressing systematic questions. The information we have at present does not support the "new systematist's" confidence in genetics, nor the confidence with which the new techniques have been, and continue to be, used. This is not to say that genetic information is of little systematic use, but only that there is no equivalent (short of time travel) of the "philosopher's stone" for studying evolutionary history. History, whether human or mammalian, recent or primordial, is studied by inference from information incomplete enough to engender not certainty, but ambiguity.

What follows is not an attempt to review the literature on the use of genetic techniques in systematics. Rather, I will examine how genetic information can be used to infer phylogeny, as well as the potential questions that may be addressed with genetic data within a phylogenetic context. By "genetic" I usually refer to the allelic variation demonstrated by electrophoretic and protein sequence methods (Ayala, 1976). While not discussed directly, karyotypic information has the same problems and benefits for phylogenetic study as do allelic data. Classical genetic studies have not been considered since the requisite breeding studies prevent them from being informative above the species level. Within systematics, I restrict attention to questions of relationship at and above the species level. Studies within species have a clear need for genetic perspective, and recent developments are reviewed in this volume and elsewhere (Endler, 1977; Throckmorton, 1977). Above the species level, there is only an embryonic equivalent of the sophisticated theory of population genetics (e.g., Kimura and Ohta, 1971, 1972; Wright, 1978), and study at this level is necessarily more empirically based than that within species.

Varieties of Genetic Relationship

The phylogeny of a group is a chronicle of its history, the temporal sequence of diversifications whereby one ancestral form is transformed into diverse taxa. Figure 1 is a diagrammatic representation of the phylogeny of a monophyletic group. Several lines of descent lead from one ancestor and branch repeatedly to give rise to the many living taxa and the multitude of forms extinct without issue. The tree form is a bare abstract of the detailed spatial and temporal distribution of individual organisms. At each point along these lines, individuals are found in separate population and breeding units and over even short stretches of time these units undergo a complex pattern of fission and fusion. For most morphological studies of phylogeny this detail is unimportant. Developmental canalization (Waddington, 1968) buffers most phenotypic traits of phylogenetic interest from change at and below the population level. Population processes are important, however, for understanding how characters change through time when, as with alleles, there is possible a large amount of intrapopulation variation. Thus before considering how genetic information can be used to infer phylogeny, it is necessary to examine the populational effects on alleles through time.

Through a simple situation: population size is consistently large and the alleles at one locus are neutral. At any point in time, there are one-to-many alleles present at this locus in a particular population. A mutational event transforms one of these alleles to a new state. The mutant will spread through the population as a function of time and will spread to other populations as a function of its frequency and rate of gene flow. If the mutation occurs after a speciation event only one of the sister species will possess the allele. If the mutation occurs before speciation, its spread throughout the population will determine its presence in one or both of the daughter species. Subsequent mutation can produce transient polymorphism and even replacement of a given allele by a newer mutant.

The temporal distribution of neutral alleles, then, depends on rates of mutation, fixation, and gene flow, the time between speciation and mutation, as well as more subtle factors in the history of populations. It

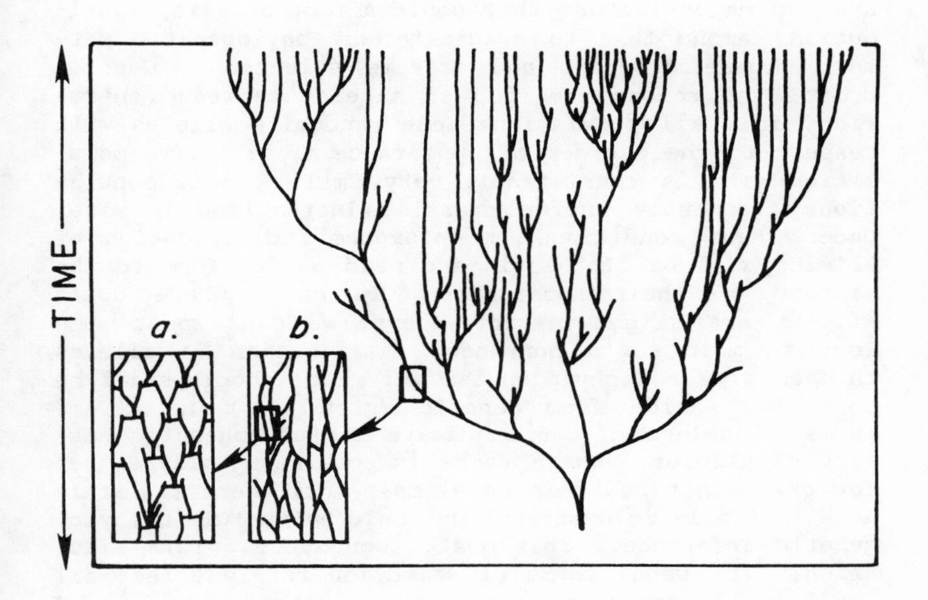

Fig. 1. Diagrammatic representation of the phylogeny of a group. Inserts magnify portions of the branches to show the temporal pattern of change in family groups (a) and populations (b).

would be a formidable exercise to obtain this information to have a phylogenetic scheme for the alleles at one locus, but we can be spared this effort. For phylogenetic inference, it is not necessary to know exactly what has happened at a locus through time, it is sufficient to know that the order in which alleles arise might be reflected in the allelic complements of living taxa. Of course, some derived alleles might be lost along some lines of descent, leaving only unique or primitive alleles in subsequent generations, thus making inference incompletely determinant. This problem can be solved, however, by comparisons of many loci which have different, locus-specific, fixation rates.

Neutral alleles in large populations, then, have a temporal behavior that allows at least partial inference of phylogeny when multiple loci are considered. It is questionable, though, whether many groups have both consistently large populations and neutral alleles. There are two major factors that could affect allelic distributions among taxa to eradicate any phylogenetic pattern: genetic drift and extreme selection. Genetic drift, the random sampling of alleles between generations in small populations, can randomize alleles with respect to their order of appearance if effective population size is consistently very small, or if populations frequently suffer drastic fluctuations in size. Under these conditions, the taxonomic distribution of alleles will be stochastic and need not conform to the history of their appearance. In cases where drift effects are slight or infrequent, we can expect some loci to maintain a nonrandom distribution of alleles through time and phylogenetic inference should still be possible. Aside from genetic drift, certain extreme forms of selection can eradicate the phylogenetic history of alleles. Where there is consistent bias either for or against new alleles at most loci there is little hope that allelic distribution could be useful in phylogenetic inference. This must, though, be a rare situation. The usual forms of selection rarely affect all loci equally so consideration of many loci should avoid the problems imposed by selective bias in the historical distribution of alleles. Balancing selection, thought by some to be the predominant force acting upon allelic forms of proteins, provides an ideal situation for phylogenetic inference by preserving allelic diversity in populations.

Any type of information characterizing a group of organisms can be used to construct statements of relationship for the involved taxa. To suggest a relationship between several taxa, one needs information that they form a group distinct from other taxa. Phylogenetic inference involves distinguishing which of the available statements reflect common ancestry of genealogical relationship, as opposed to those reflecting nongenealogical associations. Statements of genealogical relationship define monophyletic groups of taxa comprising all of the decendants of a particular common ancestor. Nongenealogical associations arise by convergence, parallelism, random change or stasis, and they include taxa that do not share a common history. The central problem in systematics is the identification of those relationships that index phylogeny distinct from those that index certain of the processes operating within the phylogenetic framework.

Genetics in Phylogeny

For genetic data this problem is met with full force. As I will argue below, the usual methods of analyzing genetic data do not necessarily produce groups defined by statements of genealogical relationship, and the usual methods that might produce such phylogenetic results cannot be easily used with genetic data. This is not to say that genetic information is useless for phylogenetic inference. As the previous discussion shows, for most cases genetic information can be expected to contain some statements of genealogical relationship. Genetic data are simply more obviously difficult to analyze phylogenetically than are other types of information. All of the problems met with in analyzing genetic data apply equally well to morphological characters but the problems are rarely recognized because ontogeny and comparative anatomy provide effective and often even appropriate solutions.

The distinction between (genealogical) relationship and (nongenealogical) association has been drawn, but not always clearly, in the cladistic-evolutionary controversy of the past few years. Briefly, cladists distinguish between the two and study the genealogical branching pattern of phylogeny, while evolutionary systematists are more interested in including evolutionary causes and processes within this framework. My discussion ignores the major source of discord between the two schools, namely the way we should construct classifications. The approach in evolutionary systematics is to study the associations among taxa to understand the processes of which particular taxa are a result. This, of course, is possible only within a cladistic framework of estimated genealogical relationships. The situation is quite analogous to the study of human history, where a distinction is made between chronicle and significance (Wartofsky, 1968:396-400). To study revolutionary movements, one needs a chronicle of when and where they occurred and of the events leading to and resulting from them. But, as "isolate and accidental particulars of mere chronicle...do not constitute history" (Wartofsky, 1968:400), the historian will use this framework to study general processes: cause motivation, success, impact, and meaning. Evolutionists find cladistic theory frustratingly narrow; it does not address the question of how phylogenetic change has occurred. The cladistic critique of evolutionary systematics, though, is partly valid. The canons of empirical evidence for genealogical inference were previously only vaguely stated. Further, many systematists have tried to estimate both genealogical and evolutionary aspects of phylogeny at the same time and even from the same data. Studies in this mode usually produce scenarios that come too close to subjectivity and circular speculation to be comfortably accepted. Cladists rightly insist that genealogical relationships be clearly established before evolutionary processes are investigated. But evolutionary systematists are justified in their reticence to simply ignore nongenealogical information. Such information, after all, is what we would use to examine the processes creating adaptive diversity as well as the mechanisms controlling those processes. Yet, as long as systematics is based upon analysis of morphological characters, there is little that can be done for interpreting the causes of nongenealogical associations. There is only a poor theoretical understanding of how complex phenotypes change in an evolutionary context (Lande, 1976). Genetic information, whose evolutionary behavior is well understood theoretically (Wright, 1969) will probably be invaluable for illuminating the processes affecting phyletic change.

Distance or Allelic Analysis?

Most systematic studies utilizing biochemically demonstrable genetic data have used distance analysis to assess relationships; there are significant exceptions (Hubby and Lewontin, 1966; Throckmorton, 1978; Wake et al., 1978). Initially, this use was motivated by the argument that, since the loci sampled were a random subset of the genome, they could be used to estimate the proportion of their genomes that two taxa shared. This assumption has since been greatly weakened (King and Wilson, 1975) and most attempts now are restricted to assessing the relationships by the available data. The distance approach was strengthened by development of biologically meaningful metrics (Nei's D; Nei, 1971), the observation that distance is roughly monotonic with time (Nei, 1975; Sarich, 1977), and by the general agreement of resulting phenograms with preexisting systematic opinion (Avise et al., 1974a).

To the extent that surveys of genetic variation are restricted to comparisons of very closely related populations, a distance-and-phenogram analysis is probably appropriate. Where distances are small (D less than about 0.3) divergence is small enough that, at most loci, populations share the same allelic complement and at only a few loci are there major allelic differences (Avise, 1974). In these cases distance measures may give a more useful assessment of relationship than would a detailed analysis of the alleles present in each operational taxonomic unit (OTU). Wright (1978) has cautioned, though, that where either (1) gene flow between populations is high or (2) random drift is the major determinant of gene frequencies, the hierarchical approach of most clustering algorithms gives an over-simplified picture of actual relationships and distance analysis with a plexus may be more appropriate.

Distance analysis, though potentially useful in studying closely related populations, has serious drawbacks as a method of phylogenetic inference. As usually conducted, distance analysis is based upon all information available in a data set, irrespective of whether each datum represents a statement of genealogical or nongenealogical relationship. Analyzing genetic data sets with genetic distance measures maintains the confusion between these two types of relationship. Only if all characters in a data set are shared derived characters (and if each derived character state is present in all descendant taxa) will distance necessarily index the pattern of genealogical relationships. Alleles primitive at the level of analysis or unique to particular taxa do not index genealogical relationship but generally make a substantial contribution to any genetic distance estimate that is relatively large. To tease apart cladistic and nongenealogical information from genetic data sets, particularly where comparisons are made over a broad taxonomic scale (Avise et al., 1974a; Straney et al., 1979), a close analysis of allelic distribution is required rather than reliance upon summary distance statistics and clustering techniques.

Two approaches suggest themselves for identifying genealogical patterns in genetic data: Wagner tree methods (Farris, 1970; not to be confused with distance Wagner trees, Farris, 1972) and cladistic techniques deriving from Hennig (1966). Wagner tree methods are gaining wide acceptance as efficient estimating procedures for finding parsimonious trees, those with the shortest total length (measured as the sum of the number of steps along each branch). Maximum parsimony trees may be useful as first approximations, but uncritically accepting as is the shortest tree as some "best" cladistic estimate has little to recommend it in practice (Throckmorton, 1978). Figure 2 presents a Wagner tree for 14 species of phyllostomatid bats based upon the presence or absence in taxa of 92 alleles at 17 gene loci (data from Straney et al., 1979). This is not the shortest tree for the data set, differing by about six of 95 steps from the shortest tree. The difference in lengths between the shortest tree and that in Fig. 2 is small enough to be well within sampling error, so that there seems little reason to accept the most parsimonious tree as the best estimate.

Romero-Herrera et al. (1978), in their study of myoglobin amino acid sequences in mammals, present an even clearer failure of the minimum step criterion. In examining eight different possible trees for the OTU's in their study, each with fundamental differences in arrangement of major groups of mammals, they found that the two shortest trees each required 281 steps. One tree has carnivores splitting from eutherian stock after hedgehogs but before treeshrews, while the other has treeshrews and hedgehogs together forming a sister group to other eutherians with carnivores the sister group to all three. Further, the other trees leave little room for choice among the eight possible; one requires 282 steps and the remaining five have 283 steps.

Even where a minimum step criterion might be useful for choosing between alternate trees, Wagner methods may produce poor phylogenetic estimates with genetic data. One of the major results from recent sequence studies of proteins has been the discovery of rampant parallel and convergent amino acid substitutions (Romero-Herrera et al., 1978; papers in Goodman and Tashian, 1976; the same is probably true for electrophoretic data). The Wagner procedure, in forcing the shortest solution, treats much of this convergence as similarity indicative of common ancestry. Thus, biochemical data where convergence is high may not be appropriate for Wagner tree analysis; Felsenstein, (1978) presents a more detailed treatment of this point. In addition, with allelic data the Wagner procedure assumes that the probability of gaining a new allele is equal to the probability of losing an old one; that is, the mutation rate equals the extinction rate of alleles. Not only are these rates assumed

Genetics in Phylogeny

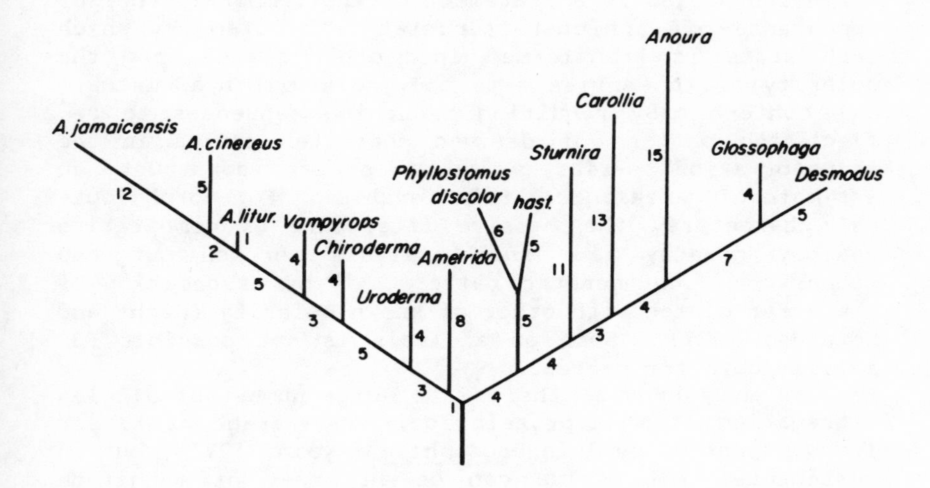

Fig. 2. Wagner tree for 14 species of phyllostomatid bats. Numbers indicate number of steps along branches. Data from Straney et al. (1979).

to equal each other, but they are assumed to be equal across loci. Clearly, these assumptions are unrealistic for most genetic data sets. The locus-specific mutation rate differs considerably among protein-coding gene loci (Dickerson, 1971). Further, where effective population size is not unreasonably large, drift effects can lead to loss of alleles at a rate much higher than the rate of appearance of new alleles. Wagner tree techniques may provide a quick analysis of the patterns of alleles over taxa, but there is little reason to believe that they will necessarily produce reasonable and complete phylogenetic estimates with genetic data. As I will discuss below, however, there are cases where techniques like Wagner methods can be useful.

Cladistic methods of phylogenetic inference differ from Wagner procedures in requiring a character by character analysis, rather than using statistical estimation. Unfortunately, currently popular cladistic methods hold little promise for analyzing genetic data. The methods stemming from Hennig (1966) are based upon constructing transformation series for character states with well-defined, primitive-derived polarity. A transformation series is a statement of the temporal order of appearance of each character state, the steps by which each state is transformed into other states, and the polarity of the series sets the whole within a historical context. By requiring branching sequences to reflect the pattern of derived character states in the transformation series, cladistics can construct an estimate of genealogical relationships. For morphological characters, the massive literature of comparative anatomy greatly aids such analysis through outgroup comparisons, ontogenetic pattern, or by association of character states with other of known polarity (Hecht and Edwards, 1977). Such an analysis is not possible for allelic data for proteins.

In many groups, there is a large number of alleles segregating at most protein loci; an average of 13 per locus is not unusual in Drosophila (Ayala, 1977), but in vertebrates this value can be an order of magnitude lower (Selander, 1976). The magnitude of within-population variation expected of allelic data is unmatched in the morphological characters usually submitted to cladistic analysis. Consider the albumin locus in Straney et al.'s (1979) phyllostomatid bat study (Fig. 3): Twelve alleles are known in the 14 taxa studied, each allele found in one (six alleles), two (three alleles), or three taxa (three alleles). Even if we arbitrarily designate one allele (e.g., Alb^{98}) as primitive, which allele is then next most primitive? Figure 3 presents only one of the very many transformation series possible for these alleles, but there is no way to tell from the albumin data, or the entire genetic data set, which series is correct.

While the distribution of alleles in living taxa no doubt indexes, at least partially, the order in which the alleles have arisen through time, there is no way to infer the exact sequence of the various transformations. This sequence could be studied only at the population level where population size, gene flow and selection will interact to produce a complex series of transformations of alleles present in a population. Random mutation must be added on top of the populational processes to produce new alleles. Further, each population not connected to others by gene flow will have its own unique transformation series. Such a pattern of charac-

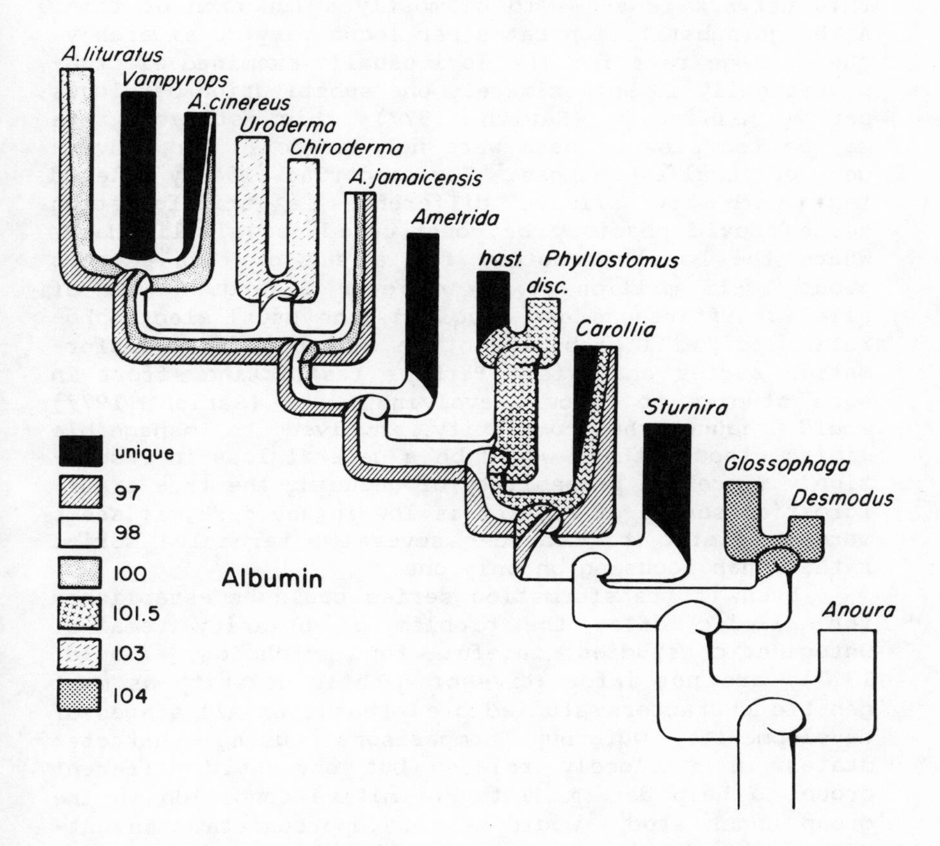

Fig. 3. Possible allelic evolution at the albumin locus in phyllostomatid bats. Terminal proportions of alleles reflect frequencies in current populations. Unique alleles are different in each taxon.

ter state transformation is far too complex to reduce to a cladistically useful series or ordination.

The mutation rate for protein alleles can be at least several orders of magnitude greater than the rate of change of morphological characters usually analyzed by cladistic methods. This, and the segregation of several to many alleles per locus within natural populations, produces a complex network for a transformation series of allelic states through time. Just how complex this network is seems to be mostly a function of time. Although substitution rates per locus vary considerably, the average rate for the loci usually examined electrophoretically is approximately one substitution per locus per 20 million yr (Sarich, 1977); this estimated rate may be too slow as data were not corrected for convergent or parallel changes. For comparing closely related taxa with few allelic differences a transformation series could possibly be constructed for allelic data. Where the basal dichotomy for a set of taxa exceeds about 10-15 million yr or wherever a large number of allelic differences are involved, the usual electrophoretic loci will probably not be suitable for transformation series analysis. Perhaps restricting effort in such studies to "slowly evolving" loci (Sarich, 1977) would reduce the complexity involved to manageable limits, though there would be a general loss in resolution. Since the probability of deducing the true transformation series of alleles is low in any case, it seems more realistic to consider several alternative series rather than focusing on only one.

Even if transformation series could be established for allelic data, the problem of polarity remains. Ontogenetic studies, useful for morphology (Gould, 1977), are not informative of genetic polarity as most genetic characters studied are present at all stages of development. Outgroup comparisons, using character states in a closely related but obviously different group to help determine the primitive condition in the group under study, would be more appropriate. An outgroup, though, is a taxon that has had an independent evolutionary history for a time longer than that since the basal split of the group under study. Given the rates of allelic substitution and chromosomal rearrangement, it is quite likely that an outgroup will share few, if any, allelic characters with the group of interest. Because outgroups have a long period of undocumented evolution relative to the group studied, shared genetic characters should not be uncritically accepted as necessarily primitive. As is shown below, primitive alleles do not all penetrate to each descendant living taxon and using only one outgroup taxon is no guarantee of finding shared primitive alleles. Also, the large amount of convergence at the molecular level (Romero-Herrera et al., 1978) raises the possibility that alleles shared with an outgroup may be simply convergent.

This is probably the case with the five alleles at four loci shared between phyllostomatid bats and the two outgroups studied by Straney et al. (1979), since the outgroups have had an independent history for probably 50 million yr.

The presence of large amounts of within-population variation, and high mutation and substitution rates, makes it unlikely that currently available cladistic methods will be overly useful in finding genealogical patterns in genetic or molecular data. The current rush to cladistic methodologies is in part due to the ease with which results are obtained with morphological data and the logic motivating their use. Nevertheless, these are not general methods, ones that can be used even with data where much variation is everywhere possible and where convergence is highly likely. Both cladistic methods and Wagner tree producers are more reasonable approaches than distance analysis for analyzing genetic data. Still, there should be some other, more general approach for inferring phylogenies from allelic patterns in living taxa. There is genealogical information contained in such data that, were we clever enough, we could winnow from convergence, parallelism, adaptation, and random change.

Towards a General Cladistic

The systematic literature is replete with different methods of genealogical inference, and each proponent of a technique is apt to view his method as the "best" estimating procedure. The "best technique" and associated "best estimate" are curious systematic concepts. The only sense with which "best" can be used in a systematic context is to describe close approximation to the actual, true phylogeny of a group. Are any of the various criteria suggested as producing "best" estimates (minimum tree length, minimum covergence, maximum likelihood, synapomorhous definition, overall gestalt, among others) actually capable of producing accurate phylogenetic estimates? Curiously, there is no answer to this question. Of course, we know no true phylogenies that would allow us to answer this question, but techniques such as computer modeling could conceivably give us a better idea than we currently have of how various proposed criteria of "best" actually index accuracy. An initial treatment of this problem is presented by Felsenstein (1978). Lacking any evidence to the contrary, I pose an extreme hypothesis: there can exist no method capable of producing necessarily accurate phylogenetic estimates. I would challenge those who would hold otherwise to demonstrate their case.

To claim that there is no necessarily accurate procedure for estimating phylogeny does not imply that phylogenetic inference is impossible. I do not construe the purpose of phylogenetic inference to lead directly to an estimate of "the" phylogeny of a group from a particular data set, but rather to identify those statements of relationship present in a data set that might be genealogical statements. The phylogeny of a group might then emerge from piecing together these possible genealogical statements. I consider it a mistake to assume that any one data set should lead to a completely accurate phylogenetic estimate. For most real data sets, both genealogical and anagenetic mechanisms are responsible for the distribution of character states in the studied taxa. Thus, the statements of relationship in the data set are apt to define some phylogenetic groups, some homoplasic groups defined by nongenealogical information, and some groups that are not really groups at all because they are defined by a combination of both types of information. It will be a rare data set that contains sufficient genealogical statements to define a complete phylogeny for a group when a data set contains no homoplasic information. Phylogenetic inference, then, is more fundamental than producing a phylogeny; it is the basis for recognizing the elements that form the phylogeny.

The theoretical aspects of phylogenetic inference have rarely been discussed, though elements of such discussion can be found scattered in the systematic literature. Pertinent general aspects have been treated by Throckmorton (1968, 1977, 1978), Nelson (1979) and Straney (1980). The discussion here is limited to genetic characters; extension to other types of characters presents no additional problems (Straney, 1980).

Several times I have referred to the statements of relationship present in a data set. Each character state or allele makes a statement of relationship concerning the taxa in which it occurs. The taxa with a given allele constitute a group, or component, defined by the presence of that allele. Following Nelson (1979) the elements (OTU's) of a component are labeled "terms." Only allelic presence is considered sufficient to define a component. Allele frequency is incidental to group definition. Absence of an allele is not useful information for defining groups (Platnick and Nelson, 1979). Copi (1972:139-140) presents the logical basis for this requirement. Only defined components will be considered. Thus, if an allele is present in taxa A, B, and C, the component ABC is defined, but subcomponents AB, BC and AC are not.

Table 1 illustrates how a data set can be analyzed into components. Reported there are the components found in Ayala's (1977) summary of frequencies of 152 alleles at 30 loci for five species of the *willistoni* group of *Drosophila*. Only about 40% of the total known allelic complement was included in the data summary, and several important taxa were omitted, but these limitations do little to the spirit of the present analysis. For five taxa, there are five possible one-term components, 10 possible two- or three-term components, five possible four-term components, and one five-term component. Of the 31 possible, 28 components are present in Ayala's data set. In all, these components can be used to define 67 completely dichotomous cladograms, not all of which are equally sound. To avoid confusion, I will use cladogram to refer to a branching diagram representing only relationships of taxa. A tree is a branching diagram where branch lengths are specified, and branch points can represent actual ancestral species. Figure 4 presents a frequency distribution of the number of defining alleles for these cladograms. The least defined cladogram (Fig. 5h) is based upon only four alleles, while the most defined (Fig. 5a) is based upon 31 alleles (alleles unique to terminal taxa have not been counted). Rather than ask which cladogram represents the phylogeny of these species, we will attempt to determine which components might represent groups of species that share a recent common ancestry.

Three of the components in Table 1 are not worth considering further, as each is represented by only a single allele. It is most likely that the terms of these components share the corresponding alleles for reasons other than recent common ancestry. If this data set does contain phylogenetic information, these components will reflect it only if the vast majority of alleles are shared because of anagenetic processes.

Table 1. Four-, three- and two-term allelic components for 152 alleles over 30 loci in 5 species for the _willistoni_ group of _Drosophila_ (Ayala, 1977). N is the number of alleles indicating a particular component (locus); + indicates presence of an allele, and - allele absence in a particular species. Species are: w = _willistoni_, t = _tropicalis_, e = _equinoxialis_, p = _paulistorum_, n = _nebulosa_. An asterisk indicates alleles discussed in text.

Component w t e p n	N	Component w t e p n	N	Component w t e p n	N
+ + + + -	12*	+ + + - -	15*	+ + - - -	4
+ - + - +	8	+ - + + -	9*	- + - - +	7
+ + - + +	7	+ + - - +	2	+ - - - +	5
+ - + + +	6	+ + - + -	2	+ - - + -	2
- + + + +	1*	+ - - + +	2	- + + - -	2
+ + + + +	41	+ - + - +	2	- - + + -	2
5 unique		- + - + +	2	- - - + +	2
		- - + + +	2	+ - + - -	1*
		10 unique		- + - + -	1*
				10 unique	

Components that are replicated in the data set contain information that may be meaningful; unfortunately, for this data set most components are replicated. Of course, there are only 31 possible components and with 152 alleles it is certain that replication would occur even if all of the components were merely a random sample. We can calculate the probability of observing a

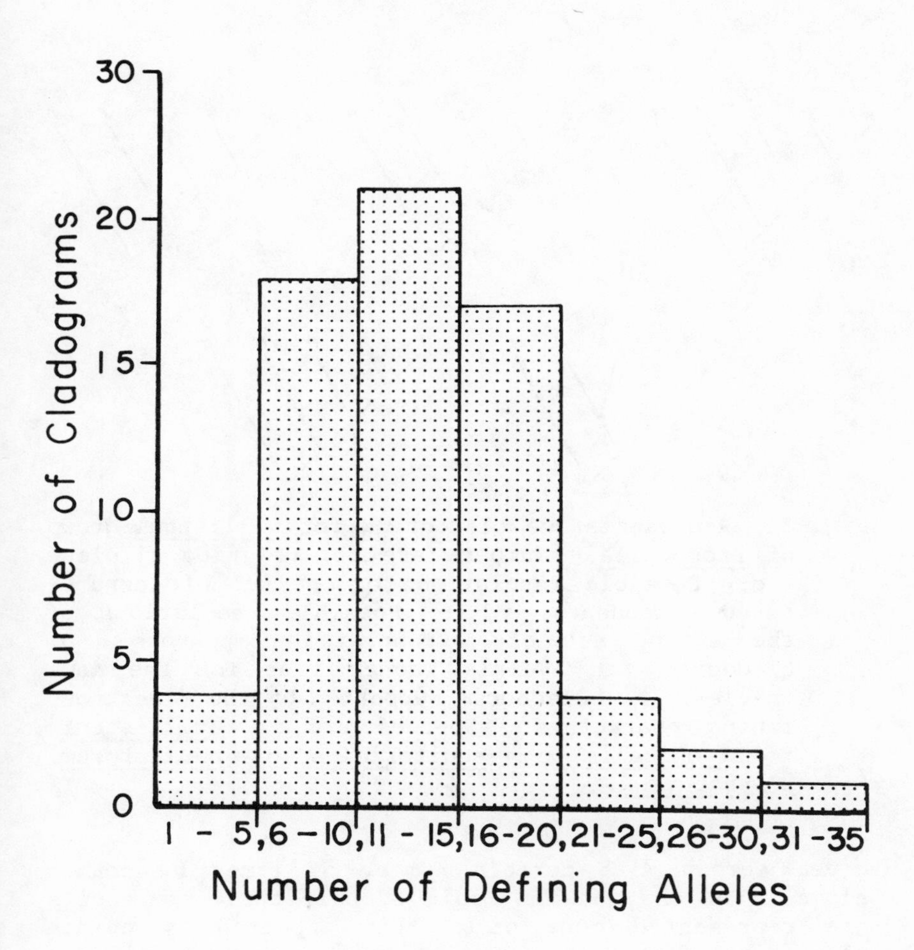

Fig. 4. Histogram of the number of cladograms defined by a given number of alleles in Drosophila willistoni complex. Data from Ayala (1977).

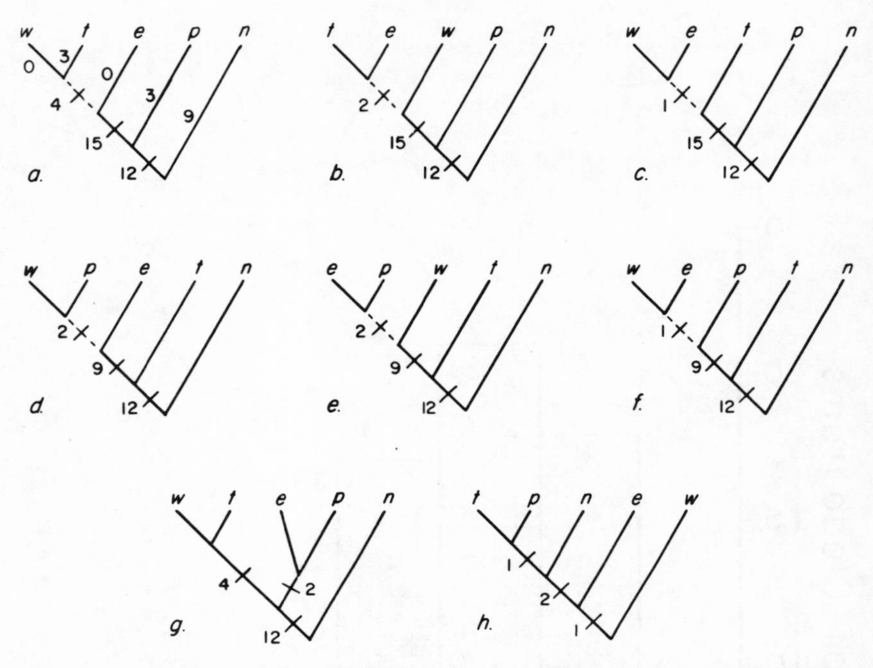

Fig. 5. Representative cladograms for *willistoni* group of *Drosophila*. Numbers are for defining alleles. a-c: Possible resolutions of one set of corroborated components. d-f: Possible resolutions of the second set of corroborated components. g: Cladogram for Ayala's Wagner tree for the same species. h: Cladogram supported by the fewest defining characters. Species are w = *willistoni*, t = *tropicalis*, e = *equitorialis*, p = *paulistorum*, n = *nebulosa*.

given component a certain number of times by chance alone. For the data in Table 1, only three components are represented more often than expected by chance alone: one four-term component is present 12 times (P = 0.0026), one three-term component is present 15 times (P = 0.0001), and one other three-term component is present nine times (P = 0.0327). The five-term component is significantly repeated but is uninformative. These three components are the only ones of Table 1 that can be identified as nonrandom associations of species, i.e., the components are corroborated.

Genetics in Phylogeny

In calculating the above probabilities, we do not imply that the data set is a random assemblage of components. There are numerous reasons why taxa might share an allele; chance is only one of these. But until chance is excluded as a potential cause, there is little reason to speculate on the importance of other causes for the pattern observed. This, at least, is the lesson of statistics. We have insufficient information on the groups formed by the majority of alleles in Ayala's data set to discuss them further. It is possible that the only genealogical information here is present in these alleles, but if so we cannot know it from these data alone.

The three nonrandom components form groups whose cause is fairly certain to be interesting. The groups formed by these components could represent phylogenetic groups or anagenetic ones. Taken together, the components specify two incompatible cladograms that both have an unresolved terminal trichotomy. Both agree in finding _D. nebulosa_ an outgroup to the other four species. They differ in whether _D. paulistorum_ or _D. tropicalis_ is the next included outgroup. Figures 5a-c give the three fully resolved cladograms representing one combination of compatible components, and Figs. 5d-f are cladograms based on the other combination. Fig. 5g is a cladogram representation of the Wagner tree resulting from this data set (Ayala et al., 1974). From the perspective of the corroborated cladograms, the Wagner cladogram illustrates tenuous patterns of relationships. The two major groups formed by the Wagner procedure (_D. willistoni_ and _D. tropicalis_; _D. equinoxialis_ and _D. paulistorum_) are based upon evidence phylogeny for these species, there is no reason to believe so from this data set alone.

What are the causes for the terms of the nonrandom components sharing alleles? Because there are two inconsistent sets of components, at least one of them must represent anagenetic groups. The three-term component repeated nine times, serving as a basis for Figs. 5d-f, is a likely candidate for representing an anagenetic group. Of the nine alleles exhibiting this component, three are alleles at the esterase-5 locus and three are alleles at the xanthine dehydrogenase locus. Thus, there are only five independent pieces of information that suggest this component. Most of the alleles representing this component probably do so because of something decidedly nonrandom occurring at these two loci. We cannot exclude the possibility that the group formed by this component is one based upon the history of selection, or possibly drift at a very few loci. Both esterase-5 and xanthine dehydrogenase show a similar pattern and may have been subjected to the same basic anagenetic forces, but it is unlikely that this pattern is produced simply by a recent common ancestry for the three species in question.

The other three-term component (Figs. 5a-c) is represented by 15 alleles that are found at 12 loci. While certain forms of selection might possibly result in a common pattern over this many different loci, it may be more reasonable to view this pattern as phylogenetic. The potential causes for this pattern are certainly fewer than for any other component in the data set, except for the four-term component represented 12 times and compatible with this one. While we cannot be certain that this pattern represents genealogical information, it is the only information in the data set that seems likely to represent phylogenetic grouping of species. Further information is necessary to test whether the cladograms of Figs. 5a-c are reasonable estimates of the phylogeny.

Ayala's Drosophila data set is unusual in the number of alleles reported per taxon studied. With 152 alleles studied in five species, Ayala has examined over five times as many character states per species as is usual in electrophoretic studies of vertebrate taxa. In these latter studies there is much less opportunity for replication of components. Table 2 presents some of the 92 components present in Straney et al.'s (1979) data set for phyllostomatid bats. For the 14 taxa studied, there are 16,383 possible components. Nevertheless, two two-term components and one six-term component are repeated: components I (four replicates), II (four replicates) and III (two replicates) of Table 2. The probability that the replication seen is due to chance alone is small; for the two-term components, $P = 3.86 \times 10^{-II}$, and for the six-term component, $P = 1.55 \times 10^{-5}$. The alleles unique to terminal taxa are also decidedly nonrandom in distribution. Forty-six of the 92 alleles in the data set are unique, but in a random collection of components only 8.60×10^{4} unique alleles are expected.

Genetics in Phylogeny

Table 2. Replicated components for allelic data from phyllostomatid bats (Straney et al., 1979). Components (numbered columns) are patterns of allelic presence (+) grouping together a given set of taxa. Each component is represented by more than one allele (individual columns; asterisks identify alleles discussed in text). Transformation from a particular allelic state to unique or other alleles is indicated by u_x and b_1 and c_1, respectively (see text for further explanation).

Species	Components					
	I	II	III	IV	V	VI
Artibeus cinereus			+ +	+ +*	+ +*	+ +
A. lituratus			+ +	+ +	+ b_1	+ +
A. jamaicensis			+ +	+ +	+ +	+ +
Vampyrops			+ +	+ u_v	+ +	+ +
Chiroderma			+ +	+ u_c	+ +	+ u_c
Uroderma			+ +	+ +	+ +	+ +
Ametrida				+ +	u_a +	+ u_a
Phyllostomus hastatus	+ + + +			+ +	+ u_n	+ u_n
P. discolor	+ + + +			+ +	+ u_d	+ c_1
Carollia					+ u_c	+ c_1
Sturnira					+ +	+ +
Glossophaga		++++				+ +
Desmodus		++++				+ +
Anoura						

Additional information can be gleaned from this data set by judicious use of transformation series analysis. I will construct transformation series strictly as hypotheses. Consider component IV, represented completely by only one allele (Table 2). The right-most allele in this component, call it a, is missing in _Vampyrops_ and _Chiroderma_, but, at the locus involved, these taxa are fixed for different unique alleles, u_v and u_c. Two transformations can be hypothesized: a into u_v and a into u_c. If true, these transformations imply that at some point allele a was present in _Vampyrops_ and _Chiroderma_ but is currently present in transformed states, u_v and u_c. Thus allele a can be considered a replicate of component IV. In like manner, hypotheses of transformation to unique alleles can be used to fill out replicates of components V and VI. Here, though, another form of transformation series is possible. The right-most allele of V, call it b, is missing in _Artibeus lituratus_, which is fixed at this locus for an allele, b_1, present also in _A_. _cinereus_ and _Vampyrops_. The taxa having b_1 form a group that is part of a previously defined group (component III), nested within the current component V. Thus, if these components reflect phylogenetic relationships, it is not unreasonable to suggest the transformation b into b_1. This transformation assumes that b_1 is derived from b. If this assumption is true, b's presence in _A_. _lituratus_ may be assumed. This hypothesis is more ad hoc than unique transformation hypotheses, for it assumes that the two alleles b and b_1 have been polymorphic for some length of time, and have differentially penetrated to living taxa. Both types of transformation hypotheses have been used to produce replication for components V and VI (four additional replicates of VI are possible, but require many transformations of the second type). I will refer to conditional components as those components, such as IV, V and VI of Table 2, that are corroborated by use of transformation hypotheses; these components are conditional upon the truth of the hypotheses made to produce replication. This is no different, though, than the usual use of transformation series analysis in morphological studies where each transformation is itself hypothetical, and the resulting components equally conditional.

The corroborated and conditional components of Table 2 are all compatible and can be used to specify a single basic cladogram. Other cladograms would result from hypothesizing other conditional components. The remaining alleles of Table 2 can be used to further resolve the cladogram by indicating possible but unreplicated components; one resolved cladogram is illustrated in Fig. 6, where the relative confidence of the components is indicated by the density of the branches. Clearly, this cladogram is a hypothesis and only the three nonterminal branches corresponding to components I, II and III have much likelihood of being parts of the true phylogeny of these bats. While this cladogram has the same basic form of any cladogram usually produced in estimating phylogenies, it is graphical representation of a more complex estimate. The illustrated confidence we can have in the various components is also a prediction of the ease or difficulty with which the various components can be falsified. While phylogenetic inference can be used to estimate phylogenies, it does so indirectly by answering other questions. For example, which groups can we reasonably expect to share a recent common ancestry, which groupings of taxa will be difficult to falsify, and therefore are most worth testing, or for which groups do we lack sufficient information to allow placing them within a phylogenetic context? Figures 5a-c and 6 are pictorial representations of answers to these questions.

For both the *Drosophila* and phyllostomatid bat data sets, all but one compatible set of components can be excluded from consideration as necessarily genealogical components. How might we test whether these single sets of components reflect the phylogeny of a group? The information present in the data sets has been exhausted and any test must come from different data for these taxa. It is curious that very few workers comparing genetic and morphological data sets have recognized that just such comparisons are the most powerful tests for phylogenetic components.

Mickevich (1978) and Mickevich and Johnson (1976) have noted that genealogical relationships are expected to emerge from the studies of different independent data sets collected from the same taxa. Indeed, this may be the only way that phylogenetic relationships can be inferred. Consider three data sets for a group of taxa, one allelic, one of morphological characters associated with feeding structures, and one of locomotor morphological characters, and suppose that each data set

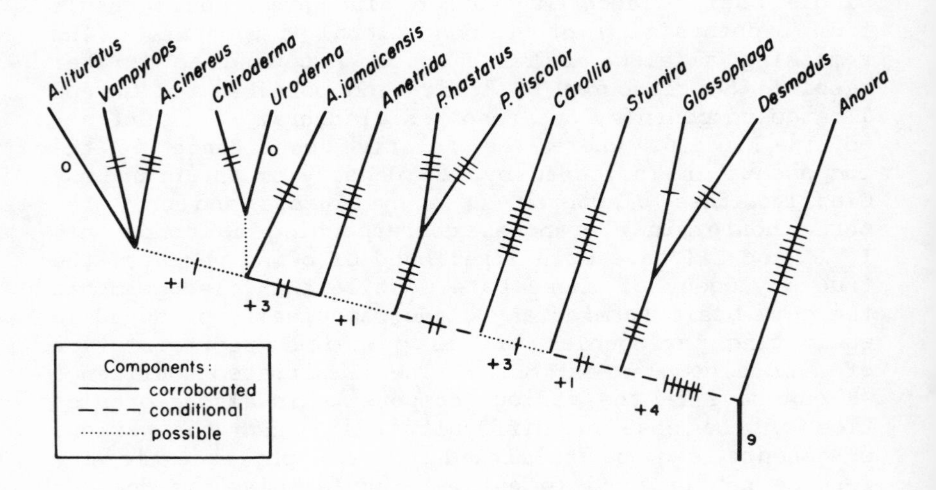

Fig. 6. Cladogram for phyllostomatid bats. Tickmarks indicate defining alleles. Numbers refer to alleles present in some but not all subsequent taxa.

yields a consistent set of corroborated, potentially phylogenetic components. If none of these components is common to two or more data sets nothing can be said about the phylogeny of the group. Each set of components could have arisen because of anagenetic processes affecting each type of character differently. Some workers, I am sure, would tend to accept the allelic components as most likely to be phylogenetic groups, but this is based upon the assumption that genetic data are somehow "better" at preserving phylogenetic information than are morphological characters (Avise, 1974). I know of no evidence that this view is anything more than genetic chauvinism, and until such evidence is available inference based upon this assumption should be avoided. On the other hand, corroborated components common to two or more independent data sets are most likely phylogenetic in origin. The two morphological data sets might not be independent if feeding and locomotor adaptations could have appeared in response to the same environmental conditions. We would, therefore, accept as phylogenetic those components shared between the morphological data sets if we can assume that the characters involved are independent. Allelic characters have the advantage of being independent of most types of morphological characters. Individual loci might be correlated with certain morphological features (Johnston, 1974), but a pattern common to a number of gene loci and a number of morphological characters is not likely to be the result of a common adaptive response. We can be reasonably confident that corroborated components repeated across independent data sets reflect elements of the phylogeny of a group.

Testing the phyllostomatid bat cladogram of Fig. 6 with reference to other data sets is not as easy as one might wish. Surprisingly, there are no morphological studies of these bats useful for testing these components. Prevailing systematic opinion is decidedly against the basic framework of Fig. 6 representing phylogenetic relationships, but this is a weak test. The current view of phyllostomatid systematics is based primarily upon aspects of feeding habits and specializations, so it would not be surprising to find that groups formed on this basis do not correspond completely to phylogenetic groups. Chromosomal data (Baker, 1979) and information from microcomplement fixation studies (V. M. Sarich, pers. commu.) are accumulating for species in this family of bats, but the available information tends to confirm current systematic opinion more than the cladogram of Fig. 6. Of the corroborated and conditional components, only the corroborated ones are also indicated in other data sets. The hypotheses used to construct the conditional components, then, do not appear to be very realistic and can be rejected.

The bat data set contains little necessarily phylogenetic information; only three components potentially represent phylogenetic relationships and one of these is certainly not phylogenetic. I suggest that this result is based upon poor sampling of character states (alleles) rather than some intrinsic inability of genetic information to reflect phylogeny. While 92 character states for 14 taxa would probably be sufficient for morphological studies, genetic characters can change at a much faster rate than do morphological ones and more character states will be required to obtain a clear pattern of genetic relationships. It will be interesting to see whether the component analysis of Ayala's *Drosophila* allelic data agrees with other data sets. Spassky et al.'s (1971) phylogeny for these species can not be accepted as a test without knowing more about the basis of its construction. The response to situations where genetic data sets provide little potential phylogenetic information should be the same as with morphologically based studies: we should gather more data rather than throw away what we have. When the former solution is impractical, though, a less rigorous approach is possible.

Nelson (1979) has concluded that comparisons between data sets are necessary for strong phylogenetic inference, but his analysis is based upon comparing phenograms from different morphological data sets. Analysis of actual data sets in terms of components is a much more trustworthy way of identifying potential genealogical components than any clustering algorithm that produces groups from any data supplied to it. Yet, realistically, there will be cases when a component analysis will not recognize sufficient corroborated components and other techniques will have to be used. Clustering techniques produce groups, no matter what the algorithm, by specifying transformation hypotheses for the available character states. Thus, clustering algorithms are no different in principle from, although vastly more efficient than, the approach used above to recognize conditional components. If the hypotheses are true, the output of a clustering algorithm is a phylogenetic estimate; if the hypotheses are false, the resulting tree consists of a number of random groups.

How might we test the transformation hypotheses implicit in a particular cluster analysis? One approach is to see whether the same conditional components appear in similar analyses of different data sets. Mickevich and Johnson (1976), for example, examined electrophoretic and morphometric characters in five species of the fish Menidia. Wagner tree estimates from both data sets show excellent agreement in the groups formed; Fig. 7a is a cladogram including all components present in both trees. Phenetic analyses of the same data sets produce phenograms sharing no components. The Wagner procedure constructs basically the same transformation hypotheses for the two data sets while the phenetic methods fail to imply the same set of hypotheses. We can confidently reject the phenetic transformation hypotheses in favor of the Wagner hypotheses, although the possibility still remains that the Wagner hypotheses are congruent only by chance. This can be checked only by finding corrobo-

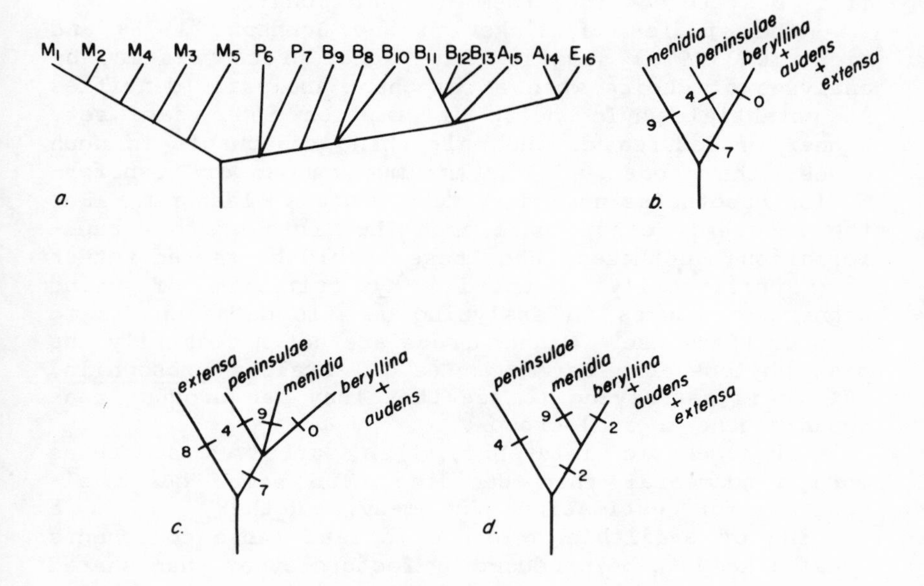

Fig. 7. a: Cladogram for populations of the fish, Menidia. Components are included in this cladogram if present in both morphometric and allelic data sets of Mickevich and Johnson (1976). Letters indicate species (first letter of specific designation) and numbers refer to populations. b-d: Cladograms for replicated components of the allelic data set for Menidia (Mickevich and Johnson, 1976). Numbers indicate defining alleles.

rated components within the analyzed data sets, or in new ones, that are elements of the Wagner tree. For Mickevich and Johnson's morphometric data, there are no corroborated components, but the allelic data set does contain nonrandomly replicated components ($P < 0.001$). Figures 7b-d illustrate the cladograms resulting from combinations of consistent corroborated components; one (Fig. 7b) is identical to parts of the cladogram common to both data sets. Thus, at least some of the groups of Fig. 7a can reasonably be viewed as phylogenetic groups. We expect that the reamining groups are as well, although until they are confirmed as phylogenetic groups it is best to consider them as conditional.

The studies of Mickevich and Johnson (1976) and Mickevich (1978) suggest that Wagner procedures may be analyses of choice when a component analysis identifies no potential phylogenetic information in a data set. Wagner procedures do the only thing reasonable in such cases; they construct the minimum number of transformation hypotheses necessary to produce a cladogram. But the result is conditional upon the truth of the transformation hypotheses, and these should be tested rather than uncritically accepted. My criticism of using Wagner procedures in analyzing genetic data applies to an uncritical use. Wagner trees are not necessarily the best phylogenetic estimates (e.g., Ayala's *Drosophila*) but it is satisfying to see that they can produce reasonable genealogical groups.

Phylogenetic inference, then, involves something more fundamental than deciding upon some "best" algorithm for estimating phylogeny. Rather, it is a sifting of available evidence to find those components least likely to be produced by factors other than shared recent common ancestry. In short, we can infer a phylogenetic basis for a particular component if: (1) The component is corroborated within a data set, or, if not, simple transformation hypotheses can be formed to produce repetition; (2) we can exclude the possibilities that the component is replicated because of chance sampling or special factors affecting a very few characters; and, (3) the component is found corroborated in data sets drawn from very different aspects of the organism under study. This form of inference may not always identify phylogenetic components, and may rarely produce a fully resolved phylogeny. But it is a conservative and probabilistic form of inference. Only those aspects of a phylogeny are identified that are supported by sufficient evidence to be recognizably nonrandom. We should not expect more from our data sets than is really there.

Study of Historical Process

Component analysis of most data sets identifies only a small portion of the information present as potentially phylogenetic. Much of the remaining data have the aspect of random noise, but are all of them totally uninformative? There is, after all, more to the history of a lineage than the sequence of cladistic events, just as in any family history there is more than a bare recital of who begat whom. The challenge of evolutionary biology is to study the nongenealogical aspects of lineage histories, the processes and conditions for the changes that have occurred. We do this by examining, within the context of an inferred phylogeny, the processes by which taxa share character states. Since evolutionary processes are population processes, genetic information makes such studies more than academic exercises. While there is no theory of population morphology, there is reasonable understanding of the sufficient conditions for particular types of genetic change. Thus, the hope is that we might be able to find nontrivial explanations for particular patterns of allelic change through time.

Studies of evolutionary processes will be only as sound as the phylogenetic estimates upon which they are based. Robust phylogenetic estimates are not available for the examples used here (Ayala's _Drosophila_ and Straney et al.'s phyllostomatid bats); these discussions are meant to be heuristic. In some respects, a complete phylogeny for a group may not be necessary for studying general processes. While details will depend upon the actual tree used, we might hope to find some general patterns that emerge from various possible phylogenies of a group. Such results would be robust where the phylogenies themselves were not. Romero-Herrera et al. (1978) have used this strategy of considering multiple hypothetical phylogenies in their studies of myoglobin evolution in mammals; their results are correspondingly strengthened by being independent of the tree used. In like manner, I have examined the patterns discussed below on several possible phylogenetic trees and, while the details differ, the same general patterns emerge.

To change a cladogram resulting from phylogenetic inference into a tree necessary for evolutionary study requires only conceptual change. Internal nodes are now considered to represent actual ancestral species and branch lengths have temporal meaning. This poses problems for placing the alleles from a data set onto the branches of a tree. There is no way to know precisely when an allele arose, nor how long it was maintained in low frequencies. I have followed two simple rules in deciding when alleles arose: (1) Alleles are placed along the latest branch where they could have arisen, consistent with their distribution in sampled taxa; (2) alleles present in some, but not all, taxa on both sides of a basal split are considered to be primitive if the number of alleles per locus in the hypothetical common ancestor does not exceed the average number seen in living taxa. When this limit is exceeded alleles present in the fewest, most widely separated taxa are considered convergent. Figure 8 presents an hypothesis of allelic change at two loci in the _willistoni_ group of _Drosophila_ (Table 1, Fig. 5; Ayala, 1977). In both cases, the number of inferred primitive alleles matches closely the observed allelic diversity in living taxa; AO-1, six primitive alleles, 6.4 alleles per species, and XDH, seven primitive alleles, seven alleles per species. Illustrated are mutation and loss estimates for each allele present in living taxa, as well as probable points of shifts from one major allele to another.

The _Drosophila_ data set is interesting because Ayala has inferred from it that the major cause of the distribution of alleles in current species in natural selection. He views his alternative explanation of convergence as "a totally unlikely possibility" (Ayala, 1977:199). From a molecular viewpoint, this view is hardly tenable (Coyne, 1976; Romero-Herrera et al., 1978), but it is true that convergence is an unnecessary explanation for these patterns. We can just as easily assume that alleles are primitive when present in taxa scattered across the tree, at least for these data (possible exceptions are XDH alleles 92, 94, and 95 present in _D_. _tropicalis_ and _D_. _nebulosa_; electrophoretic convergence is well known at the XDH locus in _Drosophila_ (Throckmorton, 1977). But if convergence is unnecessary, natural selection is as well. The shifts from one major allele to another may have a selective basis, but the main pattern seen at the loci studied is gain and loss of alleles (Fig. 8). It is certainly not necessary to postulate a selective basis for origination and maintenance of alleles (Kimura and Ohta, 1971). Selection also seems unlikely as the cause for the loss of alleles. Effective population size is very large in these species (Wright, 1978), and selection would have to be either very strong or present over long periods of time to result in the amount of loss required to produce

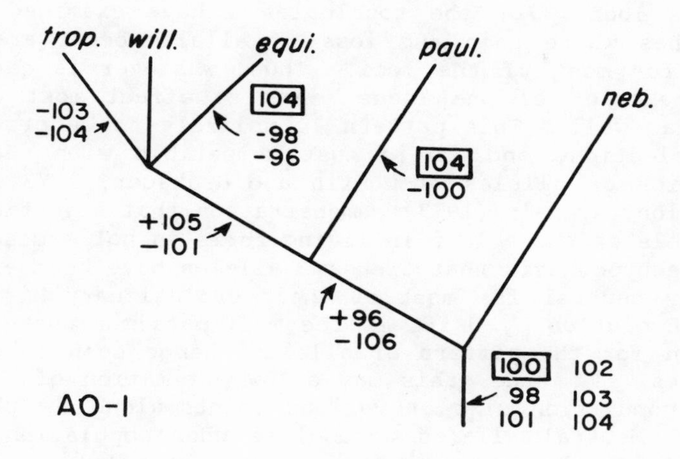

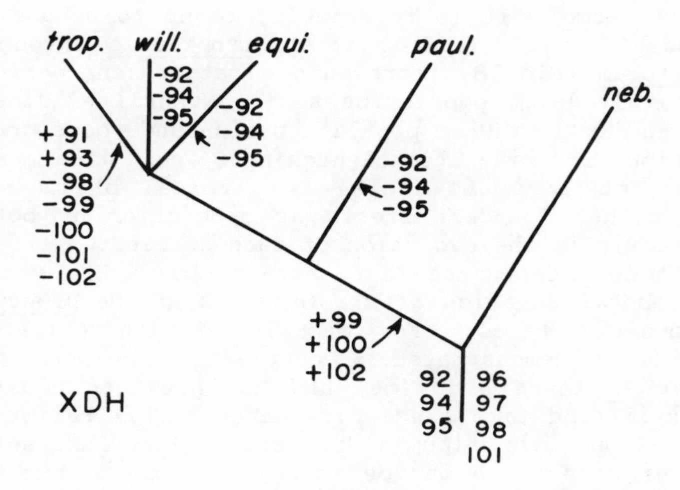

Fig. 8. Possible evolutionary histories of alleles at AO-1 locus and XDH in the willistoni group of Drosophila. Designation of the common allele is boxed; other allelic changes indicate gain (+) and loss (-) of alleles along a given path. The primitive complement of alleles is indicated at the root.

currently observed allelic complements. More important, though, is the pattern of gain and loss of alleles across loci. For the topologies I have examined, the branches where gain and loss of alleles occur are the same for most of the loci. Thus, whatever is causing the behavior of one locus seems to affect most other loci as well. This pattern is unlikely to be produced by selection, and it is most compatible with neutral behavior of alleles (Lewontin and Krakauer, 1973). In addition, Ayala's (1977) demonstration that selection is possible at these loci in living forms is not a disproof of the hypothesis that the same alleles have been effectively neutral for most of their evolutionary history.

Evolution by drift is the most parsimonious explanation for the pattern of allelic change seen in these species. If true, this may allow estimation of effective population size at various points along the phylogeny. Neutral alleles accumulate when population size is effectively large, and the number of alleles arising is a function of the amount of time population size remains large. It is reasonable, then, to assume that the basal branch leading to _D. tropicalis_ through _D. paulistorum_ (Fig. 8) represents a rather long period of effectively large population size. Neutral alleles are most readily lost when population size undergoes drastic reduction in size (bottlenecking), and the general pattern of loss of alleles in terminal branches may index either founder effects at speciation or bottlenecks early in the evolution of each species.

It is unfortunate that there is no way to estimate the temporal duration of the branches of the _Drosophila_ phylogenetic trees. For mammals, Sarich (1977) has empirically demonstrated a correlation between Nei's genetic distance and time, and has provided equations for estimating time since divergence. This calibration is based upon the albumin "molecular clock" and suffers whatever bias there may be in that timepiece, but using Sarich's approach should at least give reasonable relative time estimates. Relative estimates of the temporal lengths of branches are all that is necessary for examining rates of change.

Temporal branch-length estimates can be used to calculate allelic "substitution" rates along the various portions of the tree. "Substitution" is operationally defined as the point of appearance of an allele that is found in all later taxa, except where a later arising allele has become fixed. Sarich's calibration has been used to estimate times of divergence for the phyllostomatid bat cladogram (Fig. 9). Substitutions per branch are indicated in Fig. 6; this cladogram is only used for convenience. The average overall rate of substitution, across loci, from the basal common ancestor to each living species is 0.33 alleles/million yr, with a range of 0.21 to 0.42. Rates along each branch of the cladogram can be expressed as deviations from this average rate, as illustrated in Fig. 10 where, for simplicity, branch rates within the observed range of overall rates are not considered deviations.

Considerable heterogeneity is evident in substitution rates along branches in this phylogeny. This heterogeneity of single branch rates compared with the relative homogeneity of overall rates is a result common to several recent studies (Romero-Herrera et al., 1978). Terminal branch rates reflect substitution of unique alleles and these are particularly heterogeneous. The differences in rates possibly reflect differences in population structure during the histories of these species. Most of the internal branches have substitution rates much higher than average. For example, the branch underlying the split of _Phyllostomus_ species from the remaining taxa has an estimated temporal duration of zero million yr (Fig. 9), which may be a very short period of time for two alleles to undergo substitution. On average, two substitutions require about 6 million yr. The number of alleles studied in this case are too small to have much confidence in this anomaly, but were it true it would indicate very rapid change along this particular branch, perhaps evidence that selective forces were acting. The temporal duration of this branch poses some problems in interpretation, but if, with better sampling, we find alleles at several loci undergoing substitution along this branch a strict selective interpretation becomes less likely, and we might look additionally to small population effects for possible causes.

These two examples illustrate how allelic data can be used to examine aspects of the evolutionary history of a group. These data will be most useful in addressing questions of rates and effective population size during evolution, though attention needs to be paid to analytical techniques for such studies. We know very little about the way these factors mold adaptive respon-

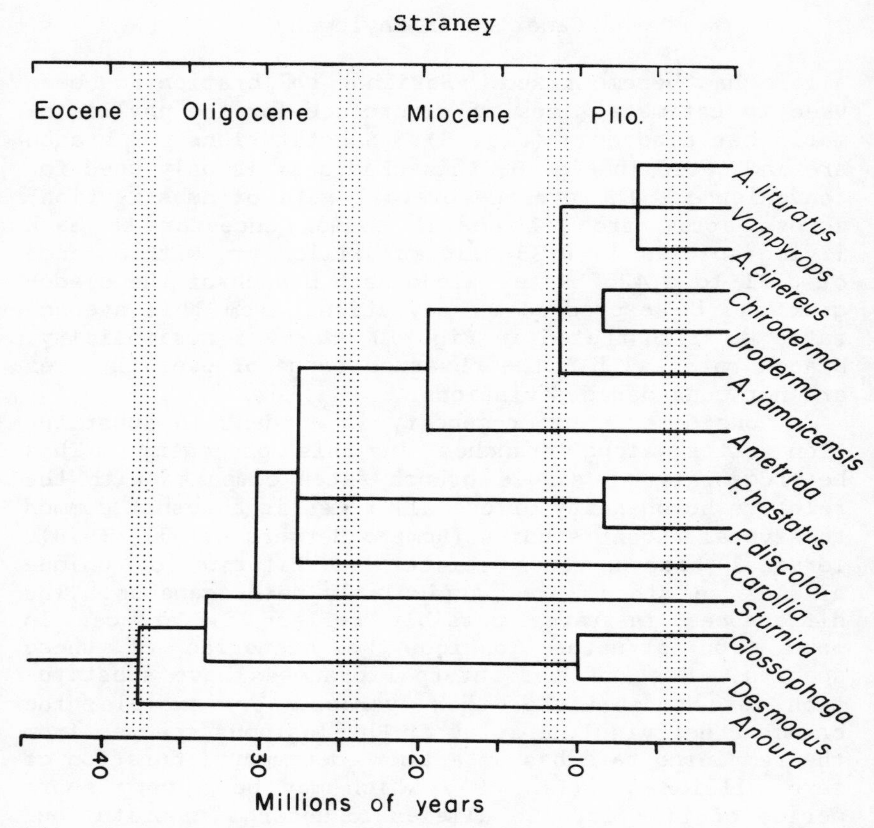

Fig. 9. Topology of Fig. 6 placed upon a time scale based upon Sarich's (1977) conversion of genetic distance to time.

ses of organisms, although some theoretical relationships have been postulated (Eldredge and Gould, 1977). With more and better genetic data sets, we may gain an insight into the complex processes involved in phyletic evolution.

Hidden Variation

The preceding discussions have implicity assumed that homology is easily established for genetic characters. In theory this is true, but recent developments argue that in practice it is more difficult than most have believed. In the past few years it has become quite apparent that electrophoretic alleles are not homogene-

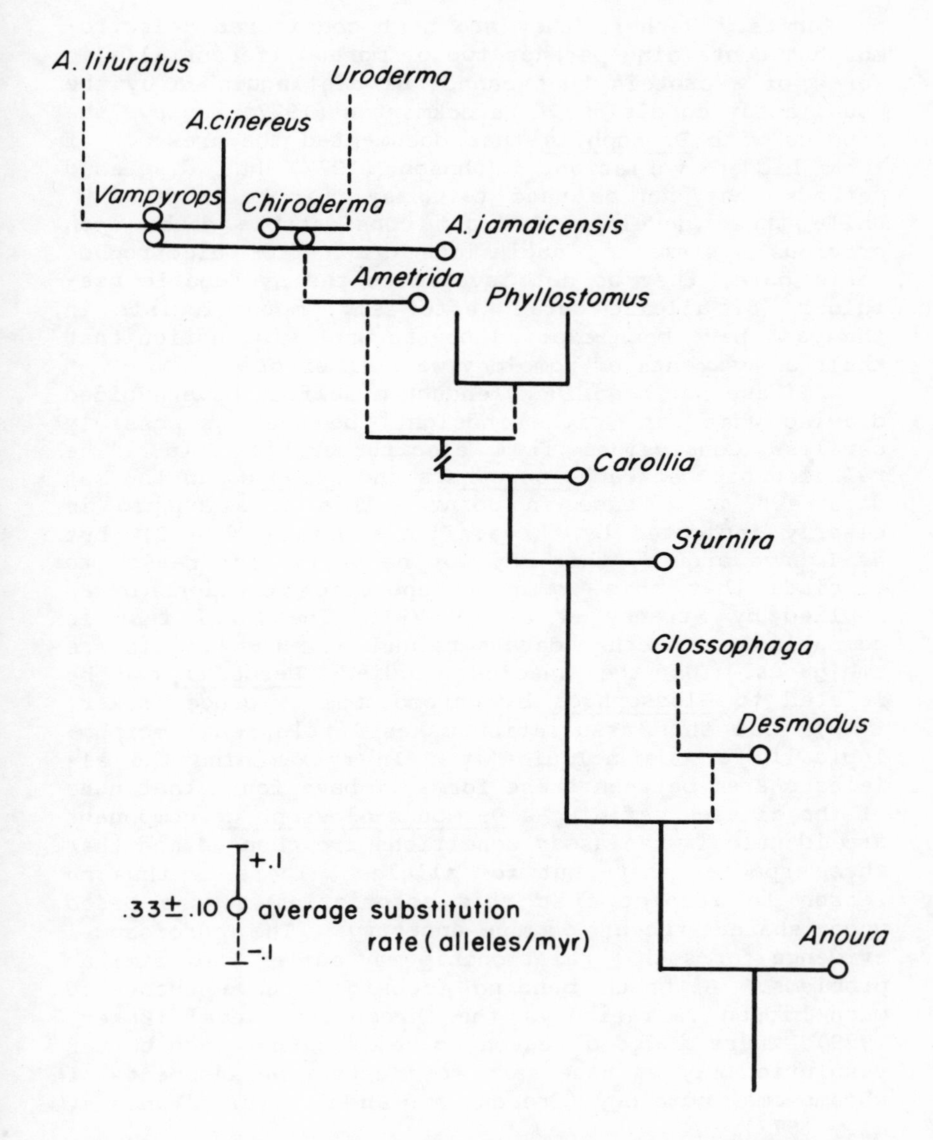

Fig. 10. Rates of allelic substitution in phyllostomatid bats. Branch lengths reflect the positive (solid) and negative (dashed) deviations in rate from the overall lineage mean. Circles represent taxa where the rate of substitution falls within the range observed for overall rates.

ous units. Rather, they are best considered "electromorphs" containing perhaps two or more different allelic forms of a protein that cannot be distinguished by the usual assay conditions. Throckmorton (1977) reviews the studies with *Drosophila* that documented the presence of this hidden variation. Johnson (1977) has discussed methods that can be used to assay this heterogeneity. While these developments cast considerable doubt upon previous systematic conclusions drawn from electrophoretic data, they do not invalidate the systematic usefulness of allelic data. After all, morphologists in the past have been exposed to the unnerving notion that their assessments of homology were in error.

I have not been lucky enough myself to have avoided drawing what can, with hindsight, be seen as possibly careless conclusions from electrophoretic data. The relationship between *Glossophaga* and *Desmodus* in the bat data set is a case in point. This relationship is clearly indicated by the available data (Table 2), but as I have argued, this may not be sufficient reason to conclude that this is a phylogenetic relationship as implied by Straney et al. (1979). The final test is comparison with other data sets and here the results are ambiguous. Of the species studied, *Desmodus* can be related to *Glossophaga* by chromosomal evidence (Baker, 1979), but this association makes little sense morphologically or immunologically. In reexamining the alleles shared between these forms, I have found that none of the alleles defining a *Desmodus-Glossophaga* component are identical when assay conditions are changed and that these species share but few alleles. There is thus no reason to suspect that this association is one based upon shared recent common ancestry. The chromosomal evidence for such a relationship may suffer from similar problems. Although banding techniques have uncovered much hidden variation at the chromosomal level (Baker, 1979), there is good reason to believe that much better resolution may be necessary for accurate assessments of chromosomal homology (Creagan and Ruddle, 1977; Yunis et al., 1977).

Techniques of nucleotide sequencing of codons will sooner or later replace electrophoresis as "the" way to document allelic variation. These techniques are currently costly and time consuming, but as it becomes possible to assay sequence variation quickly it will be possible to take population level variation into account for systematic comparisons. The utility of sequence information has been nicely detailed by Romero-Herrera et al. (1978) in their study of mammalian myoglobins. This protein is particularly appropriate given the present state of the art, as population variation of myoglobin alleles appears to be quite low (only four alleles, three rare, are known in human populations). Sequence information has only the advantage of accuracy over the less sensitive electrophoretic techniques, and information on sequences is not necessarily more useful than other ways of assessing genetic relatedness in resolving systematic questions.

CONCLUSIONS

Genetic information has been directly used in systematic studies for a little more than a decade and the discovery of hidden variation presents a convenient opportunity to assess how such information should be handled in studying the history of lineages. As we succeed in obtaining more accurate means to establish allelic and chromosomal homology, it will be necessary to have a better idea of how such information can be used to determine phylogenetic relationships. The outline of phylogenetic inference presented here may be a useful start in that direction. It may seem from my discussion that allelic data are of questionable use in phylogenetic study, but this is not the case. Allelic data simply require more intensive sampling of character states and greater care in analysis than the usual data of systematics. Certainly, genetic information will continue to be useful in formulating and testing phylogenetic hypotheses. In the long run, genetic information will have its most important use in studying the processes by which evolutionary change is effected. Indeed, genetic information will be necessary for such studies, since more is known theoretically about the evolutionary behavior of alleles than about any other type of character. This area of evolutionary biology has been infrequently studied in the past, and we should expect greater insight in the future. Hopefully, it will become possible to formulate testable hypotheses for mechanisms of phyletic change. It may never be possible to map completely the stream of heredity that is phylogeny; too much of the past is completely closed to us. But genetic information provides a new perspective for viewing the past, and there is reason to be optimistic that the view will be an interesting one.

Acknowledgments--This paper was, in a very real sense, made possible by the freedom of discussion several people extended to me: J. E. Cadle, S. L. Coats, J. C. Hafner, M. S. Hafner, J. L. Patton, and D. B. Wake. J. Hanken and G. J. Nelson deserve special thanks for helping me to see through particularly difficult problems. E. D. Pierson, V. M. Sarich, S. S. Sweet and two anonymous reviewers offered useful comments on various versions of the manuscript. To these, and others, my sincere thanks.

SOCIAL SUBSTRUCTURE AND DISPERSION OF GENETIC VARIATION IN THE YELLOW-BELLIED MARMOT (*MARMOTA FLAVIVENTRIS*)

Orlando A. Schwartz and Kenneth B. Armitage

Abstract--Yellow-bellied marmots located in the East River Valley and North Pole Basin of Gunnison County, Colorado were the subject of population studies for 16 years and 5 years, respectively. The detailed knowledge of individual life histories, demography, and colonial substructure of these animals provided an ideal background for a study of the significance and dynamics of electrophoretic variation in this species. Thirty specimens were collected from each of three localities at 2680, 3170, and 3660 m elevation, 203 blood samples were obtained from the East River Valley (2900 m), and 88 blood samples were obtained from North Pole Basin (3400 m). Allozyme variation was found at 8 of 20 loci examined. Significant positive results included an association between TRF genotype and aggressiveness, a correlation between LAP gene frequency and density, gametic disequilibrium between three pairs of loci, and significant heterogeneity in the dispersion of genetic variation among the colonies of marmots. No association between gene frequencies or heterozygosity and altitude, habitat, age, sex, survivorship, litter size, and a suite of behavioral variables was found. Recent mutation was excluded as an explanation of the distribution of genetic variation. Identical selective forces or sufficient gene flow among the five study sites could account for their similarities in gene frequencies and heterozygosity. Deterministic forces were not of sufficient magnitude to prevent significant drift acting within the spatial and temporal structure of marmot colonies.

INTRODUCTION

Population structure, substructure, and their interactions with other deterministic and stochastic forces are important considerations of population genetic theory. Endler's (1977) models for clines in gene frequency had population structure as an essential element. Group selection theory, particularly that part dealing with kin selection, requires that drift acts within small, substructured populations (Wilson, 1975:107). Mutations in large, panmictic populations face a high probability of loss after only a few generations (Spiess, 1977:374). Hence, Wilson et al. (1975) deduced that the social substructure of mammalian populations (bottlenecking) allowed the rapid incorporation of chromosomal mutations into populations and the rapid evolution of mammalian taxa. Sampling errors in the association of gametes and inbreeding in small substructured populations greatly increased the probability of fixation of a mutation through drift, conditions that also lead to genetic heterogeneity between subgroups in the population.

Genetic heterogeneity among small, social groups, and by analogy the establishment of the effective action of kin selection in populations, is enhanced under the following conditions (McCracken and Bradbury, 1978): (1) preferential recruitment of juveniles into their natal colony; (2) the restriction of mate selection to those within the social group; and, (3) a low exchange rate between members of groups. McCracken and Bradbury (1978) found no heterogeneity in three allozyme loci in social colonies of a phyllostomid bat. All juveniles dispersed from the natal colony, and recruitment into harems was random. They concluded that sociality in this species was not due to kin selection.

The yellow-bellied marmot (*Marmota flaviventris*) is a large, herbivorous, diurnal, ground squirrel living in the mountainous regions of western North America. Sociality in a population of marmots located in the East River Valley (ERV) of Gunnison County, Colorado has been the subject of 16 yr of study by Armitage. Marmots were trapped, ear tagged, and marked with a black fur dye for individual recognition. Social relations within colonies were observed for 250 hr or more each summer. Marmots generally occupy a habitat of large rocks associated with nearby meadows; such habitat is patchy in the Transition Zone vegetation of the East River Valley. Smaller patches termed satellite sites are occupied by one or a few marmots; whereas in larger patches socially structured colonies of one or more polygynous groups occur. Satellite marmots are characterized by rapid turnover in occupancy, poorer reproductive success, and lack of social structure when compared to colony animals (Svendsen, 1974). The typical polygynous group consists of a territorial male, a harem of two or three females, yearlings, and young of the year (Downhower and Armitage, 1971). Six colonies (in this paper colony is synonymous with polygynous group) in the present study (Nos. 1, 4, 6, 7, 8, and 10 in Fig. 1) were observed annually as typical marmot societies (Armitage and Downhower, 1974) though population density and social behavior varied from year to year (Armitage, 1975, 1977). Four additional areas, not under close observation, were designated as colonies for this study because they contained numbers of individuals and the stability of occupancy typical of colonies.

Sociality in marmots was also studied for the last five years (Johns and Armitage, 1979) at a subalpine site in North Pole Basin (NPB) of Gunnison County, Colorado. The methods of study were the same as for the ERV colonies. In NPB and similar alpine areas suitable habitat was more continuously distributed than in ERV, resulting in colonies being more continuously distributed with the boundaries of adjacent colonies defined by social mechanisms rather than the boundary of a suitable patch. Existing detailed knowledge of the life histories, population processes, and social substructure of marmots from the two study sites provided the ideal background to describe the relationship of genetic variation to these variables. To provide a broader ecological context in which to view allozymic variation in ERV and NPB populations, we collected additional marmots from elevations below the East River Valley, above North Pole Basin, and between the two areas.

MATERIALS AND METHODS

Blood samples were taken from 203 marmots in the East River Valley, elevation 2900 m and from 88 marmots in North Pole Basin, elevation 3400 m, of Gunnison County, Colorado during the summers of 1975 to 1977. A trapped animal was restrained in a handling sock and blood was collected from the femoral vein into a Vacutainer. Each animal was released at the site of capture. Blood samples were processed in the field and laboratory as described by Selander et al. (1971).

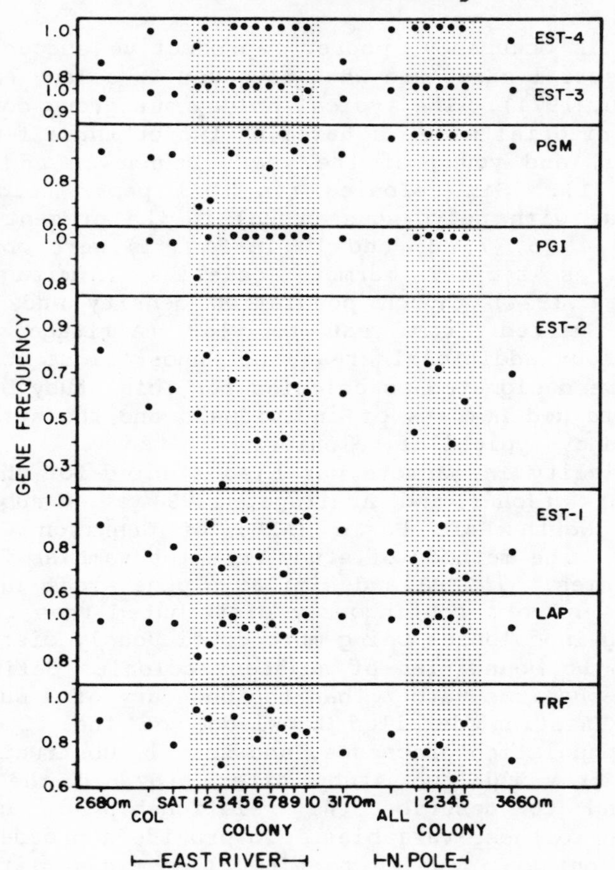

Fig. 1. Gene frequencies from eight blood allozyme systems from five study sites. The average gene frequencies from colonies (COL) and satellites (SAT) in the East River Valley study site are reported separately since they differ demographically. Gene frequencies from each colony in the East River Valley and North Pole Basin are in the shaded portions. Sample size for the three sites designated by altitude was 30. Sample sizes for TRF, LAP, EST-1 and 2 systems were: COL, 136; SAT, 67; ERV colonies 1 to 10: 8, 8, 7, 28, 8, 16, 8, 16, 11, 34; NPB all, 88; NPB colonies 1 to 5: 8, 15, 10, 10, 20. Sample sizes for the PGI, PGM, EST-3 and 4 systems, as for the above areas, were respectively: 118, 27, 6, 8, 0, 23, 8, 12, 8, 17, 8, 28, 58, 5, 15, 6, 5, 5.

Social Substructure and Genetic Variation

Ninety additional marmot specimens were collected in 1976 and 1977 by trapping or shooting from the following localities of Gunnison County, Colorado: 30 specimens from along the courses of Brush Creek and Cement Creek located 3.4 km and 7.5 km south of the town of Crested Butte at an altitude of 2680 m; 30 specimens from Schofield Park located 18.3 km north of Crested Butte at an altitude of 3170 m; and, 30 specimens from upper North Pole Basin located 18.4 km north, 9.8 km west of Crested Butte at an altitude of 3660 m. Samples of blood, heart, liver, and kidney tissues were taken from each specimen and frozen at -20 C. We attempted to randomize our sample of specimens by collecting at many different localities at the above sites. The five study sites represented an altitudinal gradient of approximately 1000 m and had marmots living in the Upper Sonoran, Transition, Canadian, Krumholtz, and Alpine life zones.

Horizontal starch-gel electrophoresis using a solution of 13% Electrostarch (Otto Hiller, Madison, Wisc.) was used for the analysis of all tissue proteins. The stains and buffer systems described by Selander et al. (1971) were used unless otherwise indicated; the phosphate buffer was described by Engel et al. (1970). Proteins were obtained from the following sources:

(1) Plasma--transferrin, two general protein systems, albumin, leucine aminopeptidase, and the two most anodal esterase systems. These esterase systems were scored by inhibiting the most anodal system with eserine sulfate (Selander et al., 1971).

(2) Hemolysate--6-phosphogluconate dehydrogenase (using buffer 4), hemoglobin (two systems, buffer 3), the two most anodal esterase systems (buffer 1) and the following systems using phosphate: phosphoglucose isomerase, phosphoglucomutase (two systems were scorable), lactate dehydrogenase (coded by two loci), α-glycerophosphate dehydrogenase, glucose-6-phosphate dehydrogenase, and NAD-dependent malate dehydrogenase.

(3) Other Tissue--glutamate oxalate transaminase (two systems for liver), sorbitol dehydrogenase (from liver using buffer 5 and stain 1 with sorbitol substituted for ethanol as the substrate), xanthine dehydrogenase (from liver), and the following systems using phos-

phate buffer: alcohol dehydrogenase (from liver), isocitrate dehydrogenase (two systems from heart and liver), α-glycerophosphate dehydrogenase (from heart), phosphoglucomutase (liver), and acid phosphatase (liver; Shaw and Prasad, 1970).

From 1963 to 1974 Armitage periodically collected and froze blood plasma samples for studies of serum hormones. These 185 samples had biochemically active leucine aminopeptidase, though the other plasma systems were not scorable. Our total sample size for leucine aminopeptidase was 567. Since we collected only plasma in 1975 and were unable to recapture some animals in 1976 or 1977, plasma system sample size was 203 for ERV marmots and 88 for NPB marmots, whereas hemolysate system sample sizes were 145 and 58, respectively.

RESULTS

The variable systems found in the blood were transferrin (TRF), leucine aminopeptidase (LAP), phosphoglucose isomerase (PGI), phosphoglucomutase (PGM), and four esterases (EST-1 to 4), with two alleles at each locus. The gene frequencies and observed Hardy-Weinberg expected genotype frequencies are in Table 1. The variable tissue systems were α-glycerophosphate dehydrogenase (α-GPD) and phosphoglucomutase (PGM) with the variant allele found in only 1 of the 90 specimens for each protein (not the same animal). Tissue α-GPD and PGM were systems that were either not present or not scorable in hemolysate.

The two plasma esterase systems were in close proximity on the gel with the slow band (S) of EST-1, the most anodal of the two systems close to the fast band (F) of EST-2 so there was some difficulty in scoring these systems. Among 76 young in 31 litters in which both parents were known there were five phenotypes in the EST-1 system and two in the EST-2 system that were not consistent with a Mendelian interpretation of inheritance and these were regarded as scoring errors in a Mendelian system. The other six variable systems supported a Mendelian interpretation of inheritance without exception.

The proportion of polymorphic loci, loci with the frequency of the alternative allele greater than 0.01,

Table 1. Gene frequencies (p_1 and p_2) and observed (O) and Hardy-Weinberg expected (E) numbers of genotypes for blood allozymes for all marmots sampled.

System*	p_1	p_2	p_1p_1	p_1p_2	p_2p_2
TRF					
O	.831	.169	254	127	1
E			263.8	107.3	10.9
LAP					
O	.952	.048	515	49	3
E			513.3	52.3	1.3
PGI					
O	.989	.011	288	4	1
E			287.0	5.9	0.0
PGM					
O	.928	.072	257	30	6
E			252.5	39.0	1.5
EST-1					
O	.797	.203	230	149	3
E			242.7	123.6	15.7
EST-2					
O	.637	.363	138	211	33
E			155.2	176.6	50.2
EST-3					
O	.981	.019	283	9	1
E			282.1	10.8	0.1
EST-4					
O	.956	.044	267	26	0
E			267.6	24.9	0.6

*Abbreviations in text.

estimated from the blood systems was 0.40 (8/20) and from the blood and tissue loci together was 0.27 (8/30). Average heterozygosity ($H = \Sigma (1 - \Sigma p_i^2)/20$) was 0.0749, where p is the frequency of the ith allele at each locus. The average proportion of heterozygous loci per individual was 0.0872 (SE = 0.0034) calculated over 20 loci. To facilitate comparisons between marmots from the five study sites the rare tissue variants were not used in any calculations. Later in this paper the proportion of heterozygous loci was calculated as a fraction of the eight variable systems, thus removing a constant factor of 12 from these calculations, and these proportions were arcsine transformed to normalize their distributions (Sokal and Rohlf, 1969:386).

There were no significant differences in gene frequencies and heterozygosities for marmots at the five study sites shown in Figs. 1 and 2. The gene frequencies of ERV colonial and satellite marmots were reported separately to demonstrate their differences in demographics and sociability. To test for differing gene frequencies between study sites 95% confidence intervals were drawn around each gene frequency. Frequencies that did not overlap other confidence intervals were compared with a t-test of proportion. Gene frequencies for marmots from five study sites, compared two at a time for eight loci, dictated a probability level of $P < 0.012$ for the 80 comparisons to avoid a Type I statistical error (Sokal and Rohlf, 1969:156); no significant case at this level was found. A one-way analysis of variance (ANOVA) of individual heterozygosities showed nonsignificant differences among the five populations ($F_{4,377} = 1.31$; $P = 0.24$). Three loci, TRF, EST-1, and EST-2, showed large excesses of heterozygotes, possibly due to gametic disequilibrium among these systems (discussed later); the small excess of homozygotes shown in four systems indicated a low magnitude of Wahlund effect from pooling the gene frequencies for marmots from the five study sites (Table 1).

There were no significant differences in gene frequencies or heterozygosity between colonial and satellite animals in the East River Valley. However, genetic heterogeneity existed among colonies in both the East River Valley and North Pole Basin, as reflected by divergent gene frequencies (Fig. 1) and heterozygosities (Fig. 2).

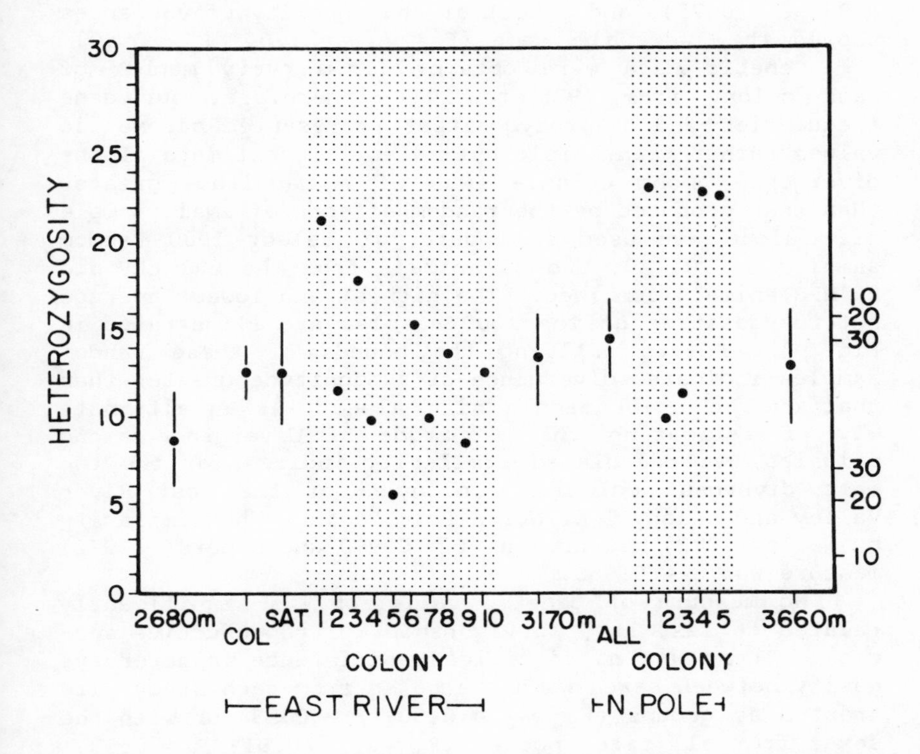

Fig. 2. Average individual heterozygosity from eight blood allozyme systems from five study sites. The legend for the X axis is as described in Fig. 1, and sample sizes were averages of those in Fig. 1. The highest and lowest heterozygosities from 1000 random samples of 10, 20, and 30 marmots are shown on the right. Vertical lines represent two standard errors around the mean heterozygosities.

The simplest prediction that could be made from a genetic comparison of colonies and satellites would be that colonies would be more homozygous with a greater variance than the approximately panmictic satellites because of drift from small effective population size, nonrandom mating, and the possibility of inbreeding. Such was not the case. The average heterozygosity of the two groups was not significantly different ($F_{1,201}$ =

0.05; $P > 0.75$), and a test of the equality of variances showed they were the same ($F_{62, 141} = 1.04$; $P > 0.40$).

Genetic data were obtained from every member of each colony from 1975 to 1977; therefore, our gene frequencies and heterozygosities represented parametric values rather than sample estimates. To estimate if the divergence among colonies was of a magnitude greater than that produced by the random effects of small sample size alone, we used a computer to select 1000 random samples of 10, 20, and 30 animals from the 203 ERV and 88 NPB animals combined. The highest and lowest average heterozygosities of the 1000 samples are illustrated in Fig. 2. Points outlying the bounds of these random samples indicated divergence of a magnitude greater than that due to small sample size alone. As an alternate way of estimating the magnitude of divergence among colonies, we calculated similarity indices for the two most divergent colonies, colony 5 of the East River Valley and colony 1 of North Pole Basin. The similarity by Nei's (1972) measure was $I = 0.990$ and Rogers' (1972) measure was $S = 0.963$.

No measure of genetic variation was significantly related to sex, age, survivorship, or reproductive success. There was no significant difference in heterozygosity between sexes with the sexes from each study site treated as groups ($F_{9,372} = 0.90$; $P = 0.93$) or with the sexes from all sites pooled ($F_{1,380} < 0.01$; $P = 0.93$). There were no significant differences in gene frequencies between the sexes at the five study sites.

An ANOVA of the heterozygosities of marmots grouped as 0, 1, 2, and 3 or greater years of age showed no significant differences among these age classes in the ERV and NPB populations ($F_{7,283} = 1.8$; $P = 0.09$). In addition, an ANOVA of the data for the age classes pooled over the two populations showed nonsignificant differences ($F_{3,287} = 0.14$; $P = 0.93$), as did a test for overlapping confidence intervals of gene frequencies among age classes.

Survivorship from 0 to 1 yr of age was tested for association with genetic variation in ERV marmots born in 1975 and 1976 treated as a cohort. Data from the NPB populations were insufficient to replicate this analysis. Thirty-seven of 53 ERV young survived to one yr of age and were the basis for this analysis. The mortality rate from 0 to 1 yr of age, q_x from life table statistics, calculated for 507 ERV young born since 1962 was

0.609; hence survivorship through this age class is a critical period in a marmot's life history. A Chi-square test of the contingency of survivorship on genotype, number of heterozygous loci, and animals with low heterozygosity or high heterozygosity did not show significant association between the variables. For the analysis of high and low heterozygosity the sample was divided into two nearly equal groups, those with low heterozygosity (0 or 1 locus) and those with high heterozygosity (2, 3, or 4 loci).

The number of heterozygous loci (1 to 4 loci) of each of the 15 ERV female marmots that produced 28 litters was used to group litter sizes for an ANOVA. There was no significant difference in litter sizes among the groups ($F_{3,24} = 2.61$; $P > 0.07$). Data were insufficient to test for significant differences between genotypes for the female marmots or for the effects of parity, and too few territorial males fathered all the litters to allow a similar analysis for males.

There was little relationship between genetic variation of ERV marmots and their behavioral characteristics derived from mirror image stimulation (Svendsen and Armitage, 1973). The first five axes of a factor analysis (BMDP4M program; Dixon, 1975) of the frequencies of 22 behaviors recorded while each marmot was exposed to mirror image stimulation accounted for 56.7 % of the observed variation. These axes were identified as I. Sociable; II. Aggressive; III. Avoidance; with factors IV and V not named. One-way ANOVA's were calculated using genotypes, the number of heterozygous loci, and animals with high and low heterozygosities (as described for survivorship) as groups. The factor scores on the five axes were used as dependent variables. The only significant relationship showed transferrin-FF and -FS genotypes grouped on the second axis ($F_{1,172} = 6.25$; $P = 0.02$).

Animals heterozygous (FS) at the transferrin locus scored more negatively on the second axis than those with the FF genotypes; this relationship suggests that marmots with a heterozygous locus are more aggressive. We had no behavioral data on animals with SS genotypes. Similarly, absence of suitable sample size within groups precluded analysis of behavior and genotypes for PGI and EST-3 and 4. These 42 ANOVA dictated a probability level of $P < 0.025$ for significance to avoid a Type-I error.

There was a positive correlation between the minimum number of marmots alive and the frequency of the most common allele of LAP ($r = 0.48$; $0.05 < P < 0.06$; Fig. 3) in the East River Valley. Although the correlation was of marginal statistical significance, we decided to report this unique relationship since this same correlation was reported in cyclic microtine rodents (Gaines et al., 1978; Tamarin and Krebs, 1969). This density cycle can be partially dampened by the explanation of artificial mortality that occurred in this marmot population. Part of the first increase in density was due to new colonies being discovered and trapped. Approximately 10 animals were lost during the decline in density when marmots proximal to occupied cabins were removed at the request of the cabin owners.

The change in density was reflected in both the numbers of colonial and satellite animals trapped. To test the hypothesis that the cycle was a function of trapping success, we calculated a Jolly (1965) stochastic estimate. The upper bound of that estimate is at the top of the vertical lines in Fig. 3. The change in density was reflected in the Jolly estimate, colony and satellite portions of the population, and average colony size. Hence, the density change has a real element as well as the artificial ones discussed above.

Even though there was an artificial element of density change, its variation with the frequency of the Lap-F allele was not so readily explained. The frequency of that allele was relatively lower in three of the seven colonies trapped during the decline phase. Kohn and Tamarin (1978) and Tamarin and Krebs (1969) analyzed such a correlation with a plot of change in gene frequency Δp vs gene frequency p. They suggested, citing Li (1955), that a significant negative slope of Δp on p was a statistical test for a stable polymorphism with the population being perturbed and returning to its equilibrium gene frequency. Our evaluation of this technique led us to think it was an unsuitable test for a stable polymorphism. Nevertheless, our regression of Δp on p produced a significant positive slope ($m = 0.33$; $t = 2.42$; $P < 0.04$). Li's (1955) theory was developed for nonoverlapping generations, hence our regression violated that condition.

Significant gametic disequilibrium, nonrandom association of alleles in gametes, occurred between the TRF/EST-1 loci, TRF/EST-2 loci, and EST-1/EST-2 loci

Social Substructure and Genetic Variation

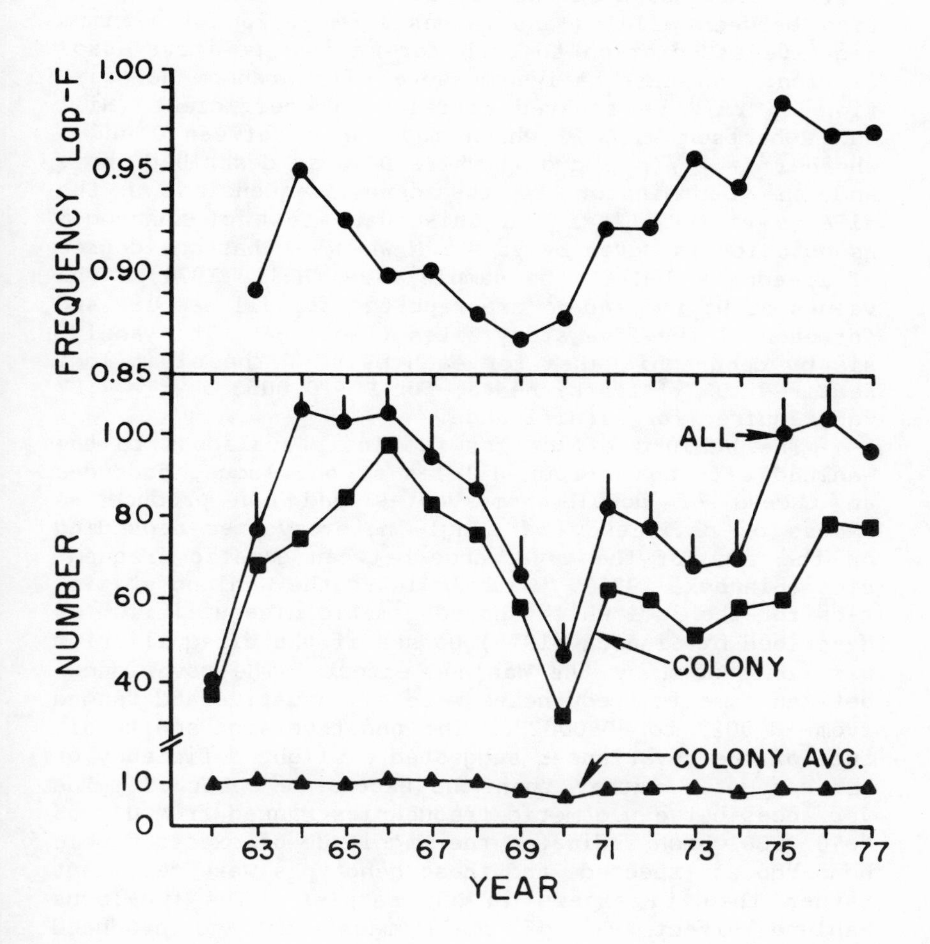

Fig. 3. Change of marmot density and the frequency of the Lap-F allele. The curves labeled colony and all (colony and satellite) represent minimum numbers alive, and the curve labeled colony average is the minimum number alive per colony. The vertical lines represent the upper bound of a Jolly (1965) stochastic estimate of numbers.

(Table 2). The linkage disequilibrium statistic, D, (Hill, 1974) measured the deviation from random association between alleles and ranges from -0.25 for maximum negative association to 0.25 for maximum positive association. An alternative measure of nonrandom association is r^2, the squared correlation coefficient (Hill and Robertson, 1968), which may range between 0 and 1 where $r^2 = D^2/(p_1 p_2 q_1 q_2)$, where D is as described above and the denominator is the gene frequencies of the alleles at both loci. A Chi-square test of nonrandom association is given by $\chi^2 = r^2 N$ where χ^2 has one degree of freedom and N is the sample size (Hill, 1974). The values of D, r^2, and χ^2 are reported for all samples and for each of the five study sites (Table 2). This analysis by geographic area for each pair of the eight loci required 196 χ^2 tests; hence, for $P < 0.005$, a $\chi^2 > 7.87$ was required for significance.

The Wahlund effect for two loci, analagous to the Wahlund effect for four alleles at one locus, produces an excess of double homozygotes and can produce an excess or deficiency of single heterozygotes depending on the sign of the covariance between gametic frequencies (Sinnock, 1975). We calculated the Wahlund statistics for the loci which showed gametic disequilibrium as described by Sinnock (1975) to see if the disequilibrium was confounded by the Wahlund effect. The covariances between gametic frequencies were all negative and ranged from -0.0027 to -0.00023. The negative sign and magnitude of the covariances suggested a slight deficiency of heterozygotes rather than the excess we observed. The variances between gametic frequencies ranged from 0.0053 to 0.0086 which indicated the magnitude of excess double homozygotes expected, and these genotypes were deficient rather than in excess in our samples. The two-locus Wahlund effect was of small magnitude and produced effects opposite of those necessary to account for our observations.

We have no way of causally explaining the observed gametic disequilibrium, but the genetic divergence due to drift within the colonies suggested that drift may also have given rise to the observed disequilibrium. We divided the ERV population into two sections, with those animals living in the southeast end of the valley as one group and those living in the northwest sections as the other group, and then calculated the measures of gametic disequilibrium for those samples (Table 3). The distri-

Table 2. Measure of linkage disequilibria (D), squared multiple correlation coefficients (r^2), and χ^2 tests for significant gametic disequilibria were calculated for marmots from the five study sites and for all sites pooled.

Systems*	Measure	2680 m	E. River	3170 m	N. Pole	3660 m	Total
TRF/EST-1	D	-0.0040	0.0227	0.0080	0.0390	0.0051	0.0222
	r^2	0.0024	0.0236	0.0036	0.0700	0.0007	0.0222
	χ^2	0.07	4.47	0.61	6.22	0.02	8.47***
TRF/EST-2	D	-0.0152	0.0388	0.0397	0.0395	-0.0104	0.0181
	r^2	0.0196	0.0488	0.0396	0.0699	0.0026	0.0104
	χ^2	0.06	9.91***	1.20	6.22	0.08	3.95
EST-1/EST-2	D	0.0047	0.0648	0.0632	0.0401	-0.0104	0.0800
	r^2	0.0020	0.1006	0.1554	0.0515	0.0031	0.0634
	χ^2	0.06	20.42***	4.66	3.70	0.09	24.21***
Sample Size		30	203	30	88	30	382

*Abbreviations in text; ***$P < 0.005$.

Table 3. Measures of gametic disequilibrium calculated after subdividing the marmot samples from the East River Valley into southeast and northwest sections. Systems notations are the same as in Table 2.

Systems	Measure	East River Valley	
		Southeast	Northwest
TRF/EST-1	D	0.0204	0.0175
	r^2	0.0176	0.0185
	χ^2	1.92	1.11
TRF/EST-2	D	0.0623	0.0254
	r^2	0.1141	0.0292
	χ^2	12.44***	1.74
EST-1/EST-2	D	0.0519	0.0989
	r^2	0.0694	0.2166
	χ^2	7.56**	12.99***
Sample Size		109	60

** $P < 0.01$; *** $P < 0.005$.

bution of the rare alleles of PGI and PGM which were largely found in southeast colonies suggested this division. Large sample size is typically required to detect gametic disequilibrium (Brown, 1975a), thus preventing a finer scale of analysis. The results of this division suggested that local population parameters were an element of the observed gametic disequilibrium. Between TRF/EST-1, disequilibrium was significant only in the southeast colonies. Also, the magnitude of the EST-1/EST-2 disequilibrium was different in the two groups.

Social Substructure and Genetic Variation

DISCUSSION

The significant positive results of this study included an association between transferrin genotype and aggressiveness, a weak correlation between Lap-F gene frequency and density, gametic disequilibrium between three pairs of loci, and significant heterogeneity in dispersion of genetic variation through the colonies of marmots. There was no association between genetic variation and altitude, habitat, age, sex, survivorship, litter size, and a suite of other behavioral variables. The presence of each allele of the eight loci at each study site (except for the rare PGI missing from the 3660 m sample) indicated that recent mutation was not an explanation for the distribution of genetic variation. The similarity of gene frequencies among study sites could be accounted for by identical selective forces in each area or a sufficient rate of gene flow among areas.

Our analysis was statistically conservative compared with other papers attempting similar correlations between genetic variation and life history and environment. Such conservatism was justified considering that in excess of 1000 statistics were calculated in the analyses presented. A model by Dickinson and Antonovics (1973) showed that such correlations may provide evidence of the action of selection on finite populations, but stochastic elements can significantly vary the observed correlation through time. Clarke (1974) considered ecological methods valuable in the detection of selection. The general lack of such correlations in our study may be accounted for by theories that consider the relationship of genetic variation and environmental "grain" (Soule, 1976; Selander and Kaufman, 1973a). The theory suggested that organisms rely less on genetic mechanisms to cope with environmental change as they become larger and more capable of homeostatic control, i.e., the organism experiences the environment as fine-grained. The marmot is a relatively large mammal, hibernates for seven months of the year, and is generally restricted to a habitat of rocky meadows; hence, it probably experiences the environment as fine-grained. The theory of grain also predicted that fine-grained organisms should tend toward a single phenotype (Selander and Kaufman, 1973a), but the level of variation in marmots was typical of that of rodents (Selander, 1976). Selander and Kaufman (1973a) suggested that there was a threshold size beyond which heterozygosity does not decrease; perhaps the marmot is beyond that limit.

Gaines et al. (1978) reviewed the literature relative to the commonly reported covariation of gene frequency and population density in mammals. Such correlation, apart from a direct causal relationship, could be due to the linkage of an allozyme locus to loci subject to selection. Models by Charlesworth and Giesel (1972) accounted for such covariation as side effects of demographic change.

Linkage between a neutral and selected allele would be subject to rapid decay, and nonrandom association due to drift would vary in its direction and magnitude through time (Hedrick et al., 1978). Hence, linkage and drift do not satisfactorily account for this phenomenon for many species in widely scattered geographic areas. The simplest hypothesis that would explain our observation would be chance variation of gene frequency in the same direction in several marmot colonies through the population processes relative to each colony. A more complex hypothesis would be that of density-dependent or frequency-dependent selection acting on semi-isolated colonies over the 5 km that separate the most distant colonies.

The positive relationship between aggressiveness and transferrin genotype could be causal or linked, as discussed for density, or a spurious statistic. We have no way of choosing among these hypotheses. Our study of the relationship of genetic variation and behavior was generally comparable to that of Garten (1976) on the old-field mouse, *Peromyscus polionotus*. Numerous behavioral acts were correlated with heterozygosity in geographic samples. Garten's suggestions, that attack time, food control behavior, and social dominance are more than 96% predictable from the variation found in soluble tissue proteins, were unsatisfying to us. We suggest that these correlations were artifacts of his statistical analysis. Specifically, Garten regressed average heterozygosity as the independent variable on the average behavioral score for each of five study sites. Linear and polynomial regressions, fitted to five pairs of variables which were averages, could account for the reported predictability. A regression model with more than one value of Y per X would be more appropriate (Sokal and Rohlf, 1969:430). In this model an ANOVA is calculated for the groups of Y variables; given a significant ANOVA, a regression of X on the Y variables (rather than the average of Y's for groups) may be calculated.

An explanation of gametic disequilibrium found in nature is frequently difficult (Clarke, 1974). Disequilibrium may be caused by migration, selection, mutation, and drift in substructured populations (Nei and Li, 1973). Population substructure was implicated as an element of the gametic disequilibrium in ERV marmots. Gametic disequilibrium was reported between esterase loci in barley (Kahler and Allard, 1970), fruit flies (Baker, 1975), and salamanders (Webster, 1973). Chromosomal mapping of the involved esterase loci in barley and fruit flies showed they were closely linked and that the association probably arose by gene duplication. Webster also hypothesized gene duplication due to the similar relative mobilities of the allozymes on starch gel. The marmot EST-1 and EST-2 alleles were also in close proximity on starch gels. Hence, gene duplication and population substructure may be elements of the gametic disequilibrium we observed. The observed excess of TRF, EST-1, and EST-2 heterozygotes may be related to this nonrandom pattern of association.

Though the above correlations and gametic disequilibrium may represent the action of deterministic forces on genetic variation, the magnitude of these forces was insufficient to prevent significant drift within colonies. This heterogeneity between colonies was the significant finding of our study. The relationship of drift to demographic processes among colonies is detailed elsewhere (Schwartz and Armitage, 1980). Demographic processes may enhance group heterogeneity; greater than 60% of harem residents were recruited from their natal colony. The Mendelian pattern of inheritance of allozymes suggested no mate selection outside the social group. Only 40 of 790 marmots observed since 1962 moved between colonies, and only 15 of these movements probably resulted in gene flow. Drift among colonies was of greater magnitude than that expected due to random sampling alone. A Nei's (1972) index of 0.989, calculated for the most divergent marmot colonies, is comparable to the level of differentiation seen at the subspecies level in fruit flies (Avise, 1976). A Rogers' (1972) index of 0.963 is comparable to that seen between closely related species of _Spalax_ and _Thomomys_

(Nevo et al., 1974) and *Spermophilus* (Cothran et al., 1977), although the average of mammalian sibling species is 0.81 (Avise, 1974).

The functional significance of allozyme variation in mammalian populations is little known, hence the ecological and evolutionary meaning of the genetic divergence shown among marmot colonies is difficult to assess. Though we cannot exclude the action of selection on the observed allozyme variation, its magnitude is less than that of drift. Thus, the genetic variation found in our study functions best as marker alleles to be studied relative to population processes.

The significance of regulatory gene and chromosomal evolution in mammalian populations has received increasing attention (Cherry et al., 1978; Wilson et al., 1975; King and Wilson, 1975), and the impact through evolutionary time of heterogeneity due to the substructure of a population is perhaps best considered in the context of regulatory genes and chromosomal changes. The dynamics of genes through long periods of time require the consideration of the actions and interactions of the many deterministic and stochastic forces that affect gene frequency. A contrast in the alternate results of the forces affecting genetic heterogeneity might be found in considering the 11 subspecies of yellow-bellied marmot and the over 400 subspecies of the gopher *Thomomys* found in western North America (Hall and Kelson, 1959). We offer the following speculation on the dynamics of marmot genetics.

The models of Levin et al. (1969) relate to the conditions found among colonies of marmots. Their models for maintaining t alleles in small demes of house mice showed rapid drift to fixation for the T allele in unconnected demes. Small demes, approximating the effective population size of marmot colonies, interconnected by gene flow of 3% were effective in preventing the loss of the t allele for 200 generations. A model of drift and gene flow (Endler, 1977) between demes (of 100 individuals exchanging genes at the rate of 0.20) may represent elements relative to marmot populations if the East River Valley is considered as a genetic unit. Endler's model yielded a wide range of gene frequencies through 1000 generations with pronounced area effects and great heterogeneity between areas. The ERV contains patches of suitable habitat with structural and biological features such as cliffs, large expanses of forest, and a broad river floodplain which may serve as partial barriers to gene flow between suitable habitat patches and between the more continuous alpine habitat. Given that the substructure of a marmot population like that of the ERV is an element retarding the loss of genetic variation and that heterogeneous populations in semi-isolated areas exchange genes, the potential for retaining genetic variation in the population by drift for long periods of time is great.

The well-known social structure of larger mammals, e.g., langurs (Blaffer Hrdy, 1977) or lions (Bertram, 1977), and the increasing evidence for substructure in small mammal populations suggest the generality of the genetic findings of our study. Those considering the rate of mammalian evolution, evolutionary ecology of mammals, and social evolution may continue to look at population substructure as a valuable element of their arguments.

Acknowledgments--We thank C. Baker, M. S. Gaines, P. W. Hedrick, and R. S. Hoffmann for continued guidance throughout this study and for critically reading drafts of this manuscript. We particularly thank M. S. Gaines for generously supplying his time and laboratory facilities for electrophoresis. We also thank Keith Armitage, U. Diedenhofen, D. Johns, S. Nowicki, and A. Torres for assistance in the field or laboratory. This study was supported by National Science Foundation grant BMS 74-2193 and University of Kansas Biomedical grant 4442 to K. B. Armitage.

RELATIONSHIPS OF GENOTYPE, REPRODUCTION, AND WOUNDING IN KANSAS PRAIRIE VOLES

Robert K. Rose and Michael S. Gaines

Abstract--From May 1971 through March 1973, samples of 20 to 30 prairie voles were obtained each month by removal live trapping. Control grids provided information on population dynamics. In all, 640 voles were examined genetically and for evidence of reproduction and wounding. For three electrophoretic loci (EST-1, EST-4, and XDH), there were significant differences in reproductive contribution within a density period for individuals of differing genotypes. For LAP, EST-1, and XDH, the rarer alleles increased in frequency throughout the cycle and were positively associated with body wounding. Chitty's hypothesis that selection favors different genotypes and/or behavioral types at different stages in the density cycle was supported.

INTRODUCTION

Chitty (1960, 1967) proposed that a genetic-behavioral polymorphism may exist in fluctuating populations of microtine rodents, permitting certain genotypes or behavioral types to be selected during the phase of increasing density but others during the phase of peak or declining numbers. Chitty suggested that during the phase of increasing numbers when density-independent mortality is prevalent, genotypes with high reproductive potential will have a selective advantage, but during the peak phase when density-dependent mortality prevails, there is selection for aggressive behavior. Although aggressive genotypes may have an advantage with respect to intraspecific competition for food and space, their fitness is reduced in other ways and this change in fitness sets the stage for the decline phase.

Chitty's hypothesis is appealing because it follows so well from Darwinian theory, i.e., nature selects those organisms that are best suited for a set of environmental conditions. When conditions change, as happens when population density increases, the frequency of the favored genotype, behavioral type, morphotype, or other types may also change within the population. Constantly changing proportions of the different types within a population are thus expected and should be correlated with demographic characteristics.

Between 1970 and 1973 we monitored several aspects of the biology of prairie voles (Microtus ochrogaster) in population studies conducted near Lawrence, Kansas. These studies involved both mark-release and removal-live trapping. Publications on the population dynamics (Gaines and Rose, 1976), genetics (Gaines et al., 1978), patterns of wounding (Rose and Gaines, 1976), and reproductive cycle (Rose and Gaines, 1978) have appeared, but the relationships between the latter three sets of data have not yet been described. In this paper we (1) examine the relationship between genotype and reproductive condition (2) examine the possible relationship between genotype and the amount of wounding, and on the basis of these findings, (3) test Chitty's hypothesis.

MATERIALS AND METHODS

Three principal study areas were selected to estimate the degree of synchrony of density and reproduction between spatially isolated populations and to obtain detailed information on wounding and reproduction for individuals of known genotypes. Two of the study areas were located 11 and 14 km northeast of Lawrence, in the vicinity of grids A-D (Gaines and Rose, 1976) and the third was located 6 km west of the city. Each study area contained three or more sampling sites, all of which were oldfields dominated by perennial cultivated brome grass (Bromus inermis). Lawrence is located in northeastern Kansas near the western edge of the eastern hardwood forest at an elevation of 310 m. The average annual temperature is 13.6 C, and mean rainfall is 878 mm. Rainfall varies greatly from month to month with 70% of annual precipitation falling during the six months from April through September. The winter months are the driest, and when winter precipitation does occur as snow, it usually melts within a few days.

Within each sampling site one Sherman and one modified Fitch (Rose, 1973) trap were placed along irregularly-shaped lines at intervals of about 8 m. The trap lines were 20 m or more apart, and although we do not know how the removal of animals affected the genetic composition of the population, the trapping regime was designed to avoid the excessive reduction of local populations resulting in the fixation of alleles. After 10-13 days of prebaiting, the traps were set for four days, or until a sample of 15 to 20 voles had been obtained. Voles weighing 20 g or more were removed, while those of lighter body weights were counted and released. Removal live-trapping was begun in May 1971 and continued through March 1973 at biweekly intervals. With five sampling sites used in sequence, the voles in each oldfield were trapped on a 10-week rotation. Nine such trapping rotations were conducted.

At the time of capture, each vole was ear tagged with a numbered fish fingerling tag, weighed, and examined for external reproductive condition. Animals were returned to the laboratory, bled through the suborbital sinus and then killed. In some instances the voles were autopsied immediately, but usually the liver tissues were prepared for electrophoresis after the animals had been frozen at -20 C for up to six weeks. Blood samples were centrifuged at 10,000 rpm for 10 min in a refrigerated centrifuge. The plasma was pipetted off and frozen at -20 C. The cellular fraction was washed three times in 0.9% NaCl solution (Selander et al., 1971), after which the cells were lysed with one-half volume of distilled water plus two drops of methylene chloride. After 20 min of centrifugation, the hemolysate was pipetted into a second tube, refrigerated and electrophoresed within 24 hr. Liver tissues were homogenized in two-to-three volumes of distilled water using a sonicator. Tubes of homogenate were then centrifuged at 1000 g for 5 min and the supernatent removed for electrophoresis. Horizontal starch-gel electrophoresis was performed following the methods of Selander et al. (1971). Starch gels of 11.5% Electrostarch (Otto Hiller, Madison, WI) were used with two different buffer systems, lithium hydroxide and continuous tris-citrate I. Standard biochemical staining procedures (Selander et al., 1971) were used to survey isozymic variation at 26 structural gene loci.

Each vole was autopsied and the reproductive data noted as in Rose and Gaines (1978). For males, the position of the testes, weight of paired testes, and condition of the tubules in the cauda epididymides were recorded. Males were judged capable of breeding if the epididymal tubules were convoluted (Jameson, 1950). For females, the variables included the condition of the pubic symphysis and vaginal orifice, uterus size and weight, numbers of placental scars and embryos, and numbers of corpora lutea and corpora albicantia. Examination of these features resulted in an assessment of current female breeding condition and of parity. At the time of autopsy, the skin of each vole was also removed and examined for evidence of puncture wounds or pigmented scars. Subdermal scar tissue is darker than undamaged dermis in everted skins. Each vole was given a score on a scale of 0 to 9; voles with higher scores had more evidence of wounding and more recent puncture wounds (Rose and Gaines, 1976).

RESULTS

The population densities on the grids A-D obtained by enumeration (Figs. 1-4, Gaines and Rose, 1976) are comparable to the densities based on catch per 100 nights of live trapping from the three removal study areas (Fig. 1). Also, the minimum number alive on grids A and B showed that the period of peak density occurred in the spring and early summer of 1972 and a slow decline through the summer of 1973 (Fig. 1).

Levels of wounding were comparatively high in winter during periods of intense reproduction among males and adults. For males, although the level of wounding increased from the prepeak into the period of peak density, there was no reduction in the amount of wounding during the period of declining numbers. For females, the level of wounding was significantly reduced during the population peak. The greatest levels of wounding were observed during the months of the late decline, from October 1972 through March 1973. In neither sex was the level of wounding positively related to population density (Rose and Gaines, 1976).

There was no association between intensity of reproduction and population density. Pregnancy rates were high with brief intervals of nonbreeding in July 1971 and 1972 and in August and December of 1972 during the population decline. There was no association between maternal weight and either parity or number of embryos. Furthermore, corpora counts were not higher in heavy or multiparous females, nor did the level of pre-

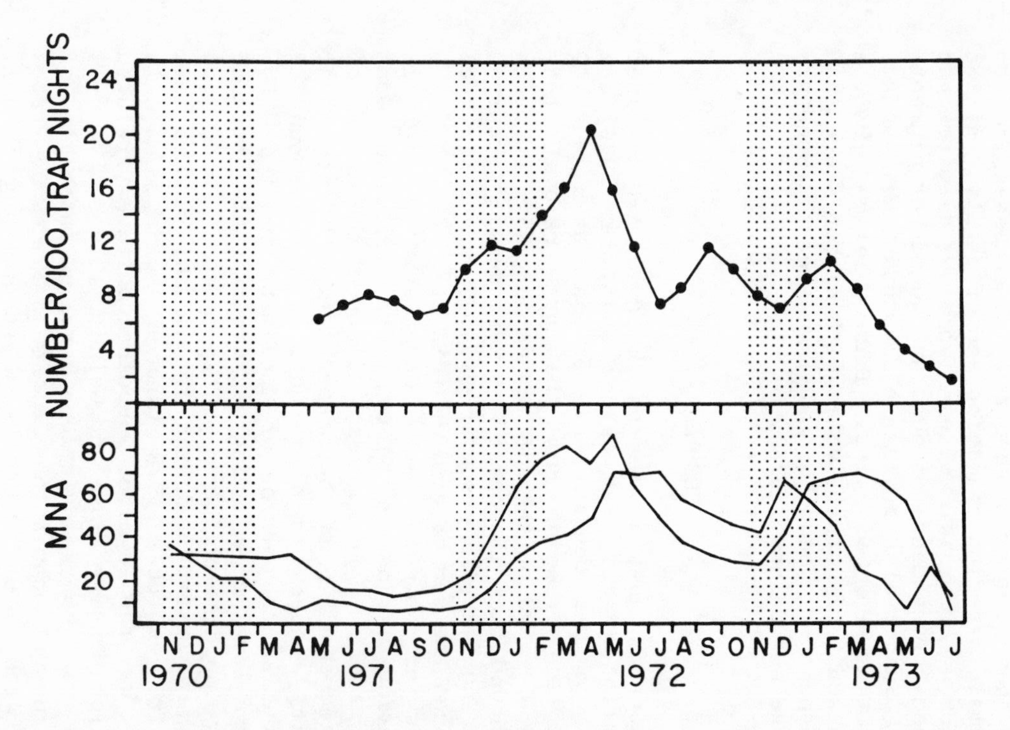

Fig. 1. The top section is a plot of the number of voles per 100 trap night by month. Below are the minimum numbers of voles known to be alive (MNA) on population grids A and B located in the main study area. Winter months are shaded. The population peak is considered to be the period including the month of greatest density, April 1972, and the two adjacent months.

natal mortality increase during the period of greatest density. Except for a few individuals taken in July and August 1971, males with a body weight of 30 g or more were judged to be fertile, as were a majority of the 20-29 g males. Body weight and testes weight were significantly correlated in fertile males but not in nonfertile males. Pregnancy rate, litter size, and adjusted testes weight showed significant increases during the March-April and September-October breeding peaks and decreases in midsummer and midwinter. The concordance of these reproductive parameters seemed to indicate that the reproductive cycle of Kansas prairie voles was much more related to seasonal than to cyclic events (Rose and Gaines, 1978).

Seven loci were genetically interpretable and polymorphic according to the 1% criterion (Lewontin, 1967): transferrin (TF), leucine aminopeptidase (LAP), 6-phosphogluconate dehydrogenase (6-PGD), 2 esterases (EST-1 and EST-4), xanthine dehydrogenase (XDH), and α-glycerophosphate dehydrogenase (α-GPD). Two other loci, EST-2 EST-3, were scored but were omitted here because neither locus behaved in a predictable Mendelian manner in test crosses. As reported by Gaines et al. (1978), some loci showed changes in the frequencies of particular alleles throughout the cycle, while for other loci no change occurred. Changes in the gene frequency for six loci are presented in Fig. 2. The frequency of the Lap^F allele decreased throughout the study, while the fast allele for EST-4 locus increased (Fig. 2a). Two common alleles that did not change in frequency throughout the cycle were $6\text{-}Pgd^B$ and Tf^E (Fig. 2b). The frequency of the $Est\text{-}1^F$ allele decreased during the late decline phase (Fig. 2c), at a time when the Xdh^F fast allele, which decreased in frequency throughout the study, fluctuated.

Effects of Genotype on Reproduction

Two methods were used to investigate the possible relationships between genotypes and levels of reproduction: (1) regression analysis of the number of embryos or corpora lutea on body weight (adjusted for pregnant females), and (2) the proportions of voles with different external reproductive characteristics (condition of vaginal opening and pubic symphysis, and testes position) were tested with the Chi-square test. Data for

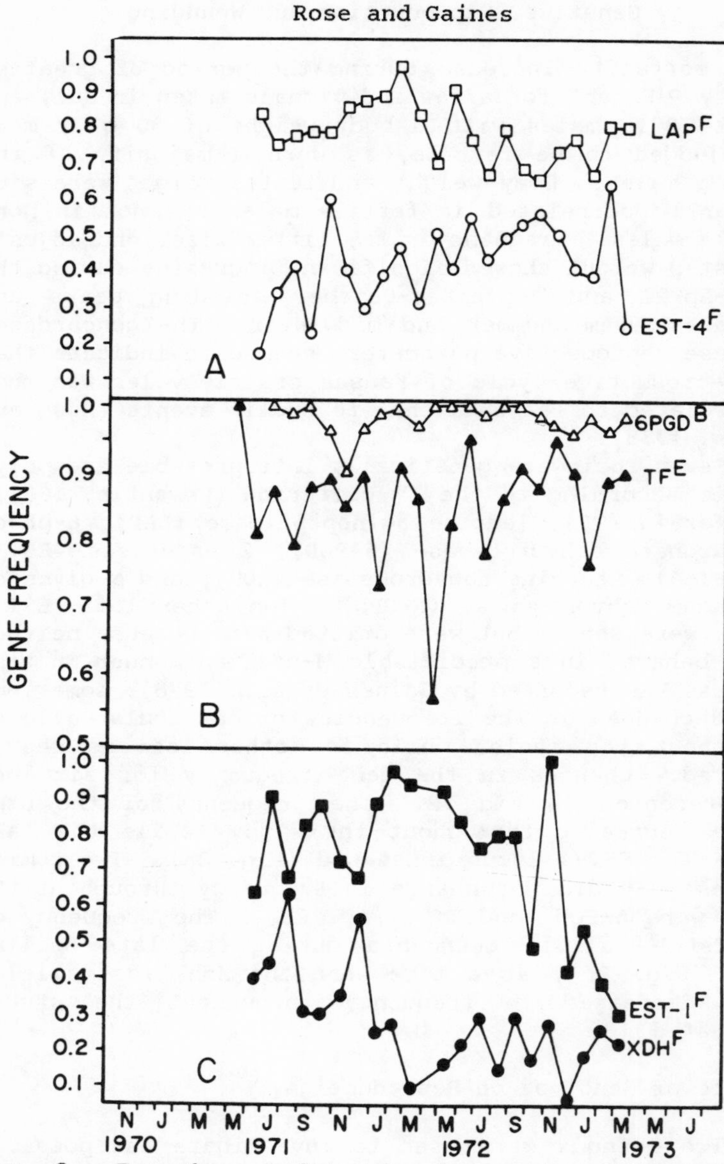

Fig. 2. For the samples of voles collected each month by removal live trapping, the frequency of the fast (F) allele for LAP and EST-4 genetic loci (a, top), the frequency of the B allele for the 6-PGD locus and of the E allele of the TF locus (b, middle) and the frequency of the F alleles for EST-1 and XDH loci (c, bottom).

voles were grouped by sex according to the density periods of capture, using a five-month period for the population peak. For the 6-PGD, three alleles and five genotypes were detected. Numbers of the less common genotypes were small, and the data from these were lumped before testing. For the 6-PGD locus, the letters A, B, and C were assigned to the three alleles in order of the increasing migration distances of their gene products on the starch gels. Thus, 6-Pgd-AA represents the homozygote for the slowest moving allele, and 6-Pgd-AB the heterozygote of the two slower moving alleles. All other loci had only two alleles. In these instances, SS represents the genotype homozygous for the slow allele, SF the heterozygous genotype, and FF the homozygote for the faster allele.

Using regression analysis, three loci (EST-1, EST-4, and XDH) showed significant differences within a density period for both corpora lutea and embryo number. The Y-variable was the number of either corpora lutea or embryos and the X-variable was body weight. EST-4 showed a trend, seen in both corpora lutea ($F_{2, 39}$ = 3.24; P = 0.05) and embryo counts ($F_{2, 39}$ = 3.82; P < 0.05) favoring homozygotes (Est-4-SS and Est-4-FF) during the increase phase and the heterozygote (Est-4-SF) in the period of peak density. The group composed of Est-1-FF homozygotes had the greatest slope in both regressions (corpora: $F_{2, 35}$ = 3.38, P < 0.05; embryos: $F_{2, 35}$ = 3.96, P < 0.05) during the months preceding the population peak, but the slopes associated with the three genotypes (Est-1-SS, -SF, and -FF) were not different in subsequent samples. Finally, for XDH the bearers of the common genotype (Xdh-SS) had the greatest slope for both corpora lutea ($F_{2, 54}$ = 3.61; P < 0.05) and embryo counts ($F_{2, 54}$ = 6.70; P < 0.05) during the population decline, the only period in which the slopes of the regression lines differed significantly from one another. Thus, for EST-1, EST-4, and XDH the genotypes differed in their reproductive contribution within a density period.

Comparisons were also made to determine whether particular genotypes made greater reproductive contributions during increase, peak, or decline populations. Again, the Y-variable was the number of either corpora lutea or embryos and the X-variable was body weight. Data from voles of similar genotypes were grouped according to density period. Two genotypes, α-Gpd-SF and

Xdh-SF, showed significant differences in the number of corpora lutea vs body weight. For α-Gpd-SF, a comparison between increase and decline populations was made with the small sample of individuals from the population peak omitted. During the population decline the slope of the regression was greater than that during increase phase ($F_{1,19} = 8.94$; $P < 0.01$), i.e., the α-Gpd-SF heterozygotes made proportionately greater reproductive contributions during the decline period. For Xdh-SF, the slope of the regression was greatest during the population peak and differed significantly from the slopes for the voles of the other two periods ($F_{2,29} = 5.38$; $P < 0.025$). No association between embryo count and body weight was detected.

Of the 16 genotypes over the seven loci with sufficient sample sizes to be tested with regression analysis, 10 were homozygous. None of the homozygous genotypes showed differences in reproductive contribution across the different periods of the cycle. However, two of the eight heterozygous genotypes did show differential contributions.

The relationships between genotype and external reproductive features, testes position, opening of vaginae, and pubic symphysis, were tested using a Chi-square analysis. For the increase phase, females grouped according to α-GPD genotype showed a significant difference in the proportion of perforate vaginae ($\chi^2_1 = 10.60$; $P < 0.005$). Individuals with the α-Gpd-SS had the lowest frequency of perforate vaginae and those with α-Gpd-FF the highest. During the population peak, there was a significant difference in the frequency of perforate vs nonperforate vaginae among females of the three LAP genotypes ($\chi^2_2 = 6.59$; $P < 0.05$). In addition, during the population decline, females with 6-Pgd-BB genotypes had a significantly higher frequency of separated pubic symphysis than that of the combined remaining genotypes ($\chi^2_1 = 4.63$; $P < 0.05$). No differences were detected among the males.

The numbers of animals of each sex within 23 genotypes were compared for deviations from a 1:1 ratio; the data were then pooled and a χ^2 test of homogeneity was performed. No comparison was significant at the 0.05 level, but for the Est-1-FF genotype there was a nearly significant excess of females during the population decline ($P < 0.10$) and a similar probability for an excess of males in the prepeak population. Thus, for

Est-1-FF there may be a trend favoring males in the increase phase and females during the population decline.

Genotype and Level of Wounding

Using several criteria, Krebs (1970) detected no differences in aggressive behavior among voles of three transferrin genotypes. We compared wounding levels in voles with Tf-EE and Tf-EF genotypes for each sex and density period. In all, 248 fertile males and 264 females, with normal or enlarged uteri, were included in the analyses. Neither sex showed a significant difference in the amount of wounding, although the analysis was almost significant for the females during the five-month period of peak numbers ($\chi^2_2 = 5.40$; $P < 0.10$). For TF, no genotype seemed associated with changes in aggression relative to density. Analyses were conducted for the other six loci testing for changes in the level of wounding among genotypic classes within density periods and throughout the entire cycle. For two loci not considered here (6-PGD and α-GPD), one genotype tended to predominate making meaningful comparisons impossible for the small samples of the rarer genotypes. In the analysis of aggression by genotype, data for fertile males and females were lumped together. The null hypothesis that wounding is not related to genotype was tested regardless of density period. For comparisons over five loci, including TF, significant differences were found only for EST-1; individuals with the Est-1-SS genotype had the highest and those with Est-1-FF the lowest level of wounding overall.

The most interesting comparisons were between individuals of different genotypes within density periods, since differences would suggest relationships between density (population phase) and individual genotypes. As a basis for comparison, data from all individuals were lumped by periods related to density. The resulting heterogeneity was significant ($\chi^2_{14} = 50.35$; $P < 0.005$) and attributable largely to reduced levels of scarring and increased incidence of puncture wounds during the population peak and to increased wounding during the decline period. Wounding levels were highest in intensity during the postpeak period.

Data from individuals of similar genotypes, grouped by density period using a 5-month period for the population peak, were then compared across periods with χ^2

tests of homogeneity. Animals with the Lap-SF genotypes differed significantly across the density cycle ($P < 0.05$), primarily because of more wounding during the peak and postpeak periods. Alternatively, for individuals with Lap-FF genotypes, significant ($P < 0.005$) differences were attributable to proportionately less wounding in the prepeak and more in the postpeak periods. Individuals with the heterozygous Lap-SF genotypes, which made up about 10% of the population during the prepeak and peak populations, increased to about 35% during the population decline. During the density cycle the frequency of the S-allele increased from 0.13 to 0.30. This trend toward increasing frequencies of slow alleles in association with increased wounding was observed for two other loci.

Different levels of wounding were shown for individuals of the three EST-1 genotypes when compared between population phases. Voles with the Est-1-SS genotype had high levels of wounding throughout the cycle and greater frequencies of wounding during the postpeak period. Alternatively, voles with the Est-1-SF genotype had a higher frequency of wounding in the period preceding the population peak, and those with Est-1-FF genotype showed reduced levels throughout, especially during the decline phase. During the study, the frequency of the S-allele also increased from 0.20 to 0.44 and was positively associated with wounding and survivorship.

Finally, voles with the Xdh-AA genotype had higher levels of wounding during the prepeak period. The trend seen for LAP and EST-1 is repeated, with the frequency of the S-allele increasing from 0.63 to 0.81 during the study, and the frequency of the individuals homozygous for the S-allele (Xdh-SS) increasing from 45 to 70% during the study. Again, there were positive associations between the frequency of the slow allele, wounding, and survival. For three loci, LAP, EST-1, and XDH, two general trends were noted: (1) an increased frequency in the S-allele during the course of the cycle, and (2) positive association between the S-allele and wounding. These observations suggest that for these three loci wounding is positively related to survival, permitting those individuals with aggressive tendencies, on the average, to survive during the decline and leave offspring compared to other genotypes. If the data are grouped by the genotype of the vole and tested for differences in mean wounding scores with a one-way analysis of variance, roughly the same results are obtained. For EST-1, voles homozygous for the S-allele had significantly ($P < 0.001$) higher wounding scores (3.62 ± 0.24) than those heterozygous (2.76 ± 0.19) or homozygous for the F-allele (2.43 ± 0.12). Individuals characterized as slow homozygotes for the α-GPD locus had significantly ($P < 0.05$) higher scores (3.16 ± 0.20) than those heterozygous (2.34 ± 0.12) or homozygous for the alternate allele (2.93 ± 0.34). Similarly, for the XDH locus, individuals homozygous for the S-allele had significantly ($P < 0.05$) higher scores (2.92 ± 0.13) than those either heterozygotes (2.48 ± 0.19) or homozygous for the alternate allele (2.37 ± 0.27).

Interaction of Reproduction, Wounding, and Density

The relationship between reproductive condition and the level of wounding has been presented in detail in Rose and Gaines (1976). In general, mature females, whether pregnant, lactating, or between periods of reproduction, have comparable levels of wounding. Nulliparous females have less wounding than parous females. Fertile males have significantly greater levels of wounding compared to nonfertile males, but throughout the study nearly 92% of the males were judged to be fertile. The relationship between reproduction and wounding was examined by determining the correlations between reproductive variables and the wounding score, shown in Table 1 for females from the increase and decline phases and Table 2 for males. The major changes from the increase to the decline phase were the higher levels of wounding during the phase of declining numbers (Rose and Gaines, 1976).

Body Weight and Genotype

Mean body weights of voles were tested using a one-way analysis of variance. Data were lumped across the cycle and tested to determine whether genotype affected body weight. There were no statistically significant relationships between maternal weight and either parity or embryo count, and the mean body weights of males and females did not differ from each other during any phase of the population cycle. Only for the α-GPD locus was there a significant ($P < 0.01$) relation-

Table 1. Product-moment correlation coefficients (r) for reproductive variables and wounding index of female *Microtus ochrogaster* during the increase phase (N = 160; May 1971 through January 1972), above the diagonal and during the phase of declining numbers (N = 158; July 1972 through March 1973), below the diagonal. *P < 0.05, **P < 0.01.

Decrease Phase / Increase Phase	Variables									
	BW	LV	PSV	VOV	US	UW	PS	LE	CL	WI
Body Weight (BW)		-.49**	.65**	-.30**	.52**	.59**	.24*	.12	.08	.28**
Lactation Value (LV)	-.30**		-.39**	.09	-.34**	-.27**	-.12	.08	.02	.03
Pubic Symphysis Value (PSV)	.69**	-.47**		-.37**	.63**	.57**	.09	-.01	-.03	.15
Vaginal Orifice Value (VOV)	-.37**	.16*	-.41**		-.40**	-.30**	-.06	.05	.05	-.21*
Uterus Size (US)	.49**	-.31**	.54**	-.44**		.46**	-.11		-.03	-.03
Uterus Weight (UW)	.50**	-.09	.51**	-.33**	.46**		-.11	.29**	.21*	.13
Placental Scars (PS)	-.03	-.09	-.08	.03	-.13	.09		.25	.33*	.40**
Living Embryos (LE)	.13	-.18	-.04	.07		-.01	.58**		.84**	-.01
Corpora Lutea (CL)	.23*	-.05	.05	-.06	-.01	-.07	.19	.82**		.02
Wounding Index (WI)	.18*	-.30**	.22**	-.15*	.22**	.02	.02	.15	.15	

Table 2. Product-moment correlation coefficients (r) for reproductive variables and wounding index of male Microtus ochrogaster during the increase phase (N = 147; May 1971 through January 1972), above the diagonal, and during the phase of declining numbers (N = 145; July 1972 through March 1973), below the diagonal.

Decrease Phase \ Increase Phase	Variables				
	BW	TW	ES	SVS	WI
Body Weight (BW)		.68**	-.40**	.57**	.16*
Testes Weight (TW)	.57**		-.43**	.71**	.33**
Epididymides Size (ES)	-.42**	-.27**		-.44**	-.22**
Seminal Vesicles Size (SVS)	.39**	.82**	-.27**		.23*
Wounding Index (WI)	-.12	.23**	-.12	.33**	

*p < .05, **p < .01

ship; voles with the slow homozygous genotype had lower body weights (37.2 ± 0.7 g) than those with either the heterozygous (40.4 ± 0.7 g) or the alternate homozygous genotypes (40.3 ± 1.3 g).

Genetic Heterozygosity

In addition to single locus effects, possible relationships between genetic heterozygosity, reproduction, and wounding were investigated by considering individuals from each density period as from one of four groups: voles with 0 to 20% of the seven loci heterozygous, and those with 21 to 40%, 41 to 60%, or > 60%. Some voles were heterozygous for five loci. We tested for group differences in reproduction by an analysis of covariance, using corpora lutea and body weight as the variables. Within the four heterozygosity groups, no significant differences were detected in the numbers of corpora lutea produced by females of prepeak, peak, or postpeak periods of density. Similar analyses examined the relationship between numbers of embryos and body weight; significance was approached ($P < 0.10$) only in the postpeak period, when the four groups varied in mean number of embryos. Based on a t-test, females in the 0 to 20% group had significantly more embryos ($P = 0.01$) per female during the postpeak period than did the females in the 21 to 40% group. The wounding index values of the four heterozygosity groups were examined with a one-way analysis of variance. There were no statistically significant differences in the levels of wounding associated with the proportion of heterozygous loci for either fertile males or parous females during any density period.

DISCUSSION

During periods of strong selection, such as might occur during a population decline, a premium is placed on those genotypes with the highest fitness, as measured by growth rate, viability, and reproduction. Of these, we have no information about growth rate. Changes in viability can be inferred from the various frequencies of different genotypes throughout the density cycle. A trend throughout the cycle toward an increase in the frequency of the S-allele was noted at three genetic loci: LAP, EST-1, and XDH. Furthermore, for these loci there was a positive association between increased frequency of the S-allele and levels of aggression as evidenced by wounding. In genetic terms, increased frequency of the rarer of two alleles increases the probable occurrence of heterozygotes, with a maximum of 50% heterozygotes resulting from random mating within a population if both alleles have equal frequencies. Individuals that are heterozygous for a large number of loci are presumed to have a higher level of vigor due to heterotic effects.

Gaines and Krebs (1971), who observed the S-allele at the LAP locus to be more common throughout the study in southern Indiana, reported males of Lap-SS genotype to be superior in viability in the increase phase and females of Lap-SS genotype to have higher survival during the phase of declining numbers. We observed similar trends with the LAP locus, and also for EST-1 and XDH loci, in the present study, although superior viability must be inferred because the sampling procedures we used do not permit the calculation of minimum survival rates. The frequency of the S-allele increased throughout the cycle from 0.13 to 0.30 for LAP, 0.20 to 0.44 for EST-1, and 0.63 to 0.81 for XDH. Only for the former two loci can the increase be related to heterosis, for reasons mentioned above.

For the LAP locus, there is an indication that the Lap-FF genotype made a larger reproductive contribution during the period of declining numbers, as deduced from the significantly higher frequency of perforate vaginae for this genotypic class. However, the slopes of the regressions of both corpora lutea and embryo counts on body weight did not differ among the three genotypes during this population phase. Gaines et al. (1978) concluded that Lap-FF females had significantly higher reproductive rates than those with other LAP genotypes during the population decline. Better evidence for greater reproductive contribution by genotype is seen for the EST-1 locus, in which the relationship for the Est-1-FF genotype had a significantly steeper slope compared to those of the other genotypes during the period of increasing numbers. Here we observe possible balancing selection during the period of increasing numbers, with the homozygotes of the S-allele becoming more common due to superior viability and at the same time the homozygotes for the F-allele having a greater reproductive contribution.

For the EST-4 and XDH loci, the slopes of the regression for both corpora lutea and embryo counts on body weight were significantly steeper for heterozygous individuals than those for the homozygous individuals only during the period of peak density. Individuals of both EST-4 homozygous classes made greater reproductive contributions than those of the heterozygous class during the increase phase. Gaines et al. (1978) also found that females with Est-4-SS genotype that were collected on the grids had significantly higher levels of reproduction during both increase and decline phases than females with the other EST-4 genotypes. These results are compatible with those of the present study. For XDH, individuals homozygous for the S-allele (Xdh-SS) produced more ova and embryos during the population decline, at a time when the slow allele was becoming common. Perhaps a combination of superior viability and higher reproductive contribution was causing the increase in the frequency of the slow allele in this instance. The critical point for Chitty's genetic-feedback hypothesis is whether differences in wounding levels, reproduction and survivorship are associated with each other and with the same genotypic classes during the same phases of the population cycle.

Genetics and Behavior

There is relatively little information about the genetic basis of behavior in natural populations of microtine rodents. Krebs (1970) found no differences in behavioral measures of aggression among voles of TF genotypes. Myers and Krebs (1971a) examined the dispersal behavior of prairie and meadow voles with different TF and LAP genotypes. Anderson (1975) examined the heritability of spacing behavior in _Microtus townsendii_. Rasmussen et al. (1977) reported that field voles (_M. agrestis_) from northern, cycling populations were more active than those from southern, noncycling populations. Since lab-reared offspring were behaviorally similar to their wild-caught parents, and hybrids produced by parents from northern and southern populations were intermediate, these authors inferred a genetic basis for the behavioral differences.

If other rodent studies are included, considerably more information is available on the genetic basis of behavior. For example, Myers (1974) examined the dispersal behavior of _Mus musculus_ with different hemoglobin and EST-2 genotypes. Garten (1976), using 31 electrophoretically detectable loci, observed a relationship between increasing heterozygosity and aggressive behavior in mainland populations of the oldfield mouse, _Peromyscus polionotus_. However, no information was presented on possible associations between a single locus and aggressive behavior. Garten (1976) reviewed the studies that have reported a relationship between genetics and behavior; these were primarily conducted with laboratory populations of murid rodents. In addition, Smith et al. (1975) postulated that the positive relationship between average heterozygosity and body weight may affect individual behavior and demographic attributes of populations.

Although cause-and-effect relationships between variants at electrophoretically detectable loci and behavioral attributes are unknown, repeated correlations could imply that at least a linkage association exists. As reviewed by Gaines (this volume), electrophoresis has not provided the numerous correlations of genotypic and demographic parameters expected by many recent investigators. Also, for no electrophoretically detectable locus is there a genotype that is always or even frequently associated with aggressive phenotypes. (A major effort has been made by several investigators to learn, as part of a test of Chitty's hypothesis (1967), whether there may be such a thing as a genotype for aggressive behavior.) Although new methods of genetic analysis are needed to increase the level of genetic resolution, even with present techniques future studies may uncover correlations between specific behavioral traits and genotypes. We agree with Tamarin (1978b) that Chitty's (1967) hypothesis cannot be discarded as having been adequately tested.

In Rose and Gaines (1976), the terms "wounding" and "aggression" were used almost interchangeably, implying a positive association. Although in some species of small mammals wounds might be inflicted predominately to subordinate animals, there is little evidence to support this argument in Kansas prairie voles. For example, the mean body weights and mean body lengths of the wounded and unwounded males and females did not differ, nor did the reproductive contributions. However, both the frequencies of the slow alleles of three loci (LAP, EST-1, and XDH) and levels of wounding increased throughout the study. In the absence of differences in reproductive contribution and dispersal, an allele can increase in frequency only if the wounded voles are surviving at high frequency. Consequently, we conclude that wounded voles are also aggressive voles, and that they are longer-lived than unwounded voles. This conclusion is consistent with the results of Turner and Iverson (1973) who found that aggressive male meadow voles had a survival advantage in natural populations. Evidence from laboratory and field studies of wounding in the meadow vole from Virginia (Rose and Hueston, 1978; Rose, 1979) reinforce the notion that wounding is common and probably occurs often within the lives of most voles. Even large, robust, and presumably dominant animals (Clarke, 1956) have significant amounts of wounding. In both prairie and meadow voles, smaller ("younger") voles have significantly less wounding than voles heavier than 30 g.

In conclusion, there are some differences in the frequency of genotypes, the level of wounding, and viability and reproductive contribution as related to the density in the population. If a balanced polymorphism such as Chitty (1967) proposed does exist, then the population would be expected to be dynamic as in our Kansas populations; the proportion of heterozygotic loci increases through a density cycle. These results are consistent with a model of the dispersion of small mammals in which at lowest densities the survivors of the population crash form small breeding units (demes), resulting in substantial inbreeding and consequently, increased homozygosity of genetic loci. As densities increase, there is increased probability of gene flow among demes, and the populations approach panmixis; hence, the highest frequency of heterozygotes should be observed at this time. However, individuals with genotypes associated with the greatest potential for population increase are not necessarily those favored during the population decline (Tamarin and Krebs, 1969; Gaines and Krebs, 1971). Chance may also play a role in determining the survivors of the population crash. The small groups that initiate the next population cycle may differ genetically to a large degree from those at the start of the present cycle. During such founding events there is a likelihood that some loci will become genetically fixed, that is, rare alleles will be lost to that population, at least until re-introduced through migration or, less likely, through mutation. Thus, both stochastic and selectional processes are important in dynamics of vole populations.

Acknowledgments--This paper is derived from a dissertation submitted to the Graduate School of the University of Kansas by the senior author. We are grateful to the University of Kansas Endowment Association for providing field facilities and to Messrs. Howard Kampschroeder and Gaylord M. Schneck for consenting to the use of their land. Financial assistance was provided by an NSF traineeship in Systematic and Evolutionary Biology to R.K.R., a grant to M.S.G. (NSF GB 29135), General Research Grant 3822-5038 of the University of Kansas, and the Kansas Biological Survey. Judith Ludington drew Fig. 2. Finally, R.K.R. wishes to thank his wife, Aleene, for her patience and support of the field and laboratory studies that resulted in this paper.

SPATIAL-TEMPORAL CHANGES IN GENETIC COMPOSITION OF DEER MOUSE POPULATIONS

Dean R. Massey and James Joule

Abstract--During 1976-77, deer mice (Peromyscus maniculatus) were sampled from six field sites representing three different habitats. Samples were collected at four times (January 1976, June 1976, September 1976, and January 1977). Dispersing animals were sampled from an additional site. Electrophoretic and morphological data were taken for all individuals captured. Significant heterogeneity among allele frequencies of the subpopulations was shown in 44% of the cases tested, while only 16% showed significant heterozygote deficiencies (Wahlund effects). Analysis of allele frequencies across loci indicated that drift was the major factor responsible for the heterogeneity. Increased mean heterozygosity ($\bar{H}$) during June 1976 was accompanied by a seasonally related breeding period, high dispersal, and a size-class composition emphasizing smaller individuals. An overall significant positive correlation between body weight and heterozygosity was found for resident male deer mice, but the increased heterozygosity of the June 1976 sample was largely due to females of lower weight classes. High heterozygosity was characteristic of dispersers throughout the study, while the resident subpopulations showed decreased heterozygosity during September 1976 and throughout the winter. The maintenance of genetic variability in local populations of deer mice appears to be related to seasonal changes in breeding structure.

INTRODUCTION

Levene (1953) described a model for maintaining intrapopulational genetic variability based on a "coarse-grained" system wherein the opposing forces of directional selection and gene flow among subpopulations maintain polymorphism. Later, fitness set theory of Levins (1968) added increased heuristic sophistication to Levene's original ideas. However, Levene's single-locus model antedated the recent disclosure of vast amounts of biochemical variability through electrophoretic techniques (e.g., Lewontin and Hubby, 1966). With this major innovation and its accompanying opportunity for investigating the genetics of field populations, new questions emerged concerning the maintenance of the observed variability not only at one locus but throughout entire genomes and gene complexes.

Upon discovery of enormous amounts of biochemical variation, the parallel assumption of extreme genetic load led to the development of the theory of neutral or slightly deleterious alleles (Kimura and Ohta, 1974). Here it is argued that the vast majority of variation detected by electrophoretic techniques is not influenced by selection, but is maintained in populations as a function of stochastic processes (e.g., drift and mutation). Mean heterozygosity ($\bar{H}$) within isolated populations can be predicted by: $H = 4N_e \mu/(4N_e \mu + 1)$, where N_e is the genetically effective population size and μ is a constant mutation rate per generation (Kimura and Ohta, 1971). Accordingly, changes in mean heterozygosity can only be caused by changes in the effective breeding size of the population (N_e).

On the other hand, mean heterozygosity (average number of heterozygous loci per average individual within a sampled cohort) appears to be correlated with increased environmental variability (Powell, 1971; Bryant, 1974a; Taylor and Mitton, 1974). These studies imply that individuals with higher heterozygosities adapt to wider ranges of environmental change. Heterosis, as a single-locus selection process, initially appeared to be the basis for explaining total genomic heterozygosity. However, heterosis has been exceedingly difficult to measure and, with the exception of the classical case of sickle-cell anemia, has never been shown clearly in natural populations (Berger, 1976). Lewontin's (1974) discussion of heterosis and overdominance still leaves us with the present lack of usable correlations between single locus and total genomic selection.

The deer mouse, *Peromyscus maniculatus*, has many desirable properties for studying the maintenance of genetic variability in the field. Hooper's (1968) cladogenic scheme defined the *maniculatus* group under the subgenus *Peromyscus* as containing only five species. Hence, speciation has not been an important outcome of genic differentiation for this group. Rather, throughout a geographic range covering most of North America (Hall and Kelson, 1959), there appears to have been a characteristic maintenance of species integrity over a wide range of environmental differences.

In Colorado, previous biochemical analyses have shown *P. maniculatus* to be a highly polymorphic species with mean heterozygosity values between 0.08 and 0.22 (Dubach, 1975; Marquardt, 1976; Avise, et al., 1979). Compared with other mammals (Nevo, 1978), these heterozygosity values are among the highest reported. Dubach, Joule and Smith (pers. comm.) studied *P. maniculatus* over a 3048 m (10,000 ft) altitudinal gradient in Colorado, measuring biochemical variation in populations distributed at 304.8 m (1,000 ft) elevational intervals (1082 to 4,115 m) during 1974, 1975 and 1976. For the same populations, Baccus, Joule and Kimberling (1979) showed that linkage between the loci studied was not an important factor in maintaining the high levels of heterozygosity measured throughout this cline. Dubach et al. (pers. comm.) also studied temporal variation by sampling populations from one location during three different seasons over a 6-yr period (1974-1979). Three findings from this geographical and temporal study stimulated the present study: (1) genetic differentiation over a broad geographic range, (2) temporal variation in mean heterozygosity, and (3) genic analyses of variation in deer mice collected from Fort Morgan at 1372 m (4,500 ft elevation) showed a relative decrease in heterozygosity accompanied by a relative increase in average number of alleles present, indicating possible allelic differences among habitat types. The area around Fort Morgan provided a field situation where spatial and temporal variation in the genetic composition of deer mouse populations could be studied over a well-defined, "coarse-grained" environment. Genic and demographic data were used to study potential mechanisms leading to the maintenance of genetic variability within local populations.

MATERIALS AND METHODS

Deer mice were live-trapped at seven field sites within a 14.5 km radius of Fort Morgan, Morgan County,

Table 1. Description of sampling locations, dominant vegetation, height and amount of cover. Habitat type was designated on the basis of the relative availability of moisture.

Site	Habitat Type	Location	Dominant Vegetation	Cover
X^{TF}	Xeric	Tormollen's farm, 9.7 km S Ft. Morgan on Co. Rd. 19.	Sage, prickly pear, herbs, and grasses	0.3 - 0.5 m sparse
X^{MC}	Xeric	MacCarteny's farm, 14.5 km N Ft. Morgan on Colo. 52.	Sage, prickly pear, herbs, and grasses	very sparse
I^{KF}	Intermediate	Kimbles farm, 4.8 km N Ft. Morgan on Colo. 52.	Tall herbs, and grasses	0.6 - 1.5 m dense
I^{AP}	Intermediate	Ft. Morgan airport, 8.1 km N Ft. Morgan on Colo. 52.	Grasses, and herbs	0.5 - 10.9 m
M^{SP}	Mesic	South Platte River, 0.8 km N Ft. Morgan on Colo. 52.	Cottonwood, willow, and tall grasses	1.5 - 7.6 m dense stratified
M^{KD}	Mesic	Kula's Dump, 2.4 km E site M^{SP}.	Cottonwood, willow, and tall grasses	1.5 - 7.6 m dense
M^{R}	Mesic	Total removal site, 14.5 km NW Ft. Morgan on Colo. 144.	Cottonwood, and willow	1.5 - 3.0 m dense

Colorado at predetermined time periods during 1976 and 1977 (Table 1). Pairs of sites represented the three major habitats commonly found in the South Platte River Basin, which were refered to as xeric, mesic and intermediate (between xeric and mesic). These habitats were distinguished qualitatively by the dominant vegetation, height and amount of cover, and relative moisture availability. Each site was separated from the others by a distance greater than 4.8 km (3 mi) and was 5 ha or more in size. Deer mouse populations of the six sites were sampled during four discrete time periods which represented the different phases of the annual population dynamics of P. maniculatus (reproductive, prereproductive, postreproductive, overwintering). Data in previously published studies, as well as those from previous studies of the Fort Morgan populations, were used in determining the precise trapping periods (Storer, et al., 1944; Jameson, 1953; Stickel, 1968; Terman, 1968; Petticrew and Sadleir, 1974; Sadleir, 1965, 1974). During each period, each population was sampled until approximately 10 deer mice were obtained, which was considered the result of a compromise between obtaining a statistically adequate sample size and maintaining the populations for further sampling. Site of capture, sex, body weight, body length, tail length, and reproductive condition were recorded for all deer mice captured. Although deer mice were of major interest, other species caught were recorded and removed from the area to prevent preemption of the habitat space.

One additional plot, Site-M^R (Table 1), where all individuals were permanently removed, was used to measure electromorphic and demographic properties of dispersing mice. This plot was chosen because of its mesic properties and adjacent xeric and mesic habitat patches. To obtain dispersing individuals prior to the establishment of a resident offspring population, samples were collected at 4- to 8-wk intervals during the reproductive season as follows: March 24-28 (resident population removed), May 26-31, July 6-10, August 10-15, and September 25-29, 1976. One additional sample was collected during January 25-29, 1977. During each sampling period, trapping continued for at least four days (used during periods of low capture rate), or until the daily number of captures became less than 5% of the total sample (used during periods of high overall capture rates).

Spatial-Temporal Changes in Deer Mice

Tissue preparation for electrophoretic analysis closely followed methods described by Selander et al. (1971). Horizontal starch-gel electrophoresis (Electrostarch lots #302 and #307, Electrostarch Co., Madison, Wisc.) was used to analyze hemolysate, serum, liver and heart-kidney tissue samples from each individual for 20 protein systems (loci). Allozymic scoring closely followed descriptions outlined by Selander et al. (1971), Mascarello and Shaw (1973), Smith et al. (1973), Dubach (1975), and Massey (1977). Allozymes for each locus were coded by numerical superscripts designating relative rates of electrophoretic migration. The most common allele at each locus was assigned an arbitrary value of 1.0 and all less common forms were coded by the mobility of their products relative to that of the common allele. Further details of the sampling methods and sites are given by Massey (1977).

RESULTS

During the four predetermined sampling periods between January 1976 and January 1977, 233 deer mice were collected from five locations, Site-X^{MF}, Site-X^{TF}, Site-I^{KF}, Site-I^{AP}, and Site-M^{SP}. Although this study was originally designed for six different resident populations, the number of animals obtained from Site-M^{KD} was extremely small and these data have been excluded from the present analysis. Preliminary analyses of morphological and genic data obtained for the animals of the five remaining sites did not reveal trends in samples paired by similar habitat type (i.e., xeric, mesic, and intermediate between xeric and mesic). Hence, samples from each population were treated as independent units. Additionally, there was no satisfactory reason to believe each sampled unit represented a complete and different population. However, there was sufficient geographical isolation by distance for use of the term subpopulation (Crow and Kimura, 1970).

One hundred thirty-two deer mice were obtained from Site-M^{R}. Forty deer mice, captured during the initial sampling period of March 1976, were considered residents. Ninety-two deer mice captured during later periods were considered dispersers. Along with each seasonal sample of the five resident subpopulations, that fraction of the 92 immigrants collected from

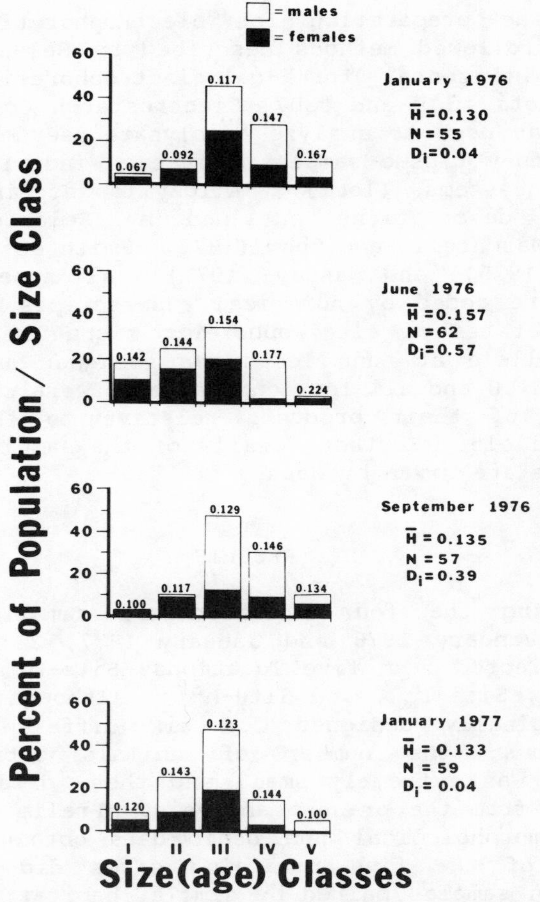

Fig. 1. Size-category distributions of pooled individuals per trapping period from the resident sites. Size classes are as follows: I. $\leq$ 13.56 g; II. 13.57-16.66 g; III. 16.67-19.76 g; IV. 19. 77-22.86 g; V. > 22.86 g. Mean heterozygosity ($\bar{H}$), pooled sample size (N) and relative dispersal index (D_i) are given for mice collected during each trapping period. Relative dispersal values were calculated as the proportion of dispersing individuals measured at Site-M^R during the particular trapping period in relation to the total dispersers taken over the entire year. Hence, D_i indicates the proportion of the total number of annual immigrants moving onto Site-M^R between sampling periods.

Site-M^R since the previous sampling period estimated the proportion of annual dispersal occurring during that period (see D_i's, Fig. 1). Using this method, 57% of annual dispersal occurred between March and June 1976 (accompanying spring reproduction); 39% occurred throughout the rest of the summer (June-September, 1976), while only 4% was recorded between September 1976 and January 1977. Since D_i was not estimated for January 1976, the estimate gained one year later (January, 1977) was also entered for this initial period (Fig. 1). During additional sampling for dispersers on Site-M^R in March 1977, no deer mice were captured. Hence, the March-June 1976 estimate gained during this study is very likely comparable to a January-June measure, with little deer mouse movement occurring before March.

Demographic Measures

The numbers of deer mice obtained from each subpopulation were not sufficient to permit a detailed demographic analysis per sampling period. Instead, seasonal data from the pooled subpopulations were tested for demographic changes. Pooling involved the assumption that many of the factors influencing population processes are climatological (e.g., Fairbairn, 1977a) and do not vary among local subpopulations. Size-class categories for the pooled data per trapping period were constructed as follows: (1) Mean body weight for deer mice over the entire study was determined ($\bar{X}$ = 18.21 g); (2) an initial size-category, one standard deviation unit (s = 3.10 g) in width, was calculated and then centered around the mean; (3) an additional four categories, with also one standard deviation widths, were placed adjacent to the central class with two on each side; and (4) data for nine individuals which fell slightly outside the five categories (five larger and four smaller) were pooled with those for the nearest category.

Direct density estimates of the resident subpopulations were not made, but indications of increasing and decreasing growth periods were shown by the relative number of individuals falling into the five size categories (Fig. 1) and by the reproductive condition of males and females (Table 2). Body weights have previously been used as estimators of relative age within small mammal populations (e.g., Joule and Cameron, 1974;

Table 2. Data on reproductive condition of deer mouse resident subpopulations pooled within each sampling season. For adult females, percent with litters in utero is given. For adult males, percent of individuals with testes descended is given. Juveniles were not included.

	PERCENT REPRODUCTIVELY ACTIVE	
Sampling Period	Females	Males
January 1976	0	4
June 1976	9	90
September 1976	19	93
January 1977**	18	67

** Based upon a subsample of 11 females and 22 males captured during this period.

Krebs et al., 1976). Hence, increased proportions of individuals in the smaller size classes indicated recent additions of new juveniles and young adults. Males tended to be slightly heavier than females and more numerous in some samples. Male reproductive condition and the presence of fetuses in utero provided the best indicators of change in reproductive activity over different seasons. The external measure of female reproductive condition (vagina open/closed) did not show differences in activity between seasons. Although Class-III was the dominant size category throughout the study, a significant shift toward the smaller sizes was measured in June 1976 and January 1977 ($\chi^2_{(12)} = 21.6$; $P < 0.001$). Contributions of each sex to the smaller size categories were equivalent during these periods ($\chi^2_{(1)} = 0.80$ for June, $\chi^2 = 0.85$ for January 1977; $P > 0.10$).

Frequency distributions of body sizes for dispersing deer mice were constructed for samples obtained from Site-M^R (Fig. 2). Data from the small samples from August 1976 (N = 7), September 1976 (N = 9), and January

1977 (N = 4) were pooled to form one distribution. The resident sample removed from Site-M^R during March 1976 showed a composition noticeably skewed toward larger sizes (58% of the individuals represented the two largest categories), while later samples of dispersers showed a significant shift toward smaller body size with only 27% representation in the two largest size categories ($\chi^2_{(12)} = 28.15$; $P < 0.01$). However, only 5% of all dispersers occurred within the smallest size category, although most were captured during the peak breeding period. While higher proportions of females occurred in the disperser sample, a test of the sex ratios of resident and disperser samples yielded a nonsignificant difference ($\chi^2_{(1)} = 0.70$; $P > 0.10$) as did a test of heterogeneity of the sex ratios between the dispersers of different trapping periods ($\chi^2_{(3)} = 5.12$; $P > 0.10$). Hence, seasonally measured shifts in body sizes were not due to significant differences in sex ratios between samples.

Genetic Analysis

Twenty isozymes, controlled by 20 loci, were analyzed for the deer mice from each location. Two loci, TO (tetrazolium oxidase) and G-6-P (glucose-6-phosphate dehydrogenase) were monomorphic for all individuals. Ten loci were slightly polymorphic, based on the criterion of the most common allele present at a frequency of 0.90 or greater in the pooled samples: LDH-1 (lactate dehydrogenase-1), LDH-2, MDH-1 (malate dehydrogenase-1), MDH-2, GOT-2 (glutamic oxalate transaminase-2), SDH (sorbitol dehydrogenase), IDH-1 (isocitrate dehydrogenase-1), PGM (phosphoglucomutase), and PGI (phosphoglucose isomerase). Eight loci were highly polymorphic (common allele frequency less than 0.90): GOT-1, ADH (alcohol dehydrogenase), TRF (transferrin), HB (hemoglobin), ALB (albumin), EST-4 (esterase-4), 6-PGD (6-phosphogluconate dehydrogenase) and PEPT-1 (peptidase-1). Allele frequencies and genotype frequencies for each subpopulation have been reported by Massey (1977, Appendices I and II). The following analyses of genetic partitioning among the subpopulations considered only the eight highly polymorphic systems.

When measurable genetic subdivision exists between populations, pooling genotype frequencies over subdivisions can lead to significant deviations from expected

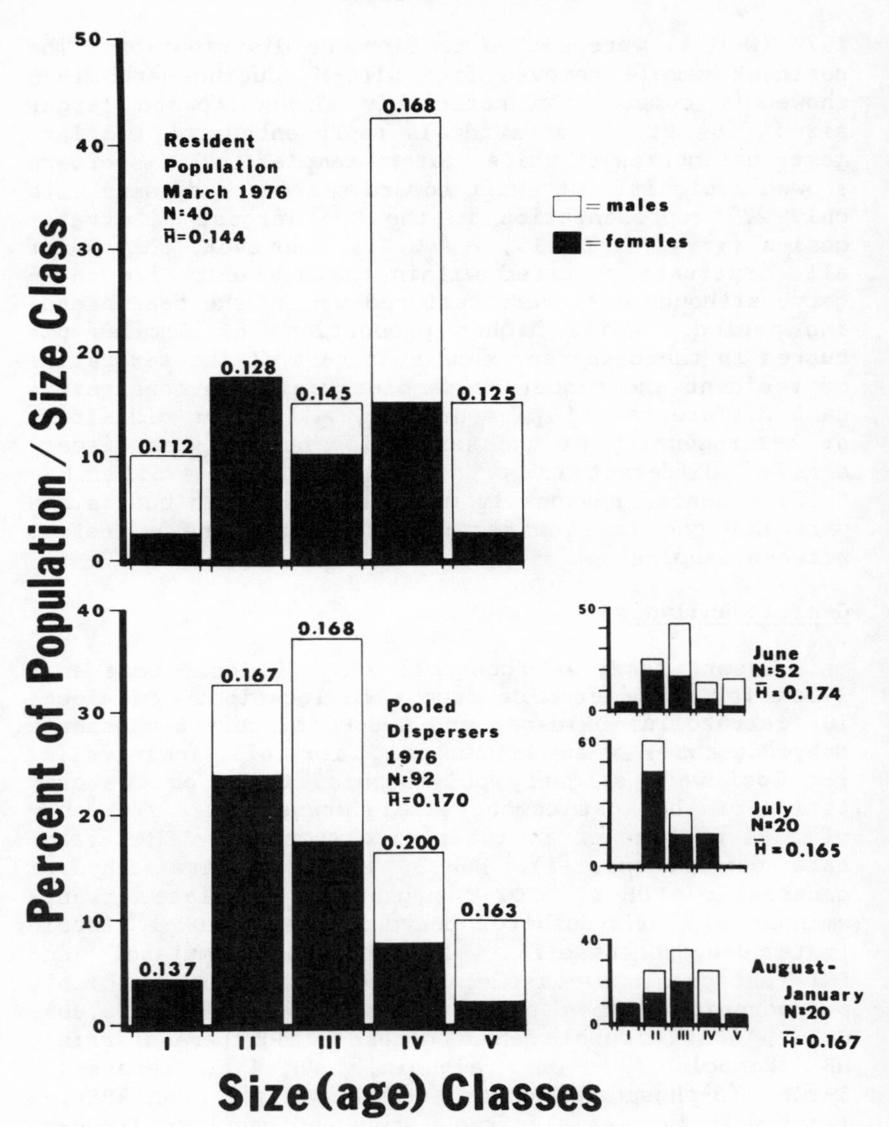

Fig. 2. Size-category distributions of initial resident and later dispersing individuals sampled through total removal of Site-M. The large dispersal distribution to the left is the pooled contribution of three smaller periodic dispersal distributions on the right. Size classes have the same ranges as described for Fig. 1.

Hardy-Weinberg proportions. Usually characterized by deficiencies of heterozygotes (i.e., Wahlund effects), these departures reflect simple statistical artifacts arising from divergent allele frequencies among subpopulations (Crow and Kimura, 1970; Sinnock, 1975). To test for subdivisional effects, genotypic frequency data from the five resident subpopulations were pooled for each season. Pooled allele frequencies were used to generate expected random mating proportions, using Levene's (1949) correction for finite populations. The null hypothesis that the observed distributions of genotypic frequencies conformed to expected distributions was tested over the eight highly polymorphic systems (Table 3). Five of the 32 tests (16%) showed significant deviations ($P < 0.05$), all showing trends toward heterozygote deficiency.

Life-history processes within populations (e.g., inbreeding, assortative mating) can have different effects upon genotypic and allele frequencies. Hence, tests for genetic differences between subpopulations must also consider differences between observed allele frequencies. As shown in Table 3, 14 of 32 tests (44%) showed significant deviations from homogeneity ($P < 0.05$), indicating that about half of the allele frequencies were heterogeneous among the subpopulations. Overall, the distribution of significant differences did not reveal seasonal trends.

Recently, Lewontin and Krakauer (L-K; 1973) developed a test of electrophoretic data which separates the deterministic effects of natural selection from stochastic processes such as drift. Basic to the L-K test is the assumption that, without selection, the F_{st} values (Wright, 1965) over all polymorphic loci of a given cohort will be distributed as Chi-square with $\bar{F}_{st} = F_{st} i/a$ (where $i = 1,2,3,...a$, and a = number of effective F_{st}'s calculated for a given cohort) and an expected variance $\sigma^2 = k\bar{F}^2_{st}/n-1$ (where $k = 2$, assuming a binomial distribution of allele frequencies and n = number of subpopulations). Tests of the distributions of the calculated F_{st} values against the expected distributions were performed by comparing variance ratios (s^2/σ^2), and showed nonsignificant deviations for all seasons (Table 4). Hence, selection does not appear to be a principal factor in maintaining heterogeneity in allele frequencies among local deer mouse subpopulations.

Table 3. Chi-square values for goodness of fit analyses of deviations from Hardy-Weinberg equilibrium for data from all populations pooled within trapping periods. Given in parentheses are the Chi-square values for tests of independence of allele frequencies testing the null hypothesis that the array of allele frequencies sampled is independent across the five sampled subpopulations.

Locus[a]	Degrees of Freedom	January 1976	June 1976	September 1976	January 1977
GOT-1	2	3.53	0.07	12.41**	4.11
	(4)	(8.14)	(12.14)*	(10.59)*	(6.44)
ADH	2	0.40	5.16	3.31	16.83***
	(4)	(13.32)**	(28.30)**	(10.68)*	(10.96)*
6-PGD	5	1.85	4.41	1.29	4.94
	(8)	(15.82)*	(32.06)***	(21.36)**	(10.81)
EST-4	5	6.94	5.87	3.73	11.97*
	(8)	(8.57)	(12.11)	(10.93)	(23.92)***
TRF	2	0.59	0.52	14.38***	3.30
	(4)	(7.01)	(16.00)*	(15.69)*	(9.52)
ALB	2	0.00	0.00	0.00	2.37
	(4)	(20.46)**	(8.08)	(3.90)	(8.62)
HB	2	19.88***	(1.95)	3.75	0.86
	(4)	(5.81)	(6.04)	(21.79)**	(9.58)
PEPT-1	2	0.21	0.00	3.31	0.35
	(4)	(4.46)	(3.44)	(6.20)	(4.81)

*P < 0.05; **P < 0.01; ***P < 0.001.

[a]Locus abbreviations are given in text.

Table 4. Lewontin-Krakauer tests for selective neutrality using F_{st} values calculated from spatial variation in allele frequencies of five subpopulations (Lewontin and Krakauer, 1973).

Locus	F_{st} January 1976	June 1976	September 1976	January 1977
GOT-1	.0901	.0843	.0980	.0470
ADH	.0481	.1335	.0855	.1270
6-PGD	.0619	.0300	.1469	.0336
	.0551	.0477	.0887	.0237
	.0985	.0077	.0953	.0171
EST-4	.0341	.0466	.0080	.1859
	.0304	.1131	.0291	.0289
	.0426	.0640	.0632	.0380
TRF	.0188	.0899	.1054	.0814
	.0084	.0826	.0224	.0789
	.0369	.0260	.2542	.0645
ALB	.1829	.0567	.0331	.0923
HB	.0576	.0466	.0808	.0889
PEPT-1	.0381	.0055	.0153	.0034
$\bar{F}_{st}$	.0574	.0603	.0804	.0648
s^2	.0019	.0014	.0041	.0024
s^2/σ^2	1.16^{ns}	$.77^{ns}$	1.27^{ns}	1.14^{ns}

$^{ns}P > .05$

Measures of Genic Variation

Mean heterozygosity ($\bar{H}$), proportion of loci polymorphic per population (P), and average number of alleles per locus (A) were calculated for each subpopulation and for the pooled subpopulations for each trapping period (Table 5). Measures of both P and A are functions of probabilities for detecting alternative alleles in the 20 loci. For the present study, P and A showed little seasonal variation with small changes following differences in sample size for each plot over seasons.

A single classification analysis of variance was used to test the null hypothesis that no significant differences in heterozygosities occurred among the subpopulations over time (Sokal and Rohlf, 1969). Prior to testing, the heterozygosity value for each individual was transformed by arcsin square root h_i. The rejected null hypothesis ($F_{(3,16)} = 3.29$; $P < 0.05$) indicated that significant variation in mean heterozygosity occurred between seasons. An a priori planned comparison between the June 1976 sample (peak reproductive) and the remaining samples (presumed low or nonreproductive) indicated the mean heterozygosity during June 1976 was significantly different (greater; $F_{(1,16)} = 8.86$; $P < 0.01$).

Mean heterozygosity of deer mice at Site-M^R continued to increase through time. The mean value for the residents was 0.14 compared to the 0.16 value for all dispersers combined. The high heterozygosity of the dispersers persisted throughout September 1976 and January 1977 in contrast to decreased heterozygosity in the resident subpopulations, indicating that immigrant mice were more heterozygous than residents.

Body Weight and Heterozygosity

Garten (1976) showed a significant relationship between body weight and heterozygosity in P. polionotus. To test the present data for this effect, product-moment correlation coefficients were calculated using a two-way frequency distribution of body weights and numbers of heterozygous loci per individual (Sokal and Rohlf, 1969). A significant positive relationship was found for males ($r = 0.26$; $P < 0.001$; $N = 141$), while females showed a nonsignificant correlation ($r = 0.01$; $P > 0.50$;

Table 5. Mean heterozygosity (H), percent polymorphism (P), and average number of alleles per locus (A). Two standard errors are given for each $\bar{H}$, one calculated per individual (reported above) and another calculated per locus.

Site		January 1976	June 1976	September 1976	January 1977
M^{SP}	$\bar{H}$	.137 ± .019	.167 ± .017	.154 ± .017	.133 ± .014
		.042	.049	.051	.036
	P	.50	.55	.50	.45
	A	1.70	1.75	1.85	1.60
	N	19	18	14	15
X^{MC}	$\bar{H}$	.133 ± .018	.138 ± .015	.135 ± .019	.179 ± .013
		.045	.042	.040	.054
	P	.45	.55	.50	.50
	A	1.60	1.70	1.65	1.60
	N	12	20	13	12
X^{TF}	$\bar{H}$	.133 ± .017	.160 ± .037	.150 ± .037	.100 ± .022
		.043	.053	.043	.045
	P	.40	.35	.45	.35
	A	1.50	1.45	1.60	1.40
	N	6	5	7	5
I^{KF}	$\bar{H}$	.129 ± .016	.178 ± .015	.120 ± .014	.135 ± .015
		.041	.057	.035	.042
	P	.45	.45	.55	.50
	A	1.55	1.60	1.70	1.60
	N	12	9	10	17
I^{AP}	$\bar{H}$	.100 ± .018	.170 ± .051	.131 ± .019	.120 ± .021
		.041	.021	.043	.049
	P	.35	.50	.45	.45
	A	1.45	1.60	1.65	1.55
	N	6	10	13	10

N = 102). When the data for the sexes were pooled a significant positive correlation was maintained due to the weighted influence of excess males ($r = 0.14$; $P < 0.05$; N = 243). Also, the mean heterozygosities per body weight category for each sampling period, as shown in Fig. 1, were ranked 1 through 5 (highest to lowest) within each sampling period and compared with the rank orders of the five weight classes of each sampling period. Spearman's rank coefficient, calculated over all periods, yielded a significant positive correlation between relative heterozygosity and body size category for the five subpopulations ($r_s = 0.92$; $P < 0.001$; N = 20).

Figure 3 shows the relationship between mean heterozygosity per individual ($\bar{h}_i$) and body weight for the residents and dispersers of each trapping period. There was an increase in heterozygosity in the resident population during June 1976. However, Fig. 3 also shows a striking difference in the relative contributions of the sexes to increased heterozygosity during June 1976, with females clearly more heterozygous over the smaller and intermediate size classes. Also, a comparison between the residents and dispersers showed higher heterozygosity for immigrating deer mice. However, the sexual differences in heterozygosity reflected by the resident subpopulations were not a contributing factor to the levels of heterozygosity observed for the samples of dispersers.

DISCUSSION

Seasonal differences in the population growth of *P. maniculatus* have been related to seasonal temperature changes (Sadleir, 1974), with overwinter breeding occurring during periods with above-normal temperatures or overabundances of food (Brown, 1945; Jameson, 1953). Using this criterion, the demographic changes recorded for *P. maniculatus* during the present study followed predictions developed from previous studies with allowances for geographic differences (Fairbairn, 1977a,b, 1978a,b). The size-category distributions for June 1976 and January 1977 (an unseasonably mild period) reflected expansive population structures, i.e., increased presence of smaller individuals (Peterson, 1969). Also, increased male reproductive activity and presence of

Spatial-Temporal Changes in Deer Mice

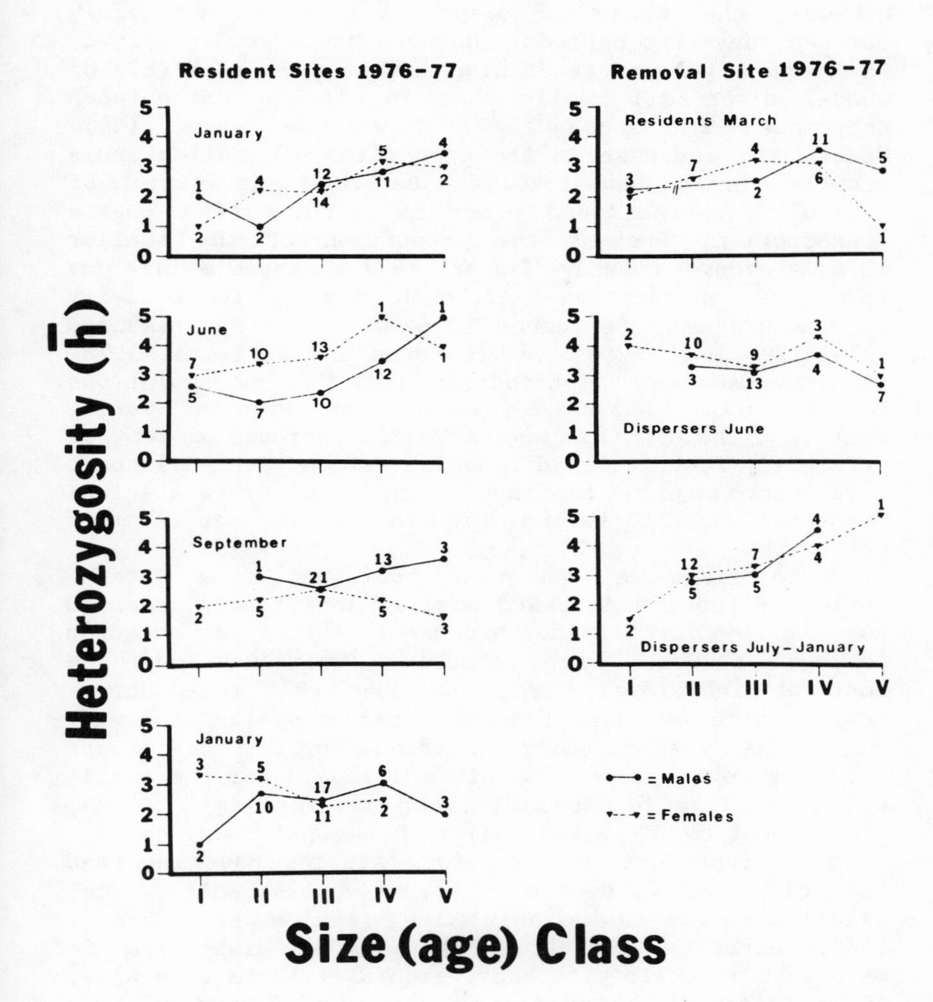

Fig. 3. Mean heterozygosity per individual ($\bar{H}$) by size-category for male and female Peromyscus maniculatus. Numbers represent individuals per size class which contributed to the class mean heterozygosity. See text for further explanation.

fetuses in utero were recorded during both periods. However, the amount of dispersal differed noticeably between these two periods. Between March and June 1976, the highest percentage of dispersal was recorded (57% of annual dispersal), while that for the period between September 1976 to January 1977 was the lowest (4%). Hence, the increase in the proportion of smaller individuals during June 1976 was due to a combination of subadult dispersers and juveniles recruited from onsite reproduction, whereas the proportion of the smaller animals recorded during January 1977 appeared mainly due to onsite reproduction. Although reproductive activity continued during September 1976 and dispersal remained relatively high (39%) between June and September 1976, the size-category distribution for the subpopulations for September showed a higher proportion of larger animals. During this period the continued growth of previously recruited individuals was important, although some reproduction continued prior to overwintering. Dispersal was curtailed for the winter after this period.

The sizes of immigrants obtained from Site-M^R reflected the trend toward smaller individuals observed as sampling continued throughout the 1976 breeding period. However, when comparing immigrants with the resident individuals from the June 1976 reproductive period which was also characterized by smaller individuals, the size-category distribution for dispersers showed a lower proportion of individuals for the smallest size class (Class I, 13.56 g and smaller). Hence, recruitment by dispersal differed somewhat from recruitment by reproduction. Previous studies have reported that dispersal in deer mice occurs when subadults reach sexual maturity during an annual breeding peak (Terman, 1968; Fairbairn, 1977a, 1978a), and that dispersers are mainly individuals with body weights of 12-13 g (Healey, 1967; Petticrew and Sadleir, 1974), or less than 15 g (Fairbairn, 1978a). After the removal of the resident deer mice from Site-M^R during March 1976, which were predominantly Class IV individuals (19.77-22.65 g), subsequent measurement of dispersal revealed predominantly Class II individuals (13.57-16.66 g).

The collecting sites for _P. maniculatus_ were selected from among locations with extreme habitat differences, largely related to differences in topography and the position of the water table relative to surface of the ground. The design of this study maximized changes in habitat factors thought to have important influences upon deer mice (e.g., moisture availability, food type, cover), thus allowing tests of genetic differences related to habitat differentiation. However, the electromorphic variation of eight highly polymorphic loci did not show significant differences related to habitat types. Alternatively, when the samples obtained from separated locations were treated as from different subpopulations, independent of habitat type, a large amount of the heterogeneity in allele frequencies was accounted for (44% of total tests). In contrast, overall genotypic frequencies remained in equilibrium in most cases (84%; Table 3). Hence these differences in trends reflected processes which changed subpopulational allele frequencies without altering the overall genotypic equilibrium. Genetic subdivision, specifically reported for several species of *Peromyscus* (*P. floridanus*, Smith et al., 1973; *P. polionotus*, Selander et al., 1971) and for *Mus* (Selander, 1970a; Anderson, 1970), seems common among small mammal species. Dubach (1975) has provided evidence of clinal differences in genic variation for deer mouse populations, but the manner in which population processes contribute to the maintenance of local variation within clinal steps remained unclear.

Dubach et al. (pers. comm.) also demonstrated temporal variation in heterozygosity for *P. maniculatus*, but again the causative factors for these seasonal changes remained unknown. In the present study, mean heterozygosity ($\bar{H}$) showed significant seasonal variation with mean seasonal values ranging from 0.13 to 0.15 and the values for specific samples varying between 0.10 to 0.17. During the period of high breeding in June 1976, a significantly higher mean heterozygosity was observed than during other sampling periods. Additionally, this increase was shown to be primarily due to the heterozygosities of the females in the resident population. To our knowledge no multilocus studies have reported females to be intrinsically more or less heterozygous than males.

Within the limitations of the present study, drift appeared to have caused the observed heterogeneities in allele frequencies among subpopulations. Achievement of divergent allele frequencies among subpopulations under drift would be analogous to the survivorship effects of the Levene model (1953). Thus, the development of heterogeneity in allele frequencies among subpopulations during nonbreeding, low-density periods, followed by panmictic breeding may explain the temporal changes in heterozygosity (Smith et al., 1978). However, there seems to be a large disparity between population size and the genetically effective population number (N_e) in P. maniculatus (Rasmussen, 1964; Dice and Howard, 1951) and also an indication that adequate panmixia may not occur. While negative assortative mating might account for the increased heterozygosity during the breeding season, Naylor (1963) has provided mathematical proof that this sort of system is theoretically impossible to maintain. Clearly, there is a need to study changes in population demography and mating structure along with their effects upon temporal changes in the genetic composition of subpopulations.

The use of size-category distributions to describe population states over different seasons contains two limitations which, while not nullifying the technique, may detract from linear interpretations between age, size, and survivorship. First, the average size of an individual at a given age will change with seasonal differences in daily resource availability and degree of harshness of the physical environment. Hence, comparing body weights for sets of individuals taken in different seasonal samples may reflect the relative condition of the adult population over seasons as well as whether recent recruitment of young has taken place. Secondly, the positive correlation between body size and heterozygosity found in the present study is similar to the interpopulational relationship found for P. polionotus over a wide geographic range (Garten, 1976). While Garten did not test for a similar intrapopulational relationship, this possibility remains. If this relationship exists, then there is no simple linear relationship between size and age because of the relationship between size and genotype and the possible interaction between these two factors. For example, some individuals will be larger than others because of genetic differences and not simply age. In the past, investigators of small mammals have satisfactorily partitioned individuals within populations into three broad categories based upon body weights: juveniles, young adults, and older adults (P. maniculatus: Healey, 1967; Sadleir, 1974; Petticrew and Sadleir, 1974; Fairbairn, 1977a, 1978a). If the mice in the present study are partitioned into these three classes, the relationship between heterozygosity and body size found using the five-class method still holds, i.e., increased size with increased heterozygosity. However, this relationship was not observed during winter 1977, a reproductive period in which mice were not dispersing. Here, dispersers did not appear to contribute to either reproduction or to direct recruitment.

In summary, deer mouse subpopulations showing differences in allele and genotype frequencies did not segregate according to habitat differences. While allele frequencies differed noticeably among subpopulations, genotype frequencies remained relatively uniform. The significant positive correlation between heterozygosity and body size appeared to be due mainly to the effects of males. Additionally, seasonally changing mean heterozygosities were highest following the spring reproductive period and the highest rate of dispersal. While an overall increase in heterozygosity with body size was shown, during winter 1976-77 this trend became inconsistent with no change in overall heterozygosity vs body size. Whether dispersal occurred prior, during or immediately after breeding could not be determined. However, the interaction between subpopulations, resulting from either outbreeding or a high degree of mixing immediately after reproduction, appeared to increase mean heterozygosity.

Acknowledgments--We thank Mr. and Mrs. Don McCartney of Ft. Morgan for their gracious assistance in developing and maintaining this field study through completion. Field assistance was provided by M. Kaufman, R. Baccus, B. A. Nelson, N. L. Gurgiolo, N. Wagner and R. A. Mills. J. M. Dubach provided the computer program for collation and analysis of the data. M. F. Stoops typed the manuscript. The study was funded in part by a Sigma Xi Grant-in-Aid to D. R. Massey. Computer time was provided by the University of Colorado. We thank the anonymous reviewers for many suggestions which improved the quality of this manuscript.

GENIC AND MORPHOLOGICAL VARIABILITY IN CENTRAL AND MARGINAL POPULATIONS OF SIGMODON HISPIDUS

Leroy R. McClenaghan, Jr., and Michael S. Gaines

Abstract--Sixteen populations of the cotton rat, *Sigmodon hispidus*, were sampled from various portions of the species distribution to provide an empirical test of the hypothesis that ecologically marginal populations of a species should have reduced genetic variability when compared to populations inhabiting more optimal environments. Populations from Kansas, Oklahoma, Tennessee, and Virginia were designated as "marginal populations", and those from the states of Veracruz and Tamaulipas, Mexico were considered as "central populations." A total of 647 individual cotton rats were obtained by live trapping. Genic variability in each population was estimated through a survey of electrophoretic variation at 23 structural gene loci. Adult cotton rats were also measured for 16 conventional external and cranial characters. Morphological variability within each population was assessed by way of multivariate coefficient of variation. The estimates of genic and morphometric variability were then compared between marginal and central populations. Marginal cotton rat populations were characterized by lower levels of genic variability as measured by both the proportion of loci polymorphic and the mean heterozygosity per individual upon comparison with central populations. However, the marginal populations reflected greater morphological variability than the central populations of *S. hispidus*. Correlation analysis demonstrated a significant negative association between genic and morphological variability. Several hypotheses are suggested to account for this relationship.

INTRODUCTION

Evolutionary biologists have long speculated about differences that might be observed when populations of a species inhabiting ecologically optimal or near-optimal environments are compared with populations of the same species inhabiting suboptimal environments (Matthew, 1915; Vavilov, 1926; Ludwig, 1950; Dobzhansky, 1951). It has become customary to refer to the former as "central populations" and to the latter as "marginal populations." However, populations have also been placed into one category or the other solely on the basis of their geographic location within the species distribution. Populations that are geographically peripheral in the species range need not be ecologically marginal. Soule (1973) has pointed out that ecologically marginal and central populations should be classified as such on the basis of population dynamics, with marginal populations being characterized by marked fluctuations in numbers and higher probabilities of extinction.

The most prevalent prediction concerning marginal and central populations is that central populations should possess greater genetic variability than conspecific marginal populations (Dobzhansky, 1951). Assuming that marginal populations are established by a few founder individuals and that they periodically pass through "bottlenecks" in numbers, inbreeding and genetic drift are important in reducing genetic variability. Selection may also be more intense in marginal environments where only a limited number of phenotypes are able to survive and reproduce. Lastly, marginal populations may be spatially isolated due to the discontinuous distribution of suitable habitat; variation that is lost will not be replenished as rapidly due to restricted flow.

In an effort to gain more insight into the question of differential genetic variability in central and marginal populations, we sampled populations of the cotton rat (<u>Sigmodon hispidus</u>) from localities along the northern limit of the species distribution in the central and eastern U.S. and from sites near the center of the distribution in eastern Mexico. <u>Sigmodon hispidus</u> seemed well suited to this study. With a species distribution ranging from temperate regions in the Great Plains of North America to tropical habitats in Central America and northern South America, populations from vastly different environmental regimes could be sampled and differences in their genetic structure measured. Also, several lines of evidence suggest that populations along the northern periphery of the species distribution are ecologically marginal. First, the fossil record indicates that the genus *Sigmodon* is of tropical origin and has only colonized more temperate regions since the Pleistocene (Hooper, 1949; Hibbard et al., 1965). Second, severe winter weather has been shown to be a source of physiological stress for cotton rats (Dunaway and Kaye, 1961). Foraging behavior of this species has been observed to be restricted in the winter (Goertz, 1964) and cotton rats are not known to store food (Schendel, 1940). Third, demographic studies of temperate populations have shown that density fluctuates markedly in a seasonal pattern, with dramatic decreases in numbers occurring in the winter (Fleharty et al., 1972; McClenaghan and Gaines, 1978), and that local extinctions following these winter density crashes may occur (Dunaway and Kaye, 1961; Gier, 1967; Fleharty et al., 1972).

In this study we assessed genic variability in central and marginal populations of *S. hispidus* through an electrophoretic survey of allozymic variation at 23 structural gene loci. In addition, adults were measured for a suite of 16 cranial and external morphometric characters to provide estimates of morphological variability. We present here the results of these analyses and a discussion of the observed patterns in genetic variability.

MATERIALS AND METHODS

A total of 647 cotton rats from 16 populations in the U.S. and Mexico were analyzed (Fig. 1). Rats were caught in Sherman live traps set under dense vegetation in oldfield habitats. Samples of blood, kidney, liver, and heart were taken from each individual and standard external measurements were made. Specimens were deposited with the Museum of Natural History at the University of Kansas.

Horizontal starch-gel electrophoresis and histochemical staining were carried out following the procedures of Selander et al. (1971). Each population of *S. hispidus* was analyzed for allozymic variation at 23 presumptive loci. A more extensive description of techniques and loci surveyed is presented by McClenaghan (1977). Loci were considered to be polymorphic if the most common allele was present at a frequency of 0.99 or less. Mean heterozygosity per individual (H) and the

Genic and Morphological Variability

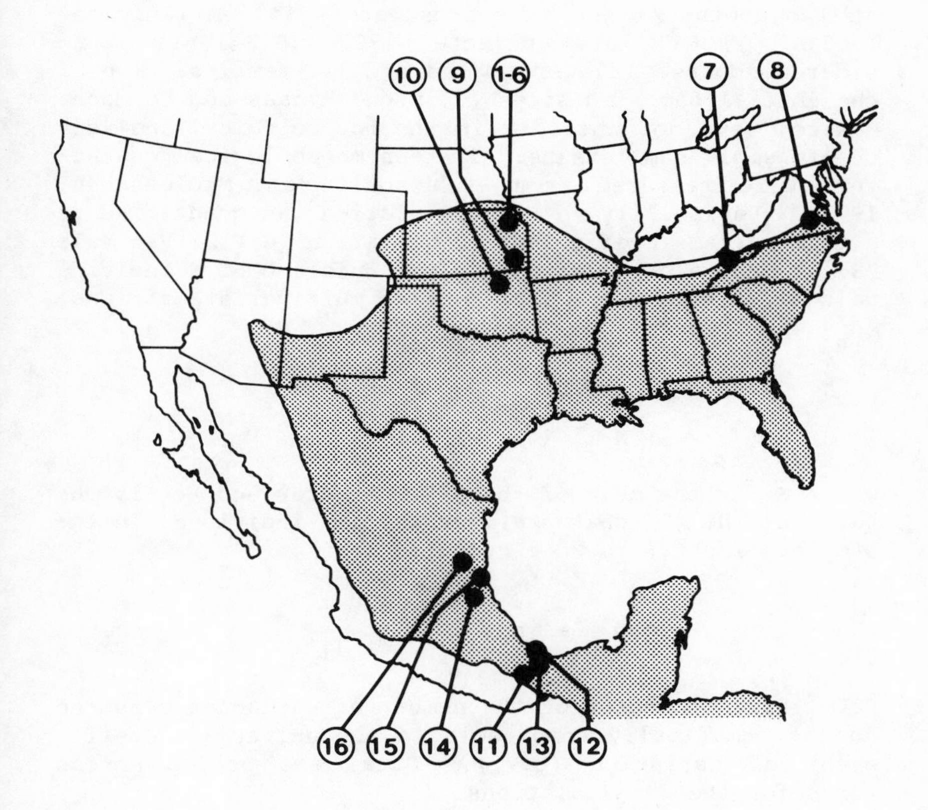

Fig. 1. Distribution of the cotton rat, Sigmodon hispidus, (shaded area) in the U.S. and Mexico. Circles represent locations of sampled populations. Sites 1-10 were classified as "marginal" and 11-16 as "central". Numerical designations of sampling sites correspond to those in Table 1.

proportion of loci polymorphic (P) were computed for each population (Lewontin and Hubby, 1966). Genetic similarities between populations were also calculated (Hedrick, 1971).

For purposes of estimating morphological variability, only adult cotton rats (body length > 140 mm) were considered. Mean body length of both males and females at two months of age is approximately 140 mm (Chipman, 1965). Of 647 rats collected, 222 (34.3%) were considered adults; 110 were males and 112 females. Two of the initial sampling sites (Clinton, Kansas and Cordoba, Mexico) were not included in the morphological analysis due to small sample size. Sixteen morphological characters were measured from each adult rat (McClenaghan, 1977). Variability in each population was quantified as a multivariate coefficient of variation ($C.V._n$; Van Valen, 1974). For a population sample consisting of N individuals measured for n characters, a multivariate variance (s_n^2) was computed as

$$s_n^2 = \frac{1}{N-1} \sum_{i=1}^{N} \sum_{j=1}^{n} (X_{ij} - \bar{X}_j)^2$$

where $\bar{X}$ is the mean of the jth character and X_{ij} is the value of the jth character for the ith individual in the sample. The $C.V._n$s were computed as

$$C.V._n = 100 s_n \sum_{j=1}^{n} \bar{X}_j^2 .$$

$C.V._n$ is independent of the number of variables measured and is numerically comparable to the univariate coefficient of variation (C.V.). Data were pooled across sexes for these calculations.

RESULTS

Populations of S. hispidus were polymorphic for six loci: leucine aminopeptidase (LAP); transferrin (TRF); 6-phosphogluconate dehydrogenase (6-PGD); two malate dehydrogenases (MDH-1 and MDH-2); and, glutamate oxalate transaminase (GOT-1). Two phenotypes were typically observed at each of these loci, with the exception of LAP which showed three. For each locus in every population, the observed phenotypic frequencies conformed to expected Hardy-Weinberg proportions.

Three populations in northeastern Kansas (Nelson, Robinson, and Ramseyer) were sampled over different months for one year. Variation in phenotypic frequencies at each polymorphic locus within each population was tested over months following the Chi-square procedure of Workman and Niswander (1970). Of 12 Chi-square tests, all loci showed nonsignificant differences through months except LAP in the Nelson population ($\chi^2_6 = 16.6$; $P < 0.025$). Samples from three sites in Mexico (Tampico, Panuco, and Ciudad Mante) were obtained in the month of July for two consecutive years (1973 and 1974). Between-year heterogeneity of phenotypic frequencies within each of these populations was tested for all polymorphic loci and none of the 18 comparisons were statistically significant.

In no instances were populations of cotton rats observed to be fixed for alternate alleles at a locus and the most common allele per locus was the same for all populations. Allelic frequencies at each polymorphic locus were weighted by sample size, pooled into marginal and central groups, and then tested for heterogeneity by contingency Chi-square analysis. Significant heterogeneity between central and marginal groups was found for all six polymorphic loci. Hedrick's (1971) probability of genotypic identity ($I_{x \cdot y}$) was calculated for the six polymorphic loci to assess relative genetic similarity between populations of S. hispidus. The $I_{x \cdot y}$ is highly correlated with other commonly used genetic similarity statistics (Hedrick, 1975). Genetic similarity between populations was very high. The average $I_{x \cdot y}$ over all pair-wise comparisons of populations was 0.937. Marginal populations were extremely similar to one another ($\bar{I}_{x \cdot y} = 0.951$), as were central populations ($\bar{I}_{x \cdot y} = 0.956$). Highest levels of divergence were observed when marginal and central populations were compared ($\bar{I}_{x \cdot y} = 0.923$).

Estimates of genic and morphological variability for populations of S. hispidus are given in Table 1. When data for individuals of all ages were used in the calculations of genic variability, the average cotton rat population was polymorphic at 19.8% of its loci. The mean P for marginal populations of S. hispidus was 17.0% and for central populations was 24.6%. Average $\bar{H}$ over all populations was 4.7%. Central populations ($\bar{H}$ = 5.54%) had significantly higher heterozygosities than marginal rat populations ($\bar{H}$ = 4.20%) when compared by a Mann-Whitney test ($U = 60$; $P < 0.001$). When only adult cotton rats are considered (Table 1), the average P for marginal populations was reduced to 15.9%, compared to

Table 1. Estimates of genic and morphological variability in central and marginal populations of *Sigmodon hispidus*. P represents the proportion of loci polymorphic; $\bar{H}$ is the mean heterozygosity per individual; and, $C.V._n$ is the multivariate coefficient of variation (Van Valen, 1974) computed from 16 morphometric variables measured from adult cotton rats. Genic indices were computed for each population based on all individuals collected and again utilizing only data from adult cotton rats.

Locality[a]	All Age Classes			Adults Only			
	Sample Size	P(%)	$\bar{H}$(%)	Sample Size	P(%)	$\bar{H}$(%)	$C.V._n$
1	155	17.4	4.14	48	17.4	3.63	9.64
2	122	17.4	3.38	38	17.4	4.00	9.40
3	44	17.4	3.95	6	13.0	4.35	7.66
4	8	13.0	4.35				
5	14	8.7	3.73	8	8.7	4.35	8.93
6	28	17.4	4.66	7	17.4	5.59	7.93
7	25	21.7	4.87	16	21.7	4.62	6.29
8	19	21.7	4.80	7	21.7	6.22	7.40
9	25	17.4	3.30	7	13.0	3.73	8.32
10	20	17.4	4.78	6	13.0	5.07	11.63
11	7	26.1	5.60				
12	21	21.7	5.74	8	21.7	7.61	5.39
13	25	26.1	6.01	14	26.1	6.21	3.42
14	39	26.1	5.69	13	17.4	5.02	4.16
15	57	26.1	5.11	27	26.1	4.67	4.91
16	38	21.7	5.12	17	17.4	5.37	6.33

[a]Specific localities: 1 = Nelson Farm, 11.3 km N, 4.8 km E of Lawrence, Leavenworth Co., Kan.; 2 = Robinson Farm, 8.8 km N, 4.8 km E of Lawrence, Douglas Co., Kan.; 3 = Ramseyer Farm, 12.9 km N, 4.0 km E of Lawrence, Jefferson, Co., Kan.; 4 = Clinton, 12.1 km W, 4.0 km S of Lawrence, Douglas Co., Kan.; 5 = Schneck Farm, 6.4 km W of Lawrence, Douglas Co., Kan.; 6 = Kampschroeder Farm, 0.8 km N, 4.8 km W of Lawrence, Douglas Co., Kan.; 7 = 22.5 km SW of Oak Ridge, Roane Co., Tenn.; 8 = 6.4 km W, 6.4 km S of Richmond, Henrico Co., Va.; 9 = 3.2 km E, 1.6 km S of St. Paul, Neosho Co., Kan.; 10 = 19.3 km N, 8.0 km E of Shidler, Osage Co., Okla.; 11 = 29.0 km E of Cordoba, Veracruz, Mex.; 12 = 16.1 km SW of Boca del Rio, Veracruz, Mex.; 13 = 11.3 km SW of Paso del Toro, Veracruz, Mex.; 14 = 19.3 km SW of Panuco, Veracruz, Mex.; 15 = 22.5 km N of Tampico, Tamaulipas, Mex.; and, 16 = 4.8 km N of Ciudad Mante, Tamaulipas, Mex.

21.7% for the central populations. However, heterozygosities in adults were slightly increased for both marginal ($\bar{H}$ = 4.62%) and central populations ($\bar{H}$ = 5.78%). Central populations still showed significantly higher heterozygosities than marginal populations (U = 36; P < 0.05).

The pattern of morphological variability is the converse of that seen for genic variability, with marginal populations of *S. hispidus* demonstrating greater morphological variability than their central counterparts in Mexico (Table 1). Marginal populations had a mean C.V. of 8.56 (range: 6.29-11.63), while the central populations had a mean $C.V._n$ of 4.83 (range: 3.42-6.33). The difference between the mean C.V.s of the marginal and central populations was significant ($F_{1,12}$ = 22.0; P < 0.001). There was no correlation between sample size and C.V. (P > 0.50). The $C.V._n$ was found to be negatively correlated with H based on cotton rats of all ages (r = -0.71; P < 0.01) and with $\bar{H}$ calculated only from adult individuals (r = -0.57; P < 0.05). These two relationships are illustrated in Fig. 2.

DISCUSSION

Estimates of genic variability in populations of *S. hispidus* (Table 1) are in close agreement with those for other species of rodents. Selander (1976) summarized data from 26 species of rodents and observed an average P of 20.2% and an average H of 5.4%. Our populations of *S. hispidus* had an average P of 19.8% and an average H of 4.7%. Johnson et al. (1972) examined genic variability in populations of *S. hispidus* from the southern U.S. and reported slightly lower estimates of variability ($\bar{P}$ = 8.16%; $\bar{H}$ = 2.14%). Differences in these results and our own probably reflect two things. First, Johnson et al. (1972) employed the 0.95 common allele frequency criterion for polymorphic loci, while we used the 0.99 frequency criterion. Second, our slightly higher estimates of heterozygosity reflect the considerable contribution of the LAP locus, while this highly polymorphic locus was not surveyed by Johnson et al. (1972). Calculations of genetic similarities between populations sampled in this study suggest very little genic differentiation in *S. hispidus*. This finding supports the contention that cotton rat populations are relatively

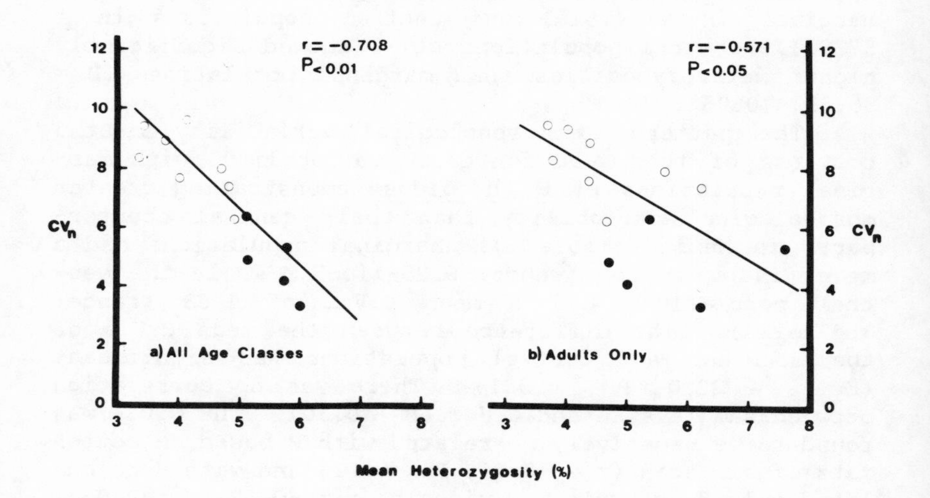

Fig. 2. Correlations of mean heterozygosity ($\bar{H}$) and multivariate morphological variability ($C.V._n$) in populations of Sigmodon hispidus. Marginal populations are symbolized by open circles, while central populations are designated by closed circles. On the left, heterozygosities were computed from cotton rats of all age classes collected at 14 sampling localities. On the right, heterozygosities were based only on genotypes of adult individuals measured for 14 morphological characters. Degrees of freedom for the two correlations are 12.

uniform with respect to geographic variation in allozymic frequencies (Johnson et al., 1972).

It is clear from the results of this study that marginal populations of S. hispidus in the U.S. have lower levels of genic variability than central populations of cotton rats in Mexico. Marginal populations were polymorphic at a smaller proportion of their gene loci and had significantly lower mean heterozygosities. Similar marginal reductions in genic variability have been reported in several other species of North American vertebrates (Dessauer and Nevo, 1969; Selander et al., 1971; McKinney et al., 1972). Several evolutionary factors could produce such a pattern. Directional selection has been considered an important force which erodes genetic variability in marginal populations (Carson, 1955, 1959; Gorman et al., 1975). To confidently invoke selection as an explanation for the patterns observed in this study would require direct evidence concerning the adaptive values of genotypes in the central and marginal populations; data of this type are not available. Nevertheless, some indirect inferences may be made. In spite of reduced proportions of polymorphic loci, heterozygosities in both central and marginal populations were slightly elevated when only adult cotton rats were considered (Table 1). This pattern is consistent with the hypothesis that allelic polymorphism is selectively advantageous in both central and marginal environments (Soule, 1973b). Heterozygosity in adults from marginal populations might be expected to decrease if these populations are subject to intense directional selection.

Reduced genic variability in marginal populations of *S. hispidus* may be a product of their history. Cotton rats have reinvaded the central and eastern U.S. since the glacial retreat in the Pleistocene. Populations in these marginal regions are thus relatively young. For example, cotton rats were not present in Kansas until 1892 (Cockrum, 1948) and were first reported from the vicinity of Lawrence, Kansas in 1941 (Rinker, 1942). Low levels of variability in these populations of recent origin may well be a consequence of genetic drift associated with colonization and subsequent bottlenecks in population size. Gene flow in marginal regions may also be restricted due to habitat discontinuities and the low vagility characteristic of *S. hispidus* and terrestrial vertebrates in general (Selander and Johnson, 1973). These factors could account for the absence in marginal populations of alleles present at low frequencies in central populations.

It is appealing on intuitive grounds to speculate that phenotypic variability and heterozygosity should be directly related. Several studies have provided evidence supporting the existence of a positive "genetic-phenetic variation correlation" (Soule, 1971; Soule et al., 1973). Marginal populations of *S. hispidus* had greater morphological variability than central populations (Table 1). In our study, morphological and genic variability were negatively correlated (Fig. 2). It should be emphasized that this correlation does not imply causation, but simply illustrates the contrasting patterns in variability observed in this study. While heterozygosity undoubtedly contributes to phenotypic variability, several other factors may also be important and should be considered. Since morphological traits are polygenic, variation in such traits is a function of both genetic and environmental variance (Falconer, 1960). Populations subjected to increased environmental variation may display increased phenotypic plasticity without an increase in genetic variability. On the other hand, increased phenotypic variability has been attributed to the breakdown of canalized developmental pathways by intense selection, resulting in a release of "cryptic" variability (Guthrie, 1965; Reeve and Robertson, 1953; Soule and Stewart, 1970). Recently, Straney (1978) has demonstrated that variation in age may account for a substantial proportion of the morphological variation observed within small mammal populations. Thus, variability in age structure of a population may also produce increased phenotypic variability.

Smith et al. (1978) proposed a conceptual model which contrasts the demographic and genetic characteristics of "primary" and "secondary" populations of small mammals. These categories are similar to the "central" and "marginal" designations employed in this paper. The model predicts that secondary (marginal) populations should be characterized as having low heterozygosities and unstable age distributions when compared to primary (central) populations of the same species. In turn, an unstable age distribution could result in increased phenotypic variability (Straney, 1978). The apparent "fit" of our data from marginal populations of S. hispidus with the model of Smith et al. (1978) is only speculative and should be viewed with caution. Difficulties in aging individual cotton rats allowed us to only classify them as "adult" or "nonadult." More precise techniques permitting us to place individuals into more specific age classes would be necessary to determine the proportion of the observed phenotypic variance that is attributable to age differences within marginal and central populations. Also, accurate demographic data from both types of populations would be required to assess differences in the stability of their respective age structures. It has been shown that temperate populations of cotton rats undergo marked seasonal changes in density and age distribution (McClenaghan and Gaines, 1978), but demographic data from populations in tropical regions are generally lacking.

Our results indicate that marginal and central populations of *S. hispidus* are quantitatively different in the relative amounts of genic and morphological variability each possesses. Further, the patterns in variability observed at these two levels of genetic organization are themselves substantially different, underscoring the complex nature of the interactions that determine intrapopulation variability. Future investigations will have to be designed to consider the dynamics of both genetic and ecological processes to unambiguously test hypotheses concerning the genetic structure of central and marginal populations of small mammals.

Acknowledgments--Financial support for this study was provided by grants from the National Science Foundation (GB-29135); the University of Kansas Museum of Natural History Watkins Fund; the University of Kansas General Research Fund (3345-5038); and, the University of Kansas Biomedical Sciences Support Fund (4076-5706). A Summer Dissertation Fellowship from the Graduate School of the University of Kansas to L.R.M. facilitated the writing of this paper. Computer facilities were generously provided by the University of Kansas Computation Center. We also wish to thank the anonymous reviewers for their valuable comments and criticisms.

MORPHOLOGICAL AND BIOCHEMICAL VARIATION AND DIFFERENTIATION IN INSULAR AND MAINLAND DEER MICE (PEROMYSCUS MANICULATUS)

Charles F. Aquadro and C. William Kilpatrick

Abstract--Patterns and levels of electrophoretic and morphological differentiation and variability were examined in six island and two mainland populations of deer mice (*Peromyscus maniculatus*) from Maine and New York. Patterns of differentiation among populations assessed by phenetic analysis of biochemical and morphological data showed essentially no concordance. Much of the biochemical differentiation, and perhaps the low correlation with morphological differentiation, appears due to stochastic shifts in electromorph frequency and absence of electromorphs from islands. Reduced levels of variability were generally observed in the insular relative to mainland populations. A positive relationship between biochemical variability ($\bar{H}$) and morphological variability (C.V.) was observed at the population level. The observed patterns of electromorph frequency shifts, the absence of unique electromorphs in any island population, and the primary importance of the square of the distance from the island to the nearest colonizing source in accounting for variation in $\bar{H}$ and C.V., suggest that genetic drift associated with reduced gene flow and founder effect has been an important contributor to the patterns of variation observed.

INTRODUCTION

Interest in the correlation between the relative amounts of differentiation and variation in natural populations as revealed by electrophoretic, morphological, karyotypic, and other studies has increased in the past few years. Knowledge of the relationship of the amount of variation within populations and differences among populations for each form of genomic expression has had a direct bearing on our understanding of selection, stochastic differentiation, and speciation. Several recent reports have documented the possibility for heterogeneity in relative rates of morphological and molecular evolution. Man and the chimpanzee are strikingly different morphologically, even when viewed from the "frog perspective," yet exhibit high biochemical similarity (King and Wilson, 1975; Cherry et al., 1978). An opposite situation exists among planaria which exhibit little morphological differentiation in the face of a fair degree of biochemical divergence (Nixon and Taylor, 1977). Other studies report independent patterns of interspecific differentiation at the biochemical and morphological levels (e.g., *Dipodomys* - Schnell et al., 1978). However, in other oganisms a fair concordance exists between the biochemically inferred relationships and those based on morphological studies (e.g., *Rattus* - Patton et al., 1975).

Analyses of morphological and biochemical variation in natural populations of both animals and plants have shown a similar diversity of results. Two hypotheses have been proposed for the relationship between genic and phenotypic variation. Lerner (1954) compiled evidence that inbreeding and concurrent reduction in heterozygosity usually resulted in an increase in morphological variability in domesticated plants and animals, presumably due to a loss of developmental homeostasis. In contrast, Soule and Yang (1973) have suggested that additive phenotypic variance should be simply and linearly related to genic variability (heterozygosity). All else being constant, overall phenotypic variance should therefore be positively correlated with genic variability.

Evidence for Lerner's (1954) hypothesis has come from findings such as Robertson and Reeve's (1952) where reduced phenotypic variance was observed in hybrids of two highly inbred stains of *Drosophila*. Studies of natural populations of killifish (*Fundulus* - Mitton, 1978), monarch butterflies (*Danaus* - Eanes, 1978), and lizards (*Uta* - Soule, 1979) also appear to support this prediction of a negative correlation between genetic and morphological variation. Other studies, however, have reported significant positive correlations between estimates of genetic and morphological variation (*Rattus* - Patton et al., 1975; *Uta* - Soule, 1971; *Anolis* - Soule et al., 1973). These latter findings suggest that both operationally independent estimates accurately reflect the same overall trends in genomic variability (Soule et al., 1973).

In this paper we examine morphological and electrophoretic differentiation and variability in insular and mainland populations of deer mice (Peromyscus maniculatus). If a broader range of genetic variance exists when comparing mainland and island populations, we should expect to see a broader range of values for testing any relationship between biochemical and morphological variability. The questions we address in this paper are as follows. Are patterns of biochemical and morphological differentiation correlated or are they independent among insular and mainland populations of deer mice? What is the relationship between biochemical and morphological variability for these populations? What best accounts for the patterns of variation and differentiation observed?

MATERIALS AND METHODS

Peromyscus maniculatus (N=129) were collected between July 1975 and August 1976 from six islands off the coast of central Maine, the adjacent mainland, and New York (Fig. 1). The six islands range in size from 3.3 km^2 to 70.9 km^2 and are located from 1.5 km to 18.5 km from the mainland. Boreal spruce and fir vegetation prevail over most of the island and mainland localities with scattered stands of northern mixed hardwoods. Deer Isle is connected to the mainland by a bridge while the other islands are not. All of the islands were connected to the mainland by land bridges within the last 10,000 yr due to a post-Pleistocene drop in sea level of approximately 60 m (Stuiver and Borns, 1975; Milliman and Emery, 1968). A following rise in sea level progressively isolated islands from about 9500 yrs B.P. (Matinicus) to circa 4500 yr B.P. (Swan's Island). Prior to about 13000 yr B.P., glaciation prevented mice from inhabiting the regions of northern New England sampled in this study (Stuiver and Borns, 1975).

The collecting localities are listed below by state and county. Numbers in parentheses refer to sample sizes for electrophoretic and morphological studies, respectively. Maine, Knox Co.: Matinicus Island (6,5), 0.2 km NE of Vinal Haven village (3,2), 4.8 km N of Vinal Haven village (7,4), SE Calderwood Neck (12,5), S Calderwood Neck (2,2), 0.5 km S of Rich's Cove, Isle au Haut (26,16), Horseman's Point, Isle au Haut (2,2),

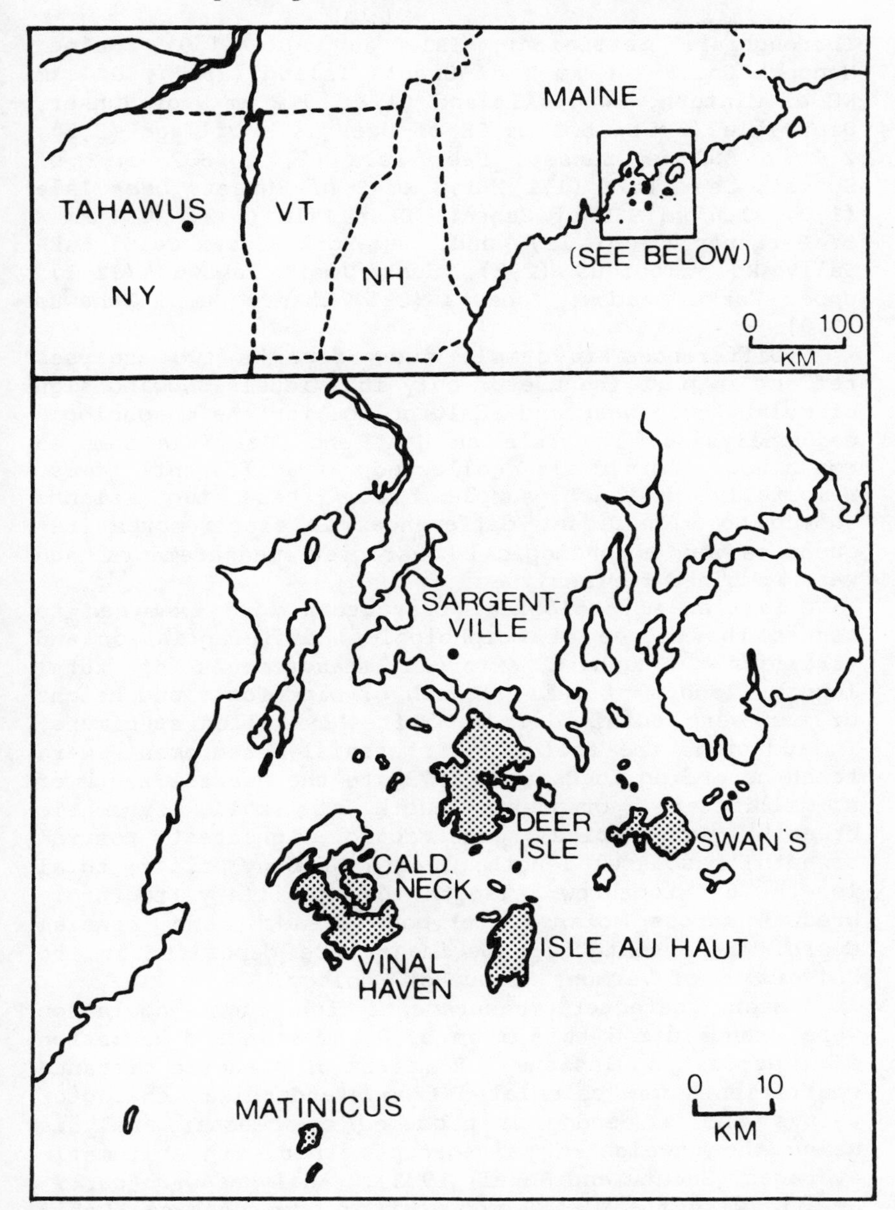

Fig. 1. Map of collecting localities for Peromyscus maniculatus from New York and Maine with an enlargement for the island localities in the Penobscot Bay region of Maine.

Thoroughfare Settlement, Isle au Haut (1,0); Maine, Hancock Co.: 1.6 km N of Swan's Island (25,7), 0.4 km NE of Minturn, Swan's Island (4,2), 1.2 km N of Sunset, Deer Isle (5,5), 1.6 km SE of Deer Isle village (3,3), 2.4 km SSE of Sunset, Deer Isle (2,2), 3.2 km S of Sunset, Deer Isle (1,2), 1.6 km S of Sunset, Deer Isle (1,0), 1.6 km SSE of Sunset, Deer Isle (5,2), 6.4 km W of Sargentville (6,11); and, New York, Essex Co.: Lake Sally Rd., Tahawus (2,2), Junk Dump, Tahawus (11,2), Upper Works meadow, Tahawus (2,1) Cheney Dump, Tahawus (3,0).

Differences in sample sizes for the two analyses reflect in part the use of only individuals showing sign of molar tooth wear and adult pelage for the morphological analysis. The Isle au Haut and Deer Isle samples contained individuals collected at different times. Temporally distinct samples from these two islands showed no significant differences in electromorph frequencies or morphological character measurements and were combined for analyses.

Fifteen morphological characters were examined to assess the degree of morphological differentiation and variation. Standard external measurements of total length, length of tail, length of hind foot, and height of ear were obtained from 75 freshly-killed specimens. In addition, the following 11 cranial measurements were taken according to Choate (1973) to the nearest tenth of a millimeter: greatest length of skull; zygomatic breadth; interorbital constriction; greatest rostral breadth; basonasal length; length of bony palate; total length of toothrow; length of maxillary toothrow; breadth across molars; pterygoid breadth; and, cranial depth. Representative specimens were deposited in the University of Vermont Museum of Zoology.

Mean character measurements for each population were standardized to a mean of 0 and standard deviation of 1 across populations. A matrix of phenetic distance coefficients was calculated from standardized character values and a dendogram produced by cluster analysis using the unweighted pair-group method with arithmetic averages (Sneath and Sokal, 1973). Analyses were carried out with the NT-SYS computer program package (Rohlf et al., 1974). Morphological variability was estimated by the mean coefficient of variation (C.V.) over the 15 characters for mice from each locality.

Morphological and Biochemical Variation

Twenty-one proteins encoded by 29 autosomal loci were examined by horizontal starch-gel electrophoresis using techniques modified from those in Selander et al. (1971), Jensen and Rasmussen (1971, 1972; albumin), and Smith et al. (1973; sorbitol dehydrogenase). The proteins (and loci) examined were: albumin (ALB-1); alcohol dehydrogenase (ADH-1); esterases (EST-1, EST-2, EST-6, EST-8); glucose-6-phosphate dehydrogenase (G-6-PD-1); glutamate oxaloacetate transaminase (GOT-1, GOT-2); α-glycerophosphate dehydrogenase (α-GPD-1); hemoglobin (HB-α_1, HB-α_2, HB-α_3, HB-β; Aquadro, 1978); hexose-6-phosphate dehydrogenase (H-6-PD-1); isocitrate dehydrogenase (IDH-1, IDH-2); lactate dehydrogenase (LDH-1, LDH-2); leucine aminopeptidase (LAP-1); malate dehydrogenase (MDH-1, MDH-2); malic enzyme (ME-1); phosphoglucomutase (PGM-1, PGM-3); phosphoglucose isomerase (PGI-1); sorbitol dehydrogenase (SDH-1); tetrazolium oxidase (TO-2); and, transferrin (TRF-1). Details of the tissue preparation, buffers, stains and loci designations used are contained in Aquadro (1978).

Genetic distances between populations were calculated from electromorph frequencies using Nei's (1972) distance coefficient (D). Populations were clustered into a dendrogram by the unweighted pair group method with arithmetic averages (Sneath and Sokal, 1973). Biochemical variability was estimated by the mean proportion of individuals heterozygous per locus ($\bar{H}$) calculated by direct count. Loci were considered polymorphic if the frequency of the common electromorph was ≤ 0.95.

The relative contribution of a variety of biogeographic variables to the observed levels of insular variability was assessed by simple regression analyses with the independent variables being island area, log island area, maximum island elevation, distance to the nearest mainland or large island colonizing source, the latter distance squared, and time since the island was last connected to the mainland by a land bridge. These variables have been discussed as possible determinants of genetic variation in island populations by a number of authors (Gorman et al., 1975; Soule, 1973a; Soule and Yang, 1973). Further details of this analysis are presented by Aquadro (1978).

RESULTS

Seventeen of the 29 loci were monomorphic in all populations sampled; two loci, LDH-2 and TO-2, each showed only a single heterozygote. Electromorph frequencies for the remaining 10 polymorphic loci are presented in Table 1. Esterases, as a group, contributed the greatest amount of variation with 63% of the populations polymorphic for EST-1 and EST-8 and 75% polymorphic for EST-2 and EST-6. PGM-3 was polymorphic in 88% of the populations examined. Other polymorphic loci included GOT-1 (75%), IDH-1 (63%), H-6-PD-1 (38%), and G-6-PD-1 (13%). The population of *P. m. gracilis* from New York was polymorphic at HB-α^3 while populations of *P. m. abietorum* sampled from Maine were monomorphic at this locus (Table 1).

Estimates of biochemical and morphological variation are presented in Table 2. Mean heterozygosities for mainland populations were 7.5% (Sargentville) and 11.7% (Tahawus) with an average of 9.6%; the proportion of loci polymorphic was 27.6% for both populations. These values are within the range reported for other continental populations of *P. maniculatus* (Avise et al., 1979a).

Reduced levels of variability were generally observed in the insular populations. Heterozygosity ranged from 0% on the small, isolated island of Matinicus to 6.6% on the larger, less isolated Deer Isle. Morphological variability (C.V.) showed the same general reduction in variability and ranged from 2.20 on Matinicus to 3.17 on Deer Isle (Table 2). Although mice from Sargentville, Maine had a C.V. of only 2.56, the C.V. for those from Tahawus, New York (3.52) was consistent with that reported by Choate (1973) for the same 11 cranial characters in *P. maniculatus* from Rutland Co., Vermont (3.71).

Insular levels of biochemical and morphological variability showed the highest correlations with the square of the distance of the island to the nearest mainland or large insular colonizing source. Correlation coefficients were -0.81 and -0.79, respectively ($0.05 < P < 0.06$). All other ecological or biogeographic variables examined, except the untransformed distance, showed no significant relationships with $\bar{H}$ or C.V. ($P > 0.10$). The coefficient of variation showed a significant positive correlation with heterozygosity (r

Table 1. Allelic frequencies at 10 polymorphic loci for Peromyscus maniculatus from Maine and New York. Sample sizes for each locality are given across the top of the table. See text for locus designations.

		Locality							
Locus	Electro-morph	Matinicus 6	Vinal Haven 10	Calderwood Neck 14	Isle au Haut 29	Deer Isle 17	Swan's Island 29	Sargent-ville 6	Tahawus 18
Hβ-α_3	100	1.00	1.00	1.00	1.00	1.00	1.00	1.00	0.34*
EST-1	100				0.53	0.16		0.17	0.44
	91		0.05			0.75		0.08	0.50
	81	1.00	0.95	1.00	0.47	0.09	1.00	0.75	0.06*
EST-2	100		0.20	0.30	0.20	0.63	0.12	0.17	0.63
	93	1.00	0.80	0.70	0.80	0.38	0.88	0.83	0.25
EST-6	100	1.00	0.68	0.61	0.63	0.76	0.43	1.00	0.73
	96		0.32	0.39	0.37	0.24	0.58		0.27*
EST-8	100	0.17	0.70	0.57		1.00	1.00	0.50	0.25
	null	0.83	0.30	0.43	1.00			0.50	0.69
GOT-1	100	1.00	0.50	0.43	0.41	0.75	1.00	0.50	0.63
	74		0.50	0.57	0.59	0.25		0.50	0.38
IDH-1	100		0.40	0.39	0.04	0.46		0.58	
	83	1.00	0.60	0.61	0.96	0.54	1.00	0.42*	1.00
G-6-PD-1	100	1.00	0.85	0.96	1.00	0.96	0.96	0.83	1.00
	91		0.15	0.04		0.04	0.04		
H-6-PD-1	100	1.00	0.90	1.00	0.96	0.96	0.98	0.92	0.94
	92		0.10		0.04	0.04	0.02*	0.08*	0.06
PGM-3	100		0.11	0.08	0.07	0.06	0.02	0.08	0.67
	92	1.00	0.89	0.92	0.93	0.94	0.94	0.33	0.33

*Alternative electromorphs found in only one population: Hb-α_3^{77}, Est-2^{87}, Est-8^{102}, and G-6-pd-1^{114}. Pgm-3^{87} was found in two populations.

Table 2. Estimates of biochemical and morphological variability for island and mainland populations of *Peromyscus maniculatus*.

Locality	Biochemical Variability				Morphological Variability		
	Sample Size	P*	$\bar{H}$**	Standard Error	Sample Size	$\overline{C.V.}$**	Standard Error
Island							
1. Matinicus	6	0.035	0		5	2.20	0.27
2. Vinal Haven	10	0.310	0.055	0.026	6	2.59	0.37
3. Calderwood Neck	14	0.207	0.058	0.026	7	2.78	0.32
4. Isle au Haut	29	0.172	0.051	0.024	18	2.99	0.32
5. Deer Isle	17	0.207	0.066	0.028	14	3.17	0.32
6. Swan's	29	0.069	0.024	0.018	9	2.59	0.29
Mean		0.167	0.042			2.72	
Mainland							
7. Sargentville, Me.	6	0.276	0.075	0.032	11	2.56	0.23
8. Tahawus, N.Y.	18	0.276	0.117	0.046	5	3.52	0.81
Mean		0.276	0.096			3.04	

* Proportion of loci polymorphic (P) defined with the frequency of the common electromorph ≤ 0.95.

** $\bar{H}$ equals mean individual heterozygosity across loci and $\overline{C.V.}$ equals mean coefficient of variation for all of the morphological characters measured.

= 0.83; $P < 0.02$; Fig. 2). Difficulty in obtaining sufficient specimens from any one area, due to extremely low population levels, made the combining of two or more sites necessary for a number of localities. Calculation of the partial correlation coefficient between $\bar{H}$ and C.V., removing any effects of differing number of sites per locality (Neter and Wasserman, 1974), gave a significant correlation of 0.79 ($P < 0.05$) between $\bar{H}$ and C.V. suggesting a relationship between molecular and morphological variability does exist.

Coefficients of Nei's (1972) genetic distance between populations ranged from 0.003 between populations from Vinal Haven and Calderwood Neck, two adjacent islands connected at low tide, to 0.128 between *P. m. abietorum* from Swan's Island, Maine and *P. m. gracilis* from Tahawus, New York. A dendrogram based on Nei's genetic distance (Fig. 3A) shows *P. m. abietorum* from Maine clustered apart from the New York *P. m. gracilis*.

Phenetic analyses of the morphological data (Fig. 3B) suggest an entirely different pattern of relationship. Two major clusters are present. One cluster included both mainland populations (Tahawus and Sargentville) and the insular populations from Matinicus and Deer Isle. The other four islands comprised the second main cluster. In contrast to their close biochemical similarity, populations from Vinal Haven and Calderwood Neck are quite distinct morphologically. Correlation and principal component analyses revealed essentially identical results and are therefore not presented here.

To quantify the differences in degrees and patterns of similarity indicated by the phenetic analyses of the electrophoretic and morphological data, correlation coefficients were calculated for pair-wise combinations of the biochemical and morphological indices from their respective matrices. Essentially no correlation exists between the relationships revealed by the two data sets ($r = 0.02$).

DISCUSSION

Patterns of biochemical and morphological differentiation in *P. maniculatus* in northern New England show a very weak concordance. Statistically, no correlation exists between phenetic assessments of morphological and biochemical relationships ($r = 0.02$). This is due, in

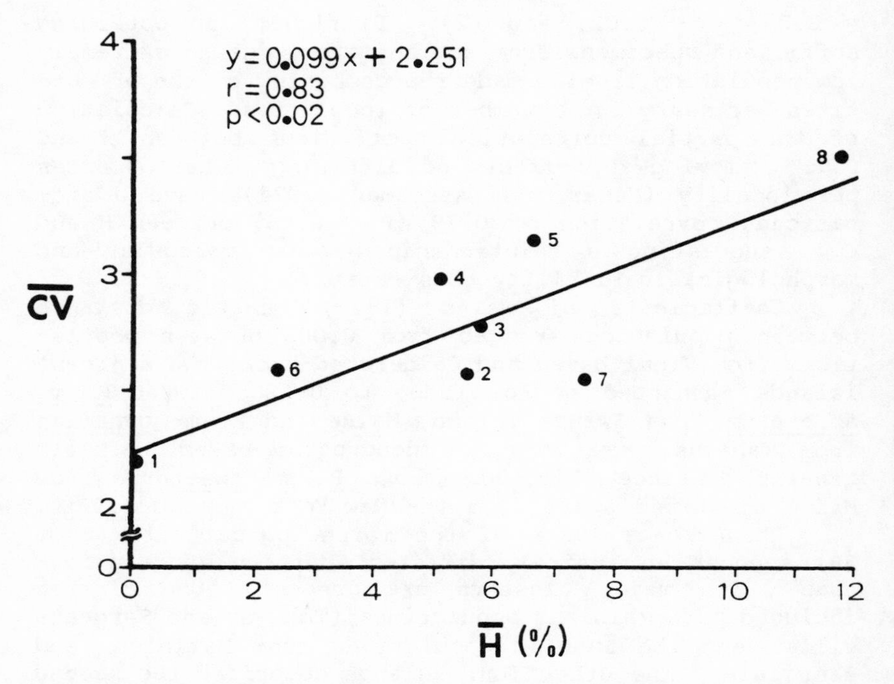

Fig. 2. Relationship between morphological variability, measured by the mean coefficient of variation (C.V.), and biochemical variability, measured by the mean proportion of individuals heterozygous per locus ($\bar{H}$), for eight populations of Peromyscus maniculatus identified by number in Table 2.

part, to the presence of two distinct clusters in the morphological phenograms (Fig. 3B) and graded clustering in the biochemical phenograms (Fig. 3A). The lack of agreement between the morphological and biochemical relationships is due to basic differences in the original distance matrices. Thus it is unlikely that other tree-forming methods which rely on these matrices would improve the correlation (e.g., Fitch and Margoliash, 1967; Farris, 1972). Matinicus Island mice, for example, are most closely allied morphologically with populations from Tahawus, New York (Fig. 3B), but clusters biochemically with populations from the geographi-

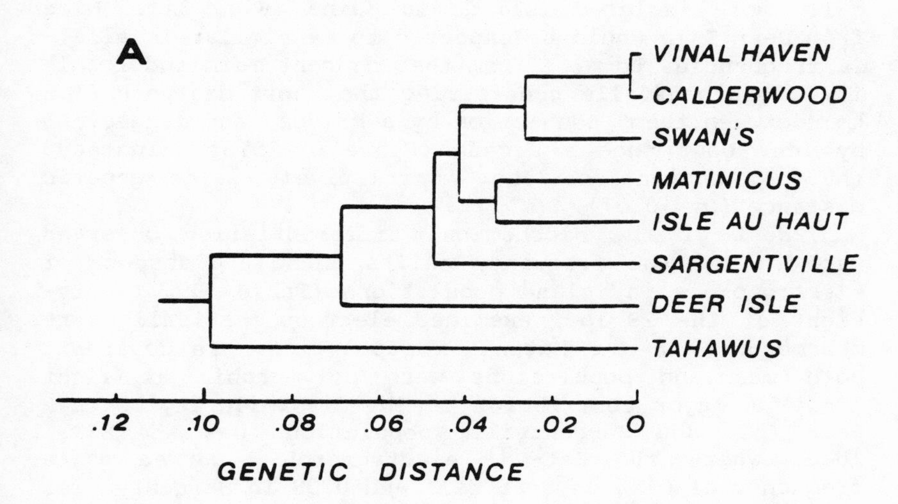

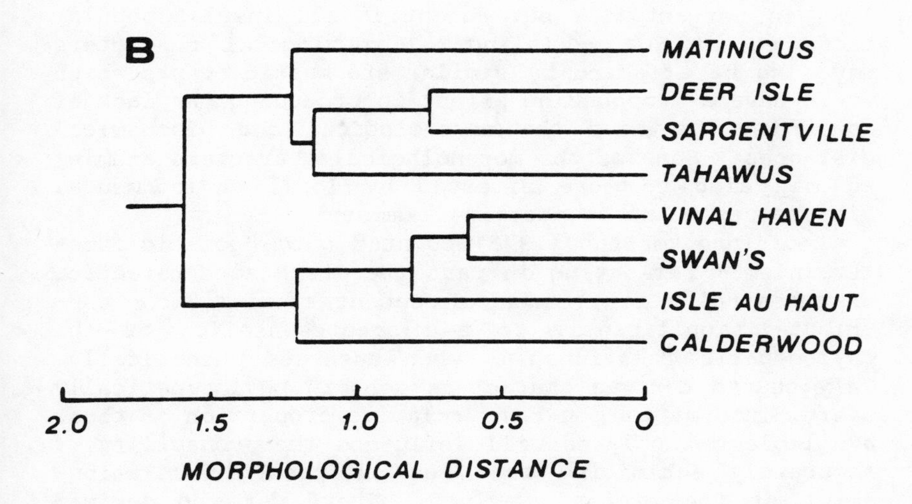

Fig. 3. Phenograms of eight populations of *Peromyscus maniculatus* showing (A) biochemical relationships based on Nei's (1972) genetic distance, and (B) morphological relationships based on coefficients of phenetic distance. The cophenetic correlations were 0.87 and 0.83, respectively.

cally more isolated islands in Maine (Fig. 3A). Mice from Deer Isle would be expected to be similar in allelic frequencies to mice from the adjacent mainland locality of Sargentville considering the short distance (1.5 km) between them, connection by a bridge, and connection by ice about once a decade (Crowell, 1973). Instead, the two populations show a relatively large genetic distance (D = 0.07).

Most of the biochemical differentiation observed was due to gene frequency shifts among and absence of electromorphs in island populations (Table 1). Twenty-eight of the 29 loci examined electrophoretically were monomorphic in the Matinicus population. In contrast, both mainland populations were polymorphic at eight loci. A major contributor to the dissimilarity between Deer Isle and Sargentville populations was the EST-1 locus where the $Est\text{-}1^{91}$ electromorph occurred at a frequency of 0.75 on Deer Isle and 0.08 in Sargentville. The distribution of the $Pgm\text{-}3^{87}$ electromorph also showed similar sharp differences, occurring at a frequency of 0.58 in Sargentville but absent in all insular populations except Swan's Island. Morphological characters may also be affected by similar stochastic events which would have a compounding effect in producing the lack of concordance between the morphological and biochemical distances. Some of the morphological characters examined may also be more affected by local environmental differences than the proteins examined.

As Throckmorton (1978) pointed out, "genetic indeterminism," reflecting in part the "haphazard retention of ancestral alleles among descedent species," can contribute significantly to a misrepresentation of the phylogenetic relationships when assessed phenetically. Only shared derived characters contain phylogenetically useful information, and the relative proportion of these available or analyzed will influence the probability of accurately estimating the true phylogenetic relationships of the groups examined. There were no derived (unique) electromorphs on any island, all of them being found in adjacent mainland populations from which the island populations were probably descended. The $Est\text{-}6^{96}$ and $G\text{-}6\text{-}pd\text{-}1^{91}$ electromorphs observed in a number of island populations were absent from the adjacent mainland locality of Sargentville (Table 1) but were observed in individuals from other adjacent mainland localities (Aquadro, 1978). Eight electromorphs appear unique to either Maine or New York populations, but it is impossible to determine if they are derived or ancestral from the data available. The morphological characters examined were mensural and cannot be used in the above manner. Given that most of the biochemical differentiation among the island and mainland populations has occurred by random electromorph frequency shifts and loss of electromorphs, the lack of agreement between biochemical and morphological relationships is not surprising. Among the conspecific insular and mainland populations of deer mice examined, genetic drift and founder effect probably had important roles in at least the biochemical patterns of differentiation and in the lack of agreement between the phenetic relationships derived from the two data sets. The effects of differential selection are impossible to assess with the present data but may also have been important.

A lack of correlation between patterns of differentiation at the biochemical and morphological levels is not restricted to the intraspecific level. Michevich and Johnson (1976) and Schnell et al. (1978) have demonstrated and reviewed numerous examples of poor concordance between morphological and allozymic data analyzed phenetically for fish (*Menidia*), rodents (*Dipodomys*), and other organisms. However, Michevich and Johnson (1976) found reasonably good agreement between the evolutionary relationships derived from their two data sets for *Menidia* when analyzed cladistically suggesting the disparity lies less with the two types of characters than with the method of analysis. We should perhaps look at differences between phenetic assessments of biochemical and morphological relationships less as conflicting phylogenies but rather as clues to the evolutionary processes acting on the different character sets examined.

In vertebrate populations with reduced gene flow due to isolation, as in the case of islands and caves, reduced levels of genetic variation have generally been observed (Nevo, 1978). Reduced levels of biochemical variability were observed among insular populations of deer mice relative to mainland populations in the present study. While directional selection (Soule, 1973b; Gorman et al., 1975) or density-dependent selection (Selander and Johnson, 1973) have been invoked to explain some patterns of reduced variability among insular populations, a reduced rate of gene flow, genetic drift, and/or founder effect seem equally applicable to many of the patterns (Avise and Selander, 1972; Browne, 1977; Schmitt, 1978; Soule, 1971).

Among the Maine populations examined, referable as *P. m. abietorum*, the pattern of decreased morphological variation on islands relative to mainland populations appears not to hold since the C.V. for Sargentville (2.56) is below the mean C.V. for the island populations (2.72). Evidence that this level may be atypically low for a mainland population comes from the C.V. observed for the Tahawus, New York population (3.52) and that reported by Choate (1973) for the same 11 cranial characters for *P. maniculatus* from Vermont (3.71). These New York and Vermont populations are referable to *P. m. gracilis*. Thus, the possibility exists that fundamental differences in levels of morphological variation exist between mainland populations of these two subspecies. Population levels on the mainland in Maine were extremely low during this study as judged by very low trap success compared to that of previous years and for the other mainland site. Since all mice were caught in the same area, the low C.V. for Sargentville may be due to analysis of only one or two families of mice.

The observed patterns of electromorph frequency shifts, the absence of unique alleles in any island population, and the primary importance of the square of the distance from the island to the nearest colonizing source in accounting for levels of heterozygosity in insular populations ($r = -0.81$) suggest that genetic drift associated with reduced gene flow and founder effect have been important contributors to the reduced levels of biochemical variation. Kilpatrick (this volume) also found heterozygosity to be best negatively correlated with the square of the distance from the colonizing source in insular *Microtus* populations. This relationship is consistent with MacArthur and Wilson's (1967) suggestion that the rate of colonization is a decreasing exponential function of distance. Insular levels of morphological variability (C.V.) were also best predicted by the square of the distance to the nearest colonizing source ($r = -0.79$). Both $\bar{H}$ and C.V. seem to reflect the same underlying genomic variability and are influenced, at least in insular populations of deer mice, by similar biogeographic variables.

Mean heterozygosity and phenotypic variation (C.V.) of a variety of mensural characters for populations of deer mice were positively correlated (Fig. 2). This correlation is as expected according to Soule and Yang's (1973) hypothesis that additive phenotypic variance should be simply and positively related to genetic variability. Do these results therefore argue against Lerner's (1954) homeostasis hypothesis which predicts a negative relationship between genic and phenotypic variability? Mitton (1978) and Soule (1979) have pointed out that the two predictions about the relationship between genetic and phenotypic variation are not contradictory but are rather at two different levels. The least intraindividual phenotypic variation is predicted for the most heterozygous individuals within a population. However, a population with high genetic diversity will have a greater additive genetic variance component and hence greater interindividual phenotypic variance than a population of low genetic diversity. Thus, in studies comparing mean heterozygosity in populations with estimates of interindividual variability (C.V.), a positive relationship is expected and has in fact been observed in rats (Patton et al., 1975), lizards (Soule, 1971; Soule et al., 1973), and deer mice (this paper).

While there is increasing evidence for a relationship between variation at enzyme loci and morphological variability, the precise nature and cause of the relationship is still unclear. It seems unlikely that all the protein loci analyzed electrophoretically would have a direct effect on developmental stability of the morphological characters although they may be closely linked to loci which do. Similarly, variation measured by $\bar{H}$ is largely, if not totally, due to genetic variation. Morphological variation, though, is controlled by genetic, environmental, and genetic-environmental interaction variance components. Unless the relative contribution of these three components to the total variance of the morphological characters is known, little can be said in the way of accounting for why the observed relationships exist.

Acknowledgments--The data for this paper represent part of a master's thesis presented to the University of Vermont by the senior author. We extend our thanks to the following people who contributed significantly in a variety of ways to this study: W. S. Aquadro, Dr. N. Bailey, Dr. and Mrs. K. L. Crowell, Dr. M. L. Kennedy,

Dr. G. L. Kirkland, S. Lindsay, and Mr. and Mrs. W. Stevens. The critical comments of Dr. J. C. Avise, R. W. Chapman, Dr. E. D. Parker, Jr., J. C. Patton, the editors, and anonymous reviewers of various drafts of this paper were extremely helpful.

BIOCHEMICAL VARIATION IN NORTHERN SEA LIONS FROM ALASKA

W. Z. Lidicker, Jr., R. D. Sage and D. G. Calkins

Abstract--Available evidence suggests that marine mammals as a group have relatively little genetic variation. If such a tentative generalization can be sustained, there are important ecological, evolutionary, management, and conservation implications that justify its exploration. We analyzed electrophoretically tissue samples from 157 individual *Eumetopias jubatus* collected over a large area of coastal Alaska. Of 32 loci studied, four exhibited rare alleles (maximum percentage of 1.6) and one (EST-3) was truly polymorphic with two alleles in approximately equal proportions. Average individual heterozygosity was 0.018. No geographic variation was apparent except for a possibly weak differentiation of the Prince William Sound population. Our results support (1) the treatment of the area surveyed as a single management unit, and (2) the hypothesis of low biochemical variation in pinnipeds in general. Possible explanations for such low variability are discussed.

INTRODUCTION

Relative to terrestrial species, marine mammals have been little studied with respect to biochemical variation. We here report on a survey of biochemical variation in the northern sea lion (*Eumetopias jubatus*) from a large area in Alaska. A total of 157 individuals was studied from an area extending from Cape St. Elias in the northern Gulf of Alaska to Chowiet Island, 168 km southwest of Kodiak Island (Fig. 1). Numerous breeding colonies of sea lions exist throughout this vast area, but it is not known to what extent gene flow occurs among colonies. Our interests were in adding to the small fund of information on biochemical variation in marine mammals as well as in identifying appropriate management units within this area.

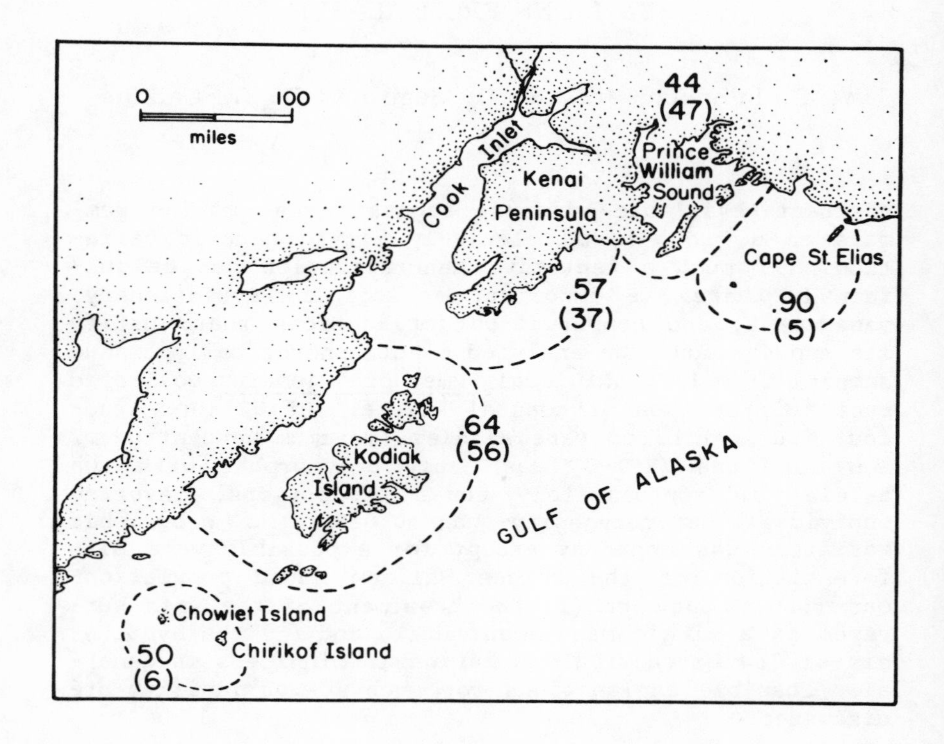

Fig. 1. Region in Alaska from which sea lion collections were made. Dashed lines enclose regional areas in which data from different collecting localities were grouped. Percentages refer to the proportion of the Esterase-3 null alleles in each region; numbers in parentheses are numbers of sea lions analyzed for this locus.

Information on genetic variation in populations is generally assumed to have important evolutionary and ecological implications. Moreover, it is of interest to know whether marine mammals as a group show patterns of variation which are similar to terrestrial species. Early studies with marine mammals (Blumberg et al.,

1960; Naevdal, 1966a,b; Shaughnessy, 1969, 1970; Seal et al., 1971) sampled only a few loci and revealed few polymorphisms. Naevdal (1966a) could distinguish populations of harp seal (*Phoca groenlandicus*) from two sides of the North Atlantic on the basis of different frequencies of transferrin alleles. However, a later study on western populations cast doubt on this interpretation (Naevdal, 1969). Shaughnessy (1970) recognized several discrete populations of southern fur seals (*Arctocephalus*) also based on transferrin variations. Published surveys of extensive numbers of biochemical loci are available only for the two species of elephant seals (*Mirounga leonina* and *M. angustirostris*). McDermid et al. (1972) found a genic heterozygosity of 0.028 in 42 individuals (18 proteins) of *M. leonina* from Macquarie Island, although this estimate was increased to 0.075 for the same population when whey proteins were included among a total of 12 loci studied (Shaughnessy, 1974). Bonnell and Selander (1974) found no polymorphisms at 24 loci in 159 *M. angustirostris*. The lack of genetic variation in *M. angustirostris* was attributed to a genetic "bottleneck" occurring when this species was reduced to less than 100 individuals in the early 1890's. Beyond this, the few other studies available on marine mammals all hint at low variabilities, although they involve too few loci or individuals for reliable estimates of genetic variation to be drawn (e.g., Heath, 1978). Of these, a study by Shaughnessy (1975) is of particular interest as he found no polymorphisms among nine loci in various Alaskan populations of the non-migratory harbor seal, *P. vitulina*. This suggests tentatively that the high vagilities characteristic of most marine mammals may not be sufficient to explain their low genetic variabilities. Finally, Selander (1976:34) refers briefly to unpublished studies indicating low variability in porpoises. Existing information, therefore, suggests the hypothesis that marine mammals as a group exhibit low genetic variability.

In addition to the important evolutionary implications of levels of genetic variation, there are also management and conservation repercussions which may emerge from such information (Smith et al., 1976). It has been suggested (Johnson, 1974) that enzyme polymorphisms enhance an individual's ability to cope with a varying environment. Moreover, Smith et al. (1975) and Berger (1976) have hypothesized that overall heterozygosity is positively correlated with viability. It would follow from these ideas, as well as from other theoretical considerations, that populations with low genetic variability might be less able to cope with rapid environmental changes. Hence they would be more vulnerable to environmental deterioration stemming from man's activities. Geographic patterns of genetic variability may also reveal the demic structure of species. Since demes are the evolutionary units of species, they are likely to be the most appropriate management units. Intelligent management was one of the motivations in Naevdal's study of harp seals (1966a), as it was in ours.

MATERIALS AND METHODS

Samples of sea lion material were collected from 1975 through 1977 from widely scattered localities throughout south-central Alaskan waters. All available age classes and both sexes were sampled, although the collecting was biased toward adult females to facilitate concurrent reproductive studies. All samples were taken on land by one of us (DGC) in the course of continuing ecological studies on the northern sea lion populations in this area. Blood samples were preserved in the field by adding acid-citrate-dextrose solution at the rate of 15 ml of preservative to 100 ml of whole blood, and refrigerated. Several grams of liver and kidney were wrapped in aluminum foil and frozen immediately. Blood and unground tissue samples from these animals have been deposited as voucher specimens in the Frozen Tissue Collection of the Museum of Vertebrate Zoology, University of California, Berkeley.

The 35 specific collection localities were grouped into five regional areas (Fig. 1). These grouped localities, the specific localities, and the number of specimens taken from each are as follows: (1) Northeastern Gulf of Alaska: Middleton Island, 4; Cape St. Elias, 2; (2) Prince William Sound: Glacier Island, 5; Port Fidalgo, 15; Perry Island, 1; between Perry Island and Dutch Group, 1; Pleiades Islands, 9; Point Eleanor, 1; Prince of Wales Passage, 1; Squire Island, 3; Montague Point, 2; Port Bainbridge, 8; The Needle, 3; (3) Kenai Peninsula: Aialik Bay, 4; Aialik Cape, 1; Harris Bay, 1; Outer Island, 11; Chiswell Islands, 9; Rugged Island,

2; Resurrection Bay, 2; Sugarloaf Island, 7; (4) Kodiak area: Shuyak Island area, Latax Rocks, 7; Sea Otter Island, 16; Marmot Island, 18; Cape Barnabas, 3; Cape Chiniak, 4; Gull Point, 3; Spiridon Point, 1; Black Cape, 1; Miners Point, 2; Cape Ugak, 3; Big Bay, Shuyak Island, 1; and, (5) Southwest of Kodiak: Chowiet Island, Semidi Islands, 1; Chirikof Island, 5.

In the laboratory, the tissues and extracts were stored at -76 C. Separate extracts were prepared of liver and kidney tissue. The tissue was minced with scissors and combined with approximately an equal volume of deionized water and centrifuged at 27,000 g for 40 min. Tissues and extracts were kept chilled at approximately 4 C during the preparative stages. Horizontal starch-gel electrophoresis was carried out using the procedures described by Selander et al. (1971) and Harris and Hopkinson (1976). Staining recipes were similar to those described in these two papers. Electrostarch (O. Hiller, Madison, Wisc.) from Lots 302 and 307 were used throughout the study. The buffer systems and electrophoresis conditions are identified in Table 1.

Interpretation of the esterase complex posed special problems for which some additional methodological comments are required. Esterases (EST)-1 and -2 stained as single-banded isozymes. Esterase-1 migrated most anodally (about 6 cm) and had strong staining activity. To stain for EST-2, the EST-1 system must be allowed to overstain while waiting for this more cathodal system to develop. This second locus migrated about 1 cm cathodal to EST-1. EST-3 appeared as a variable series of isozymes migrating about 2 cm anodally from the origin. With 4-methylumbelliferyl acetate as a substrate three bands fluoresced in some animals but not in others. When using α-napthyl phosphate as a substrate only the two slower bands stained in these same individuals. Some animals showed no staining activity with either substrate, while others showed the 2- or 3-banded isozymes depending on which of the substrates were used in the reaction. We interpreted the no-band pattern as expressing a homozygous condition for a "null" allele. It was fairly easy to classify the remaining phenotypes as weak or strong with respect to staining intensity. The difference between the two categories was particularly striking when using the fluorogenic substrate. We considered the weak staining pattern to represent hetero-

Table 1. Enzyme and nonenzymatic protein systems studied. The number of loci found for each system is given (n = 32) as well as the type of tissue and buffer system used (K = kidney; L = liver). Numbers for buffer systems refer to those given in footnote.

Enzyme Commission Number	Enzyme or Protein Name	Number of Loci	Buffer System/ Tissue*
1.1.1.1	Alcohol Dehydrogenase	1	4/L
1.1.1.8	Glycerol-3-Phosphate Dehydrogenase	1	2/K
1.1.1.14	Sorbitol Dehydrogenase	1	2/K
1.1.1.27	Lactate Dehydrogenase	2	3/K
1.1.1.37	Malate Dehydrogenase	2	2/K
1.1.1.40	Malic Enzyme	1	1/K
1.1.1.42	Isocitrate Dehydrogenase	2	2/K
1.1.1.44	Phosphogluconate Dehydro-genase	1	1/K
1.1.1.47	Glucose Dehydrogenase	1	2/K
1.15.1.1	Superoxide Dismutase	1	7/K
2.6.1	Glutamate-Oxaloacetate Transaminase	2	7/K
2.7.4.3	Adenylate Kinase	2	6/K
2.7.5.1	Phosphoglucomutase	2	5/K
3.1.1.1	Esterases	3	8/K
3.1.3.2	Acid Phosphatase	1	3/K
3.4.1.1	Peptidases	3	7/K
3.5.4.3	Guanine Deaminase	1	7/K
5.3.1.8	Mannosephosphate Isomerase	1	7/K
5.3.1.9	Glucosephosphate Isomerase	1	5/K
	Albumin	1	8/K
	Hemoglobin	2	8/L,K

* Buffers and running conditions: (1) Tris-maleic-EDTA, 100 v/4 hr (Selander et al., 1971); (2) Continuous Tris-citrate, pH 8.0, 130 v/4 hr (Selander et al., 1971); (3) Continuous Tris-citrate, pH 7.0, 180 v/3 hr (Ayala et al., 1972); (4) Phosphate-citrate, pH 7.0, 100 v/4 hr (Selander et al., 1971); (5) Phosphate, pH 6.7, 150 v/4 hr (Selander et al., 1971); (6) Succinate-citrate, pH 5.0, 100 v/3 hr (Blake, 1977); (7) Discontinuous Tris-citrate, pH 8.2 (Poulik), 250 v/3 hr (Selander et al., 1971); and, (8) Lithium hydroxide, pH 8.1, 300 v/3 hr (Selander et al., 1971).

zygotes, and the strongly staining individuals to be homozygous for the active form of this esterase.

RESULTS

Nineteen enzyme and two nonenzymatic protein systems, controlled by a presumptive total of 32 structural gene loci, were surveyed (Table 1). For most systems 126 individuals were studied, but only 90 animals were examined for alcohol dehydrogenase genotypes, while 151 were surveyed for their esterase genotypes. Most allozyme patterns indicated a homozygous condition. Except at one esterase locus (discussed below) patterns characteristic of heterozygotes were seen in a few individuals at only four other loci. Two presumptive heterozygotes for a more cathodal allele were seen for isocitrate dehydrogenase-1 (soluble form); one heterozygote pattern was observed at the lactate dehydrogenase-1 locus (heart form); two heterozygous animals were found for the most anodally migrating peptidase (dipeptidase) locus; and, four heterozygotes were found at the albumin locus. For these four loci, percentages of the rarer alleles varied from 0.4 to 1.6.

Only one esterase locus showed allozymic variability that suggested a polymorphism (EST-3). For our entire sample, the null allele at this esterase locus occurred in a frequency of 54.0%. The three genotypes were collectively distributed precisely according to Hardy-Weinberg expectations. Allelic frequencies among the five regional groupings did not differ statistically from the overall mean, and except for one sample of five individuals, the null allele ranged only from 43.6 to 63.7% (Fig. 1). However, there was heterogeneity in allele frequencies as tested by contingency Chi-square and G-statistic analyses (Sokal and Rohlf, 1969:577; $P < 0.05$ and 0.025, respectively). Although these statistics unavoidably incorporate gene frequency as well as geographical heterogeneity, they point to the Prince William Sound sample as the most deviant of the large samples. This population had a higher frequency of homozygotes for the visible allele than did the others. The difference was statistically significant when genotype frequencies of animals from Prince William Sound were compared to expected values based on all other individuals ($P < 0.005$). Moreover, among the nine heterozygotes observed for rare alleles, the one lactate dehydrogenase heterozygote and three of four albumin heterozygotes were from Prince William Sound (the fourth was from Chiswell Island along the outer coast of the Kenai Peninsula and not far from Prince William Sound). Therefore, the Prince William Sound sea lions may be somewhat different from the others.

Hemoglobin was present in both the liver and kidney tissue extracts. On the LiOH buffer system two bands of red-colored protein were visible in the unstained gel after electrophoresis. These red bands are assumed to correspond to the alpha- and beta-hemoglobin chains found in most mammalian species. In three individuals from Prince William Sound, however, variation from this pattern was observed in gels made from kidney extracts. All three individuals showed the same deviant pattern of variation together with the normal pattern on gels made with liver extracts. This unusual protein phenotype differed from the normal one in that both red bands migrated more slowly than in the common condition. One band migrated approximately 1 cm cathodally and the second appeared approximately 1 cm anodally. In the usual phenotype the bands migrated about 3 and 4 cm anodally. These forms appear "normal" in the sense that they showed sharply staining bands. To test the possibility of sample degradation, samples from other individuals were incubated in water at approximately 70 C for 30 min to stimulate protein denaturation. These treated aliquots showed the same banding pattern as observed in the untreated controls. We therefore cannot explain the three aberrant phenotypes, but do not think they have a genetic basis because of their observed tissue specificity. Naevdal (1966a) also observed aberrant hemoglobins in a few individual harp seals which he assumed to have a genetic basis.

If all variations (except for the peculiar kidney hemoglobin in three individuals) are considered, five out of 32 loci (15.6%) were found to be polymorphic. If, however, we consider only those loci in which the rare alleles occurred in a frequency of at least 5%, as is usually done, we observed only one polymorphic locus (3.1%). The average individual heterozygosity as measured by H (Nei, 1975) for the total sample was 0.018.

Biochemical Variation in Sea Lions

DISCUSSION

Northern sea lions from the region sampled show extremely low genic variability as evidenced by 32 structural gene loci. Average genic heterozygosity (0.018) is even lower than that reported for the southern elephant seal and lower than that found in most species of mammals (Selander, 1976; Smith et al., 1978). The proportion of loci polymorphic (0.031) is even more striking in being unusually low for mammals (Selander, 1976; Smith et al., 1978). Only one locus (EST-3) was found to be polymorphic, and it contained two alleles in approximately equal numbers. It is not clear from our study whether this relative genetic uniformity extends throughout the range of *Eumetopias jubatus* or applies only to the region examined; analyses of samples from other parts of the species' range will answer this important question.

If we tentatively assume that the entire species shows low genetic variability, then this case supports the hypothesis that marine mammals as a group may typically exhibit lower levels of biochemical variation than do terrestrial species. Bonnell and Selander (1974) suggest that the genetic bottleneck which occurred in the northern elephant seal can explain the lack of variability which they found in that species. While this may be true, there is no reason to suspect that such a bottleneck has occurred in *E. jubatus*, at least during the last few hundred years. Kenyon and Rice (1961), for example, estimated the Alaskan population to number 181,892 individuals. Nei et al. (1975) have suggested, however, that the genetic effects of a population bottleneck may persist for many thousands of years, especially in a species with long generation time and/or low reproductive potential. It is conceivable, therefore, that we may be observing the consequences of a bottleneck which occurred long before recorded history. The alternative that small deme size has led to low variabilities is untenable in view of the apparent large size of demes in at least some species, the impressive vagility of most marine species, and the previously mentioned report by Shaughnessy (1975) of low variability in the nonvagile harbor seal. The third possibility is that there is something about selective regimes in marine mammals which leads to low variability. Clearly we cannot escape the conclusion that our knowledge and understanding of genetic variation in marine mammals is in its infancy.

With respect to our search for natural management units within the area sampled, we find little evidence for differentiation into a demic structure within this large region. Only in the case of the Prince William Sound population is there a plausible suggestion of some geographical differentiation. The small sample from the northeastern part of the Gulf of Alaska also has a deviant ratio of genotypes at the EST-3 locus, but the sample is much too small for much to be made of this. What is impressive is the general lack of differentiation over the area surveyed. If this basically negative evidence can be supported by other lines of evidence, such as frequent movement of individuals between widely-spaced rookeries, it leads to the conclusion that the region can be considered one deme for management purposes.

Finally, we would like to speculate further that if low heterozygosity levels such as exhibited by _E. jubatus_ do in fact reduce a population's ability to cope with changing conditions, either 1) through direct effects on viability and/or behavior (Garten, 1976, 1977; Smith et al., 1975; Berger, 1976), or 2) through effects on long-term rates of evolutionary change (e.g., Haldane, 1937; Lewontin, 1958), then marine mammal species such as this may be expected to show enhanced vulnerability to rapid environmental changes perpetrated by man or occurring naturally.

Acknowledgments--Field work was supported by the Bureau of Land Management through interagency agreement with the National Oceanic and Atmospheric Administration (program managed by the Outer Continental Shelf Environmental Assessment Program office). We thank Ken Pitcher and Karl Schneider for assistance in the field work. We also express our appreciation to S. Y. Yang and Monica M. Frelow for technical assistance, and to R. E. Jones and B. J. LeBoeuf for critically commenting on early drafts of the manuscript.

Note Added in Proof--A paper on the hemoglobins of pinnipeds (Lincoln et al., 1973; Blood 41:163-170) has recently come to our attention. The typical phenotypes seen in our material are similar to those that they report for the walrus (*Odobenus rosmarus*), the fur seal (*Callorhinus ursinus*), and the California sea lion (*Zalophus californicus*). However, the aberrant pattern that we describe from kidney extracts of three individuals is reminiscent of the phenotypes they report for the grey seal (*Halichoerus grypus*), the harbor seal (*Phoca vitulina*), and the ribbon seal (*Histriophoca fasciata*)!

ALLOZYMIC VARIATION IN HOUSE MOUSE POPULATIONS

R. J. Berry and Josephine Peters

Abstract--Electrophoretic studies have been carried out on population samples of house mice from a range of geographical and ecological situations. This paper reports data on 49 loci from house mice trapped in the Central Valley of Peru. The mean heterozygosity per locus was 8%, and there was little differentiation among samples. There was also linkage disequilibrium for variants of the GPI-1, MOD-2 and HBB loci on chromosome 7. These results are compared with gene and genotypic variation in other house mouse populations.

INTRODUCTION

Until 1966 there were no accurate ways of estimating the amount of inherited variation in natural populations. Available estimates were based on either morphological variation or the frequency of some particular class of variant, such as lethal alleles or chromosomal polymorphisms (Dobzhansky, 1970; Yablokov, 1974). This situation changed drastically with the development of electrophoresis for detecting enzyme and protein variants in population samples (Harris, 1966; Lewontin and Hubby, 1966). In theory, the frequencies of allozymic variants in a random series of proteins give an accurate estimate of the variation carried by an individual; but the validity of the values obtained have been criticized on a variety of grounds (Johnson, 1974). Notwithstanding, electrophoretically-determined allozymic frequencies give the best information of genic and genotypic variety so far available, and values of the numbers of polymorphic loci and average individual heterozygosity have been produced for a wide variety of species (Selander and Kaufman, 1973a; Powell, 1975; Selander, 1976; Nevo, 1978).

The realization that a high degree of genetical variation exists in virtually every natural population of animals effectively destroyed the sophisticated mathematical basis of population genetics, and speculation about the significance of the newly revealed variation has run riot (King and Jukes, 1969; Clarke, 1970a, b; Ayala, 1976). The consequence of this is that a genetical mysticism seems in danger of arising. For example, wide claims have been made about the effects of overall levels of heterozygosity (Smith et al., 1975; Soule, 1976; Valentine, 1976). Unfortunately for these generalizations, the data supporting them are very often inadequate. In many cases a single sample is taken as characterizing a population or species, whereas the frequency of inherited traits may change markedly in space or time (Berry, 1978, 1979; Berry et al., 1979). Such genetical heterogeneity is well documented for man, some *Drosophila* species, and a range of rodent species. This paper reviews the published information on allozymic variation in the house mouse (*Mus musculus*) and summarizes data from nine population samples trapped from a small area of central Peru.

MATERIALS AND METHODS

Our data are derived from 216 mice live trapped in January and February 1976 on agricultural land and around farms near Nana, a village at a height of 549 m, 22 km east of Lima, Peru. In this area the Rimac River runs along a valley about 1 km wide and bounded by steep, dry, and barren hillsides (Fig. 1). The only rodents caught were house mice; on the valley sides this species is replaced by native species of *Phyllotis* (Pearson and Pearson, 1978). The reason for trapping in this area was that descendents of mice caught in 1961 on a farm at the center of the area produced an abnormally high incidence of deleterious mutants, including histinidaemia, a recessive lethal allele at a coat color locus, steroid upset leading to female sterility, and polydactyly (Wallace, 1971). The later trapping was part of an attempt to discover the extent and causes of this situation (Wallace and Berry, 1978).

Two hundred and ninety-nine mice were caught from nine sites, which can be regarded as linearly distributed along the Rimac valley, although sites 3 and 10 were on the opposite side of the river to the others. The river sometimes dries up during the summer. The mice were flown to England and breeding experiments carried out at the University of Cambridge by Dr. Margaret

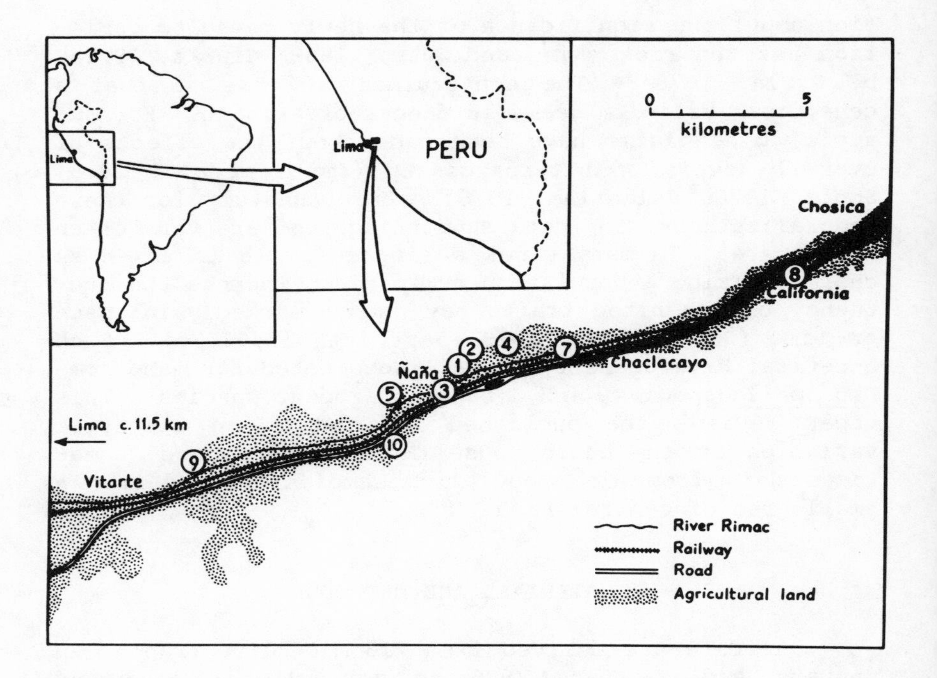

Fig. 1. Collecting sites for house mice in a valley in Central Peru. Numbers refer to trapping localities that are specifically identified in the text.

Wallace. Mice not wanted for the breeding program were killed; blood, liver, kidney, and heart were collected. Material became available from 216 animals.

Tissue samples were prepared for horizontal starch-gel electrophoresis as described by Peters and Nash (1977). Twenty-seven enzymes and proteins representing 46 known loci were screened electrophoretically. These included albumin (ALB-1), hemoglobin β chain (HBB), transferrin (TRF), sorbitol dehydrogenase (SDH-1), lactate dehydrogenase (LDH-1, LDH-2), lactate dehydrogenase regulator (LDR-1), malate dehydrogenase (MOR-1, MOR-2), malic enzyme (MOD-1, MOD-2), isocitrate dehydrogenase (IDH-1, IDH-2), phosphogluconate dehydrogenase (PGD), hexose-6-phosphate dehydrogenase (GPD-1), indophenol oxidase (IPO-1), glutamate oxaloacetate transaminase (GOT-1, GOT-2), pyruvate kinase (PK-3), phosphoglycerate kinase (PGK-1), adenylate kinase (AK-1, AK-2), phosphoglucomutase (PGM-1, PGM-2), red cell acid phosphatase (ACP-1), peptidases (PEP-1, PEP-2, PEP-3, PEP-7), guanine deaminase (GDA), adenosine deaminase (ADA), carbonic anhydrase (CAR-1, CAR-2), enolase (ENO-1), glyoxalase (GLO-1), mannose phosphate isomerase (MPI-1), glucose phosphate isomerase (GPI-1), and esterases (ES-1, ES-2, ES-3, ES-5, ES-6, ES-7, ES-9, ES-10, ES-11). Three further phenotypes (esterase-B, kidney elaidyl esterase, and liver elaidyl esterase) for which the genetic determination is unknown were assumed to be the products of three further loci. The electrophoretic methods and staining techniques have all been published (Ruddle, et al., 1969a; Nichols and Ruddle, 1973; Harris and Hopkinson, 1976; Eicher et al., 1976; Berry and Peters, 1977; Peters and Nash, 1978).

RESULTS

Twenty-eight of the 49 putative loci were monomorphic. Five loci had only rare variants, i.e., alleles occurring with a frequency of less than 0.1% (LDH-2, GPD-1, PGM-1, ES-5, ES-11), while IDH-1 had both common and rare variants. The overall frequency of rare variants was 1.12×10^{-3}. Defining polymorphism as occurring when the frequency of the most common allele at a locus is less than 99%, there were 16 polymorphic loci, and the frequencies of the alleles at these are given in Table 1 for seven population samples. The allele frequencies for samples from trapping sites 1 and 2, and 4 and 7 were not statistically different, and the data from them have been pooled. Table 1 also lists the proportion of loci, which were polymorphic in each sample, and the mean heterozygosity per individual. Genetic distances between the samples are given in Table 2 (Rogers, 1972).

When each of the segregating loci was considered separately, no significant differences from expectations were found in genotype frequencies. However, there were significant associations between three loci on chromosome 7 (Table 3). There are several methods described for estimating such associations (D: Webster, 1973; Hill, 1974). We have tested the deviation from random assortment by the method of Sinnock and Sing (1972), taking into account the sensitivity of D to allele frequencies (Hill and Robertson, 1968). There was a

Table 1. Frequency of one of the alleles for 15 polymorphic loci analyzed for samples taken north or south of the Rimac River. Percent of polymorphic loci (P), mean percent heterozygosity (H) per locus and sample size (N) are also given for each location.

	North of River					South of River	
Allele*	9	5	3	1+2	4+7	10	8
Hbb[d]	0.375	0.417	0.594	0.254	0.358	0.481	0.239
Sdh[b]	0.375	0.260	0.282	0.287	0.398	0.500	0.458
Mor-1[b]	0	0.020	0	0.088	0.027	0.020	0
Mod-1[a]	0.250	0.114	0.071	0.034	0.150	0.040	0.045
Mod-2[a]	0	0.571	0.357	0.389	0.238	0.200	0.150
Idh-1[a]	0.375	0.220	0.281	0.345	0.373	0.482	0.771
Got-2[b]	0.125	0	0.094	0	0.045	0.069	0.021
Pk-3[b]	0.125	0.036	0.107	0.125	0.262	0.077	0.100
Pgm-2[a]	0	0.020	0.120	0.073	0.018	0.121	0.125
Gpi-1[a]	0	0.542	0.469	0.306	0.391	0.483	0.283
Mpi-1[a]	0	0.023	0	0.031	0	0	0
Es-2[a]	0.707	0.447	0.433	0.183	-	0.186	-
Es-6[a]	0	0.205	0	0.110	0.117	0.229	0.381
Es-7[d]	0.875	0.545	0.333	0.429	0.343	0.556	0.545
Es-9[b]	0.125	0	0.038	0	0.022	0.167	0.024
P	21.3	28.0	24.5	30.0	26.0	27.1	24.0
H	7.3	6.7	7.3	6.3	8.2	9.3	7.4
N	4	25	16	61	57	29	24

*Locus abbreviations are given in the text.

Table 2. Genetic distances (Rogers, 1972) between samples of mice caught at a series of locations north and south of the Rimac River.

	North of River				South of River		
	5	3	1+2	4+7	10	8	Mean
9*	0.072	0.061	0.070	0.066	0.077	0.078	0.071
5*		0.027	0.035	0.044	0.043	0.059	0.047
3*			0.034	0.041	0.035	0.054	0.042
1+2*				0.029	0.039	0.039	0.041
4+7*					0.037	0.038	0.042
10						0.032	0.044
8							0.050

*North of river

association between HBB and GPI-1 variants overall, although this was statistically significant for only the three central sites (1+2, 3 and 8). The disequilibrium between GPI-1 and MOD-2 was much weaker; there was no statistically significant association ($P > 0.05$) between MOD-2 and HBB.

DISCUSSION

The house mouse is an opportunistic colonizer, subject to frequent local extinction (Justice, 1962). New populations are almost inevitably distinct from their ancestral group through the operation of the founder effect (Berry, 1964, 1977 a, b, c). Judged by the usual amount of differentiation between populations, the Peruvian samples, with the exception of the small sample from site 9, are surprisingly like each other (Table 2). Standards of comparison are provided by a mean distance of about 0.02 between successive year samples from a genetically closed population on the small Welsh

Table 3. Probable gametic frequencies and values of linkage disequilibrium (D) for alleles at beta hemoglobin (HBB), glucose phosphate isomerase (GPI-1) and malic enzyme (MOD-2).

Locations	Sample Size	Probable Gametic Frequencies				D	Probability
		$Hbb^{d}/Gpi\text{-}1^{a}$	$Hbb^{d}/Gpi\text{-}1^{b}$	$Hbb^{s}/Gpi\text{-}1^{a}$	$Hbb^{s}/Gpi\text{-}1^{b}$		
1+2	53	0.181	0.055	0.130	0.634	0.108	<0.001
5	23	0.358	0.055	0.185	0.402	0.134	<0.01
8	22	0.114	0.136	0.159	0.591	0.068	<0.01
Total	196	0.204	0.148	0.174	0.475	0.071	<0.001
		$Gpi\text{-}1^{a}/Mod\text{-}2^{a}$	$Gpi\text{-}1^{a}/Mod\text{-}2^{b}$	$Gpi\text{-}1^{b}/Mod\text{-}2^{a}$	$Gpi\text{-}1^{b}/Mod\text{-}2^{b}$		
Total	97	0.134	0.211	0.175	0.480	0.027	<0.02

*Some animals were not typed for all loci, so these sample sizes are less than the mice available from each site (Table 1).

island of Skokholm and 0.12 from samples collected from farms about 5 km apart in rural southern England (Berry and Peters, 1977). The mean genetic distance between the Peruvian samples was 0.04. This shows that there was very little differentiation between the samples trapped (without, of course implying that they are completely homogenous), despite the fact that they were caught on both sides of a normally fast-flowing river, and site 8 was 10 km from the main trapping area.

This is very different from the assumptions that are usually made about the effective population size of house mouse populations. Based on the movement of individuals, computer simulations of gene frequency dynamics, laboratory based behavior studies, and intrapopulation differentiation, it has been claimed that the population unit of the species may be as small as four individuals (De Fries and McClearn, 1972). However, feral, as opposed to commensal, mouse populations undergo considerable churning through death and replacement of territorial animals, and the Peruvian situation is not dissimilar to that found in other wild populations of this species (Berry and Jakobson, 1974; Lidicker, 1976).

Estimates of genetic variation have been determined for house mouse populations living in a wide variety of ecological and geographical situations (Table 4). The average level of heterozygosity and the proportion of polymorphic loci range from zero, for populations on two small British islands and Fugloy in the Faroe archipelago, to a maximum heterozygosity value of 11.4% of loci for one of the British mainland samples, Watten, Caithness in northern Scotland. These values indicate that house mice are relatively variable; the mean heterozygosity for all rodents is 5.4%, while for man it is 6.7% (Powell, 1975; Smith et al., 1978; Harris and Hopkinson, 1972). Most loci where segregation occurs have only two alleles. The frequency of rare alleles in the Peruvian samples was 1.12×10^{-3} compared to 1.75×10^{-3} in British mice and 1.76×10^{-3} in man (Berry and Peters, 1977; Harris et al., 1974). These alleles could have arisen by recent mutational events.

Different heterozygosity estimates are more similar than those for proportions of polymorphic loci; considerable care ought to be taken in comparing estimates of genetic variation based on different loci. There have been repeated claims that certain classes of pro-

Table 4. Amount of genetic variation as indicated by percent of loci polymorphic (P) and percent heterozygosity per locus (H) in house mouse populations.

Locations	No. of Samples	No. of Loci	P	H
North Carolina, Vermont and Alberta[1]	6	17	29.4	10.2
California[2]	10	40	35.0	10.6
Denmark[3]	5	41	41.5	7.8
Hawaii[4]	9	26	73.1	7.0
British Mainland[5]	6	22	59.1	8.6
British Islands:				
Skokholm[5]	1	22	18.5	6.0
Isle of May[5]	1	22	0	0
Inchkeith[5]	1	22	0	0
Orkney[5]	5	22	21.5	5.0
Shetland[5]	6	22	20.0	4.4
Faroe[5]	6	22	20.3	2.3
Macquarie Is.[6]	1	17	31.6	6.5
Marion Is.[7]	1	24	36.4	7.7
South Georgia[8]	1	27	7.7	3.4
Central Peru[9]	7	49	32.0	8.0

[1]Ruddle et al., 1969; [2]Selander and Yang, 1969; [3]Selander et al., 1969a; [4]Wheeler and Selander, 1972; [5]Berry and Peters, 1977; [6]Berry and Peters, 1975; [7]Berry et al., Peters, 1978b; [8]Berry et al., 1979; [9]Table 1.

teins are more variable than others (Gillespie and Kojima, 1968; Johnson, 1974). For example, Berry and Peters (1977) found that the mean heterozygosity in house mice of enzymes capable of using variable substrates was 5.3%, of regulatory enzymes was 7.0%, and of nonregulatory enzymes was only 2.5%. The problem is that the assignment of enzymes to different classes is somewhat arbitrary. Heterozygosity estimates for samples from *Drosophila* species made by Lewontin and his associates include a high proportion of glucose metabolizing enzymes and are almost always below 14%, while Ayala and his co-workers screen a higher proportion of nonspecific enzymes and all but one of their estimates are over 14% (Kojima et al., 1970; Selander, 1976). Using our normal order of scoring loci, the estimated heterozygosity in our Peruvian data after classifying 10 loci was 14.2%; after 20 it was 10.9%; and, after 40 it was 9.6%. Other laboratories will have a different order of scoring, but there is clearly danger in basing heterozygosity estimates on only a few loci.

A disequilibrium exists between the segregating alleles on chromosome 7. Classical genetical studies have established the gene order on this chromosome as GPI-1 —27— MOD-2 —5— HBB (Chapman et al., 1974). The strongest disequilibrium found was between the widely separated loci GPI-1 and HBB, and there was no association between MOD-2 and HBB ($P > 0.05$). The simplest explanation of these results would be a chromosomal inversion, such that the gene order was modified to MOD-2——GPI-1——HBB.

Linkage disequilibria may arise and be maintained by selection if there is an epistatic interaction between loci, or they may arise due to chance, in which case they will be temporary. The majority of studies on the subject have been on *Drosophila*, and most of those reported involve an enzyme locus and an inversion (Mukai et al., 1971). Since *Drosophila* species have a lower haploid number and generally more protein polymorphism than vertebrate species, linkage disequilibria are more likely to be found in them than in house mice.

It is not known whether the linkage disequilibria in the Peruvian mouse populations are transitory or are maintained by selection. The occurrence of an overall linkage disequilibrium across samples would be suggestive of selective action, but the disequilibria did not occur in all the Peruvian samples and were not found in any of the population samples studied by Berry and Peters (1977). However, if selective forces are maintaining the linkage disequilibria, these may fluctuate in both time and space. Furthermore, evidence of selection on the chromosomal segment carrying the HBB locus has been found in several mouse populations where allele or genotype frequencies change with season, social status, and age (Berry and Murphy, 1970; Bellamy et al., 1973; Myers, 1974; Berry and Peters, 1975). Allele frequencies at the HBB locus also change when selection for body size is exercised under laboratory conditions (Garnett and Falconer, 1975).

Selander et al. (1969) and Hunt and Selander (1973) studied animals from the two north European subspecies of the house mouse (domesticus and musculus) on both sides of an apparently stable zone of hybridization in Jutland, Denmark. Allozymic frequencies were different in the two races. On the grounds of only limited penetration of alleles across the hybrid zone, they concluded that there was considerable coadaptation in the genomes of the two races. There were correlations between the ES-1, ES-2 and ES-5 loci which are linked on chromosome 8, but equally strong association between the unlinked alleles $Es\text{-}1^b$ and $Idh\text{-}1^a$; $Es\text{-}2^b$ and $Idh\text{-}1^a$; and $Es\text{-}2^b$ and $Mod\text{-}1^a$. These authors concluded that linkage disequilibrium was unimportant in this situation.

The only two loci segregating in the mouse population of the South Atlantic island of South Georgia are ES-6 and GOT-2 on chromosome 8. The alleles of these loci show almost complete association (Berry et al., 1979). Finally, it has been known since at least 1962 that recombination in chromosome 17 is largely suppressed between the T and H-2 segments (Dunn et al., 1962; Forejt, 1972; Womack and Roderick, 1974). These are the only other linkage disequilibria so far reported in the house mouse, although both centric fusions and attractions are now comparatively well known in the species (Michie, 1953; Wallace, 1958; Capanna et al., 1976). As knowledge increases of this genetically and developmentally well known and ecologically tractable species, we may expect to learn more about the relative importance of different evolutionary forces than is possible from other organisms favored by ecological geneticists, including man.

Variation in House Mice

Acknowledgments--This work was supported by the Medical Research Council. Our thanks are due also to Professor C. A. B. Smith for statistical advice; Mr. H. R. Nash for technical help; Mr. A. J. Lee for drawing the figure; Mr. John Coppock for trapping the animals; and, Dr. Margaret E. Wallace for stimulating our interest in the Peruvian mice.

HERITABILITY OF MORPHOMETRIC RATIOS IN RANDOMBRED HOUSE MICE

Larry Leamy

Abstract--Estimates of heritabilities and maternal effects were made for three separate suites of morphometric ratios in randombred house mice. The suites consisted of 15 osteometric and two external metric characters expressed over skull length (A), body length (B), and body weight (C). Heritabilities were calculated from regressions of male and female offspring of three separate age groups (1, 3, and 5 months) on both male and female 5-month parents. The mean heritabilities for male and female 5-month offspring, respectively, were 0.63, 0.52 (A), 0.18, 0.36 (B), and 0.17, 0.46 (C). The general level of heritability achieved in each suite of characters was explained primarily in terms of the properties of the individual numerator and denominator variables. Intraclass correlation coefficients obtained from full-sib analyses of variance also were quite variable among the age groups and sexes, but yielded mean estimates of maternal effects for the 5-month mice (sexes averaged) of 0.05 (A), 0.13 (B), and 0.21 (C). The results indicated that certain morphometric ratios, such as the A types, are highly heritable and thus should prove particularly useful in mammalian taxonomic and paleontological studies.

INTRODUCTION

Ratios of assorted morphometric traits have been used in a wide variety of studies as measures of relative size or shape. They are to be found especially in systematic studies since it has long been assumed that absolute size measures are too much influenced by environmental forces to qualify as useful taxonomic criteria (Gould, 1966; Oxnard, 1968, 1969; Mayr, 1969; Corruccini, 1975). In the mammalian literature, ratios also have been employed in allometric (Gould, 1966), paleontological (Coombs, 1975), anthropological (Ashton et al., 1976) and animal breeding (Turner, 1959) studies and currently are becoming increasingly popular as input variables for a variety of multivariate investigations (Wilson, 1973; Gipson et al., 1974; Mares, 1976). The interpretation of ratios sometimes is complicated by their statistical peculiarities (Pearson, 1897; Atchley et al., 1976) and/or factors such as allometry, but their general usefulness as fundamentally important characters often outweighs such complications (Simpson et al., 1960; Corruccini, 1975; Dodson, 1978). The use of ratios as alternative raw variables therefore undoubtedly will continue, and this seems legitimate, especially if their practical and/or theoretical constraints are given due consideration in individual applications (Simpson et al., 1960; Sutherland, 1965; Salthe and Crump, 1977).

Given the pervasive use of ratios, it is important especially for the systematist or paleontologist to know the extent of genetic variation, i.e., the heritability, in these kinds of characters. The heritability of any character expresses the proportional contribution of the additive genetic variance to the total phenotypic variance and is a prime determiner of the expected response of the character to either artificial or natural selection (Falconer, 1960). The widely held assumption that morphometric ratios are moderately to highly heritable is supported primarily by indirect evidence, i.e., the known differences in body types among breeds of domestic animals, and the extent of response exhibited from artificial selection for anatomical ratios or indices (Cock, 1966). Direct estimates of the heritability of feed efficiency (feed consumed/gain in body weight) and similar potential selection criteria have been made for several economically important animals (Turner, 1959; Sutherland, 1965; Eisen, 1966), but these are of little relevance to mammalian systematists who conventionally employ other morphometric variables. What is particularly needed is a comparison of the heritabilities of ratios of each of a suite of morphometric traits in which several different and more appropriate measures of size in the denominators are chosen.

Recently, the heritabilities of 15 osteometric and three external metric traits were estimated in randombred house mice of three different ages (Leamy, 1974). In this study, parent-offspring regression techniques are used to calculate the heritabilities of the ratios of each of these traits over three different size measures (skull length, body length, and body weight). Full-sib correlations also are used in an assessment of the effects of nongenetic maternal influences on each of these ratios.

MATERIALS AND METHODS

The Population and Variables

Mice used in this study originated from randombred strain CV1, a derivative of inbred strain 101 (Leamy, 1974). Sufficient numbers of single-pair random matings were made such that 200 pairs were successful in producing over 1000 offspring (first litters only). All litters were reduced to six individuals except for those litters with fewer than six in which all offspring were retained. In each litter, three sublitters of two mice each were sacrificed, one at each of three ages: 1 month (35 days), 3 months (90 days), and 5 months (150 days). An attempt was made to equalize the sex distribution within sublitters, the result being that nearly all of the 1-month sublitters consisted of one male and one female, whereas most of the 5-month sublitters consisted of two mice of the same sex. The 5-month offspring also were smallest in total numbers compared to the 1-month or 3-month offspring since not all litters contained six offspring. All parents were sacrificed at 5 months of age. Other details may be found in Leamy (1974).

After being sacrificed, each mouse was weighed (W = body weight), its tail length (Ta_L) and body length (B_L= total length minus tail length) measured, and its skeleton prepared. A total of 25 separate measurements were made on each skeleton, 10 of which were paired, the remaining five (indicated by asterisks below) being unpaired. Only the mean of the two sides for the paired variables is used here, however, so that a total of 15 osteometric variables are generated. The 15 variables are as follows (detailed descriptions are given by Leamy, 1974): Sk_L*, skull length; P_L*, palate length; ZF_L, zygomatic fenestral length; M_L, mandible length; Sk_W*, skull width; Z_W*, zygomatic width; IO_W*, interorbital width; In_L, innominate length; Il_L, illium length; OF_L, obturator foramen length; Sc_L, scapula length; F_L, femur length; Ti_L, tibia length; H_L, humerus length; RU_L, radioulna length.

Although a large number of ratios conceivably could be generated with this many variables, it was decided to form ratios of the 15 osteometric characters as well as body length and tail length, over skull length (A), body length (B), and body weight (C). These three characters differ considerably in the magnitude of their phenotypic variation and heritability, and this was presumed to be an advantage for later comparison of the levels of heritability generated among the three suites of ratios. More specifically, in this population of mice, skull length is a character with low phenotypic variation but a high heritability; body length exhibits moderate variation and a low heritability; and, body weight has a relatively high magnitude of variation and a moderate heritability (Leamy, 1974). In addition, skull length is a denominator size measure used often of necessity by paleontologists, body length is widely employed in mammalian taxonomic studies, and body weight is conventionally used in studies of growth and allometry.

Statistical Methods

Heritabilities (h^2) of each of the ratios were estimated by the technique of parent-offspring regression, as was done previously (Leamy, 1974) for the individual characters themselves. The simple linear regression coefficient of the offspring on either parent estimates 0.5 h^2, and if both regressions are statistically equal, regression of the offspring on the mean of the parents (midparent) is a valid direct estimate of heritability (Falconer, 1960). In some cases common environmental influences, especially maternal effects in mammals, may act to inflate the regression on female parent and thus on midparent, however, in which case regression on the male parent is the most reliable. Significant sexual dimorphism in the means and/or variances of the characters also may occur and, if so, it is generally advisable to obtain heritability estimates for the separate sexes (Turner and Young, 1969). Parent-offspring regression provides a true estimate of heritability only when parents and offspring are of the same age; however, regression involving different ages may give approximate heritability estimates if the repeatability of the heritabilities of the characters at different ages is high (Turner and Young, 1969).

Atchley et al., (1976) recently examined the statistical consequences of creating ratios and concluded, among other things, that distributions of ratios are seldom normal and that spurious correlations, and by inference, regressions, may be generated between a ratio and its numerator or denominator or between two different ratios. The latter finding is irrelevant here since, in this study, regression of offspring on parent is used in each case on the same ratio. The issue of normality is more serious, however, for this is a basic assumption of regression. Plots of representative ratios in all three suites (A, B, and C) of the morphometric characters in the 5-month mice were made, and they did not show significant departures from normality. ($P > 0.05$)

Since in the present design there are two individuals per sublitter, it is possible to compute for each ratio character the intraclass correlation coefficient (t) of full sibs. The full-sib correlation estimates the proportional contribution to the total phenotypic variance of one half of the additive genetic variance, one fourth of the dominance variance, and any variance due to common environmental effects, i.e., $t = (1/2V_A + 1/4V_D + V_{Ec})/ V_p$ (Falconer, 1960). Thus if dominance effects are considered negligible, and if heritabilities previously have been calculated from parent-offspring regression, these t values may be used to estimate common environmental effects. These effects result from all environmental factors which act to inflate the between-litter variance, but in mammals they are primarily associated with maternal effects (Falconer, 1960). Estimates of maternal effects for all three suites of characters were made by analyses of variance of male-male, female-female, and male-female sib combinations.

RESULTS

The means, coefficients of variation, and sample sizes of the A, B, and C characters for all 5-month mice (pooled parents and offspring) of each sex are given in Table 1. These basic statistics are not presented for the 1-month or 3-month offspring since, strictly speak-

Table 1. Means ($\bar{X}$), coefficients of variation (C.V.), and sample sizes (N) for the A (numerator variables divided by skull length), B (numerator variables divided by body length), and C (numerator variables divided by body weight) characters in all 5-month male and female mice. Asterisks (placed after the larger value in each case) denote significant differences between the sexes.

	A/Sk_L						B/B_L						C/W					
	Males			Females			Males			Females			Males			Females		
	$\bar{X}$	C.V.	N	$\bar{X}$	C.V.	N	$\bar{X}$	C.V.	N	$\bar{X}$	C.V.	N	$\bar{X}$	C.V.	N	$\bar{X}$	C.V.	N
Sk_L							.251	2.73	299	.250	2.99	351	.700	9.40	302	.780**	9.60	352
P_L	.513	1.33	303	.513	1.36	346	.129**	3.10	302	.128	3.36	349	.359	9.59	305	.400**	9.76	350
ZF_L	.360	1.49	290	.363**	1.34	349	.090	3.13	294	.091**	3.26	348	.252	9.54	297	.283**	9.80	349
M_L	.388**	2.00	301	.382	1.97	351	.097**	3.13	301	.095	3.41	354	.272	9.27	304	.298**	9.63	355
Sk_W	.432**	1.86	298	.428	1.74	351	.108**	3.18	294	.107	3.45	349	.303	9.49	297	.334**	9.76	350
Z_W	.538	1.51	301	.539	1.58	352	.135	2.86	297	.135	3.16	351	.376	9.33	300	.420**	9.48	352
IO_W	.171	2.85	302	.171	2.69	353	.043	3.60	299	.043	3.87	351	.119	9.68	302	.133**	9.81	356
In_L	.826	1.61	249	.906**	2.13**	258	.207	2.87	246	.225**	3.32*	262	.581	9.44	248	.713**	9.61	262
Il_L	.475	1.86	269	.518**	2.38**	308	.119	3.04	266	.129**	3.24	311	.332	9.43	269	.406**	9.56	312
OF_L	.236	2.99	260	.270**	3.14	284	.059	3.85	259	.067**	4.30	287	.166	9.97	260	.212**	10.35	287
Sc_L	.495**	1.81	237	.488	1.73	260	.124**	3.18	236	.121	3.25	263	.351	9.85	288	.381**	9.91	264
F_L	.688	1.63	278	.698**	1.75	330	.172	2.94	274	.174**	2.92	332	.483	9.33	277	.545**	9.36	333
Ti_L	.758	1.59	278	.766**	1.68	306	.190	2.81	274	.191*	3.03	307	.533	9.42	277	.600**	9.53	307
H_L	.543**	1.65	257	.537	1.80	307	.136**	2.89	259	.134	3.15	312	.383	9.32	261	.420**	9.51	313
RU_L	.620**	1.37	208	.615	1.47	241	.156**	2.83	209	.152	3.29*	262	.445	9.43	210	.481**	10.08	242
B_L	3.991	2.69	299	4.010*	3.01*	351							2.785	8.62	324	3.125**	8.81	372
Ta_L	4.213	3.62	299	4.220	3.63	351	1.056	4.86	324	1.053	4.79	372	2.940	9.77	324	3.290**	10.14	372
Mean C.V.		1.99			2.09			3.19			3.42			9.46			9.69	

* = $P < 0.05$; ** = $P < 0.01$

ing, these individuals may not be used for heritability estimation. The sample sizes vary considerably among the characters because of missing and/or broken bones and the necessity of having both sides present for the paired variables. Also, the sample sizes for females are greater than those for males, in all cases, since the number of female-female sib combinations exceeded the male-male combinations in the 5-month progeny (Leamy, 1974). The coefficient of variation (standard deviation/mean x 100) of each character was chosen over other measures of variability since it more readily permits a direct comparison of the relative variability among the diverse characters represented here (Simpson et al., 1960). Asterisks denote statistical significance between the sexes in either the means (t tests) or coefficients of variation (F tests) in each of the three suites of characters. These asterisks, where present, are placed after the larger of the two values in each case.

Sexual dimorphism in the means of the characters (Table 1) is widespread (41 of 49 possible comparisons reach significance). In both the A and the B characters, the males have the larger means for Sc_L, two skull (M_L and Sk_W) ratios, and the two anterior limb (H_L and RU_L) ratios, whereas females are larger in the pelvic girdle measures (In_L, Il_L, and OF_L) and in the two posterior limb (F_L and Ti_L) ratios. All 17 C characters exhibit significantly greater means for females. The C.V.s for females are greater than those for males in 42 of the 49 comparisons although only five individual comparisons (all in the A and B suites) reach statistical significance. However, the mean C.V. values clearly show that overall the females are more variable, these means for the B and C characters being significantly different in paired t tests. Overall, the A, B, and C characters exhibit low, moderate, and high levels, respectively, of variability.

Because of this extensive significant sexual dimorphism, especially in the means of the various characters, heritabilities first were estimated from regressions of each of the separate sexes of offspring on both the male and female parents. Rather than present all four estimates of heritability for each of the 49 characters, however, it is worthwhile first to examine the means of each of these estimates over each suite of characters (Table 2). Although heritability estimates

Heritability of Morphometric Ratios

Table 2. Mean heritability estimates within each of the A (numerator variables divided by skull length), B (numerator variables divided by body length), and C (numerator variables divided by body weight) suites of characters for male and female mice from regressions on male and female progeny.

Characters	Ages Months	Male Progeny Male Parent	Male Progeny Female Parent	Female Progeny Male Parent	Female Progeny Female Parent
	1	0.61	0.49	0.57	0.32
A	3	0.74	0.58	0.51	0.59
	5	0.67	0.58	0.51	0.51
	1	0.38	0.14	0.18	0.12
B	3	0.25	0.29	0.27	0.37
	5	0.21	0.13	0.28	0.41
	1	0.10	0.82	-0.10	0.39
C	3	0.33	0.33	0.44	0.34
	5	0.20	0.14	0.56	0.39

for the individual characters are given below (Table 3) only for the 5-month offspring, means of the heritability estimates for the 1-month and 3-month offspring are included in Table 2 for purposes of general comparison and for completeness. Heritabilities at these ages were calculated from regressions involving one randomly chosen individual from each sublitter. The method was not used for the 5-month offspring, however, since the prevalence of like-sexed sublitters at this age would have meant less than optimal use of the data. Instead, regressions for each character were computed from the weighted mean of two separate regressions, one involving single individuals from male-female sublitters and one involving the mean of the two individuals from like-sexed sublitters, the weights being the reciprocal of the sampling variance of each estimate (Falconer, 1963).

The mean heritability values in Table 2 show considerable variation over the three suites of characters, age groups, and sexes. The repeatability among ages is

Table 3. Heritability estimates ± standard errors of the A (numerator variables divided by skull length), B (numerator variables divided by body length), and C (numerator variables divided by body weight) characters for the 5-month male and female mice from pooled regressions on male and female parents. Number of pairs in each case is indicated by parentheses.

	A/Sk_L		B/B_L		C/W	
	Males	Females	Males	Females	Males	Females
Sk_L			0.09 ± 0.16 (144)	0.15 ± 0.14 (195)	0.16 ± 0.13 (149)	0.39 ± 0.14 (196)
P_L	1.11 ± 0.13 (151)	0.79 ± 0.10 (190)	0.13 ± 0.16 (140)	0.35 ± 0.13 (191)	0.17 ± 0.13 (151)	0.38 ± 0.14 (196)
ZF_L	0.68 ± 0.13 (143)	0.35 ± 0.13 (190)	0.01 ± 0.18 (136)	0.41 ± 0.12 (189)	0.14 ± 0.13 (151)	0.50 ± 0.13 (190)
M_L	0.54 ± 0.16 (147)	0.60 ± 0.14 (196)	0.03 ± 0.16 (144)	0.52 ± 0.13 (199)	0.15 ± 0.13 (149)	0.42 ± 0.14 (200)
Sk_W	0.77 ± 0.18 (143)	0.59 ± 0.15 (194)	0.19 ± 0.14 (136)	0.62 ± 0.11 (191)	0.21 ± 0.14 (141)	0.49 ± 0.13 (192)
Z_W	0.80 ± 0.12 (147)	0.83 ± 0.08 (197)	-0.13 ± 0.17 (140)	0.50 ± 0.12 (195)	0.12 ± 0.13 (149)	0.41 ± 0.15 (196)
IO_W	0.71 ± 0.14 (150)	0.46 ± 0.13 (197)	0.43 ± 0.19 (145)	0.48 ± 0.13 (200)	0.10 ± 0.13 (150)	0.40 ± 0.14 (201)
In_L	0.21 ± 0.22 (80)	0.59 ± 0.40 (63)	0.22 ± 0.21 (75)	0.42 ± 0.26 (66)	0.20 ± 0.16 (78)	0.20 ± 0.28 (58)
Il_L	0.37 ± 0.13 (115)	0.42 ± 0.10 (129)	0.39 ± 0.15 (110)	0.52 ± 0.13 (132)	0.22 ± 0.14 (115)	0.58 ± 0.15 (132)
OF_L	0.25 ± 0.17 (85)	0.75 ± 0.25 (91)	0.18 ± 0.18 (74)	0.22 ± 0.18 (95)	0.19 ± 0.14 (85)	0.61 ± 0.17 (95)
Sc_L	0.07 ± 0.19 (55)	-0.05 ± 0.24 (71)	0.50 ± 0.25 (51)	0.84 ± 0.20 (76)	-0.12 ± 0.35 (54)	0.60 ± 0.17 (76)
F_L	0.98 ± 0.14 (117)	0.58 ± 0.13 (161)	0.22 ± 0.18 (111)	0.04 ± 0.18 (162)	0.09 ± 0.15 (115)	0.31 ± 0.15 (163)
Ti_L	0.99 ± 0.11 (111)	0.41 ± 0.13 (149)	0.12 ± 0.21 (105)	0.03 ± 0.13 (151)	0.01 ± 0.16 (109)	0.34 ± 0.17 (151)
H_L	1.07 ± 0.21 (82)	0.82 ± 0.12 (132)	0.10 ± 0.27 (79)	0.39 ± 0.13 (140)	0.29 ± 0.21 (82)	0.48 ± 0.16 (140)
RU_L	0.84 ± 0.28 (29)	0.26 ± 0.16 (58)	-0.10 ± 0.35 (25)	-0.30 ± 0.34 (58)	0.55 ± 0.32 (27)	0.81 ± 0.39 (58)
B_L	0.11 ± 0.16 (144)	0.13 ± 0.15 (195)			0.08 ± 0.12 (167)	0.41 ± 0.13 (222)
Ta_L	0.64 ± 0.14 (144)	0.68 ± 0.12 (195)	0.31 ± 0.12 (167)	0.27 ± 0.14 (220)	0.17 ± 0.12 (167)	0.42 ± 0.12 (222)
Means	0.63	0.52	0.18	0.36	0.17	0.46

fair for the A and B characters but poor for the C characters. The correspondence of the two estimates, from regressions on either parent, within each sex of offspring is somewhat erratic for the 1-month offspring but tends to stabilize at the later ages. The worst average discrepancy in the 5-month progeny in this respect occurs in the female progeny and for the C characters, but even here none of the individual comparisons were significantly different ($P > 0.05$ in t tests of the regression coefficients). Heritability estimates for the 5-month mice from regression on female parent also do not appear to be inflated by maternal effects (in only one of six cases is the difference in this direction). Thus, pooling of the two heritability estimates for each sex is permissible.

Heritability estimates for all 49 characters for the 5-month male and female mice are presented in Table 3. The estimates were derived from pooling separate regressions on each parent, the standard errors of these pooled estimates being calculated according to Falconer (1963). Pooling of the separate regressions was favored over regression on midparent since it was assumed that the extensive sexual dimorphism for many of the traits would tend to inflate the midparent variance and thus bias (decrease) these latter estimates. Preliminary estimates of heritability from midparent regressions did in fact produce mean levels lower than those seen in Table 3 and, in addition, did not substantially improve on the magnitude of most of the standard errors. The standard errors in Table 3 partially reflect the sample sizes involved; they are highest for those characters (such as RU_L) with small samples and are generally lower for females than for males. Sampling variation also accounts for the fact that two heritability estimates in Table 3 exceed the theoretical limit of unity.

As seen from the means (negative values counted as zeros) in Table 3, the A characters exhibit relatively high heritabilities, whereas the overall level for both the B and C characters is low to moderate. Male progeny generally have greater heritabilities than the female progeny for the A characters, but the reverse is true for the B and C characters. In both sexes for the A characters, the skull and limb dimensions tend to exhibit rather high heritabilities whereas the four girdle measures (In_L, Il_L, OF_L, Sc_L) are generally lower. The heritability values for the B and C characters are far more uniform; the male estimates are low and the female estimates are moderate in magnitude. For the two external dimensions, heritability estimates for tail length exceed those for body length.

The means of the intraclass correlations of male-male, female-female, and male-female full-sib combinations for each suite of characters in each of the three age groups are presented in Table 4. One-way analyses of variance were used to calculate these values for the like-sexed sublitters whereas two-way analyses of variance (sex being the second factor) were used for all male-female sublitters. A significant sex-litter interaction in any of these two-way analyses would produce a downward bias in these t estimates (Leamy, 1974); in general they do not exhibit such interactions for their overall levels seem comparable to, if not greater than, those derived from the one-way analyses (Table 4). In addition, so few like-sexed sublitters were available for the 1-month offspring that they provide the only estimates of intraclass correlations for this age group. Repeatability of the estimates for the 3- and 5-month offspring, much like the heritability estimates, is fair for the A and B characters but poor for the C characters. The means of the t values for like-sexed sublitters in the 5-month mice are similar in overall magnitude for the three suites of characters, although those for males are greater than those for females for the A and B but not the C characters.

The common environmental variance (V_{Ec}) for each ratio, previously interpreted for the individual characters themselves as primarily maternal effects (Leamy, 1974), was computed by the subtraction of 0.5 h^2 from each t estimate. These estimates of V_{Ec} are shown in Table 5 for each of the 49 characters in male and female 5-month mice. Heritabilities used in these calculations were obtained from pooled estimates of the regressions of the means of like-sexed sublitters only on each parent. Male-female sib combinations were not used because of their smaller numbers and their less reliable t values. The use of these different regressions plus individual sampling variation accounts for the discrepancy in the mean levels of V_{Ec} seen in Table 5 from that expected on the basis of the heritability values of Table 3. In general, the A characters exhibit very low levels of maternal effects, the B characters exhibit

Table 4. Means of intraclass correlations for the A (numerator variables over skull length), B (numerator variables over body length), and C (numerator variables over body weight) characters over male-male, female-female, and male-female full-sib combinations in each age group.

Characters	Ages Months	Male-Male	Female-Female	Male-Female
	1			0.38
A	3	0.26	0.33	0.32
	5	0.33	0.20	0.26
	1			0.20
B	3	0.36	0.27	0.42
	5	0.37	0.17	0.25
	1			0.42
C	3	0.71	0.31	0.41
	5	0.22	0.42	0.62

moderate amounts for males but not females, and the C characters show moderate amounts for both sexes.

DISCUSSION

The properties of the individual denominator variables (Sk_L, B_L, W) go a long way towards explaining the differences, seen throughout the results, among the three suites of ratio variables. The means of these three variables as well as most of the other variables, for example, showed significant differences between the sexes of the 5-month progeny in the original heritability study (Leamy, 1974); this would account for the widespread sexual dimorphism in the means of the ratios of this study. Furthermore, the direction of this dimorphism in the A and B ratios generally parallels that shown by the individual numerator variables (Table 1 in Leamy, 1974). Body weight is so much larger, pro-

Table 5. Common environmental (maternal) effects for the A (numerator variables over skull length), B (numerator variables over body length), and C (numerator variables over body weight) characters in 5-month male and female mice. Character abbreviations are given in text.

	A		B		C	
	Males	Females	Males	Females	Males	Females
Sk_L			0.19	0.00	0.20	0.27
P_L	0.00	0.01	0.46	0.00	0.28	0.34
ZF_L	0.00	0.04	0.26	0.05	0.23	0.28
M_L	0.03	0.01	0.18	0.00	0.24	0.27
Sk_W	0.00	0.08	0.31	0.02	0.33	0.29
Z_W	0.00	0.00	0.29	0.00	0.25	0.26
IO_W	0.00	0.07	0.00	0.02	0.25	0.25
In_L	0.00	0.00	0.30	0.09	0.05	0.38
Il_L	0.00	0.03	0.17	0.05	0.12	0.00
OF_L	0.13	0.00	0.18	0.00	0.04	0.00
Sc_L	0.12	0.00	0.00	0.00	0.52	0.00
F_L	0.00	0.05	0.02	0.05	0.27	0.30
Ti_L	0.09	0.07	0.32	0.06	0.26	0.23
H_L	0.32	0.00	0.21	0.05	0.26	0.11
RU_L	0.00	0.00	0.60	0.00	0.00	0.15
B_L	0.17	0.00			0.27	0.14
Ta_L	0.20	0.07	0.01	0.16	0.22	0.20
Means	0.07	0.03	0.22	0.03	0.22	0.20

portionately, in males compared with females than is either skull length or body length (Leamy, 1974), however, that it overrides the effect of the numerator variables, all 17 C ratios showing significantly greater means for females. The pattern for the variability of the ratios also reflects that for the individual characters (Leamy, 1974); i.e., few individual comparisons exhibit significant sexual dimorphism although overall the females tend to be more variable than males. Lande (1977) has shown that the coefficient of variation of a ratio ($V_{x/y}$) depends upon the coefficient of variation of the numerator (V_x) and denominator (V_y) variables as well as the correlation between the two variables, the maximum value ($V_x + V_y$) occurring with complete negative correlation and the minimum value ($|V_x - V_y|$) with perfect positive correlation. Since most of the individual characters are positively but moderately correlated with the three denominator variables (Leamy, 1975), the general mean levels of coefficients of variation of the three suites of ratios are between these limits, being in fact quite similar to the coefficients of variation of the respective denominator variables.

The sexual dimorphism mentioned above also was manifested in the heritability and maternal effects estimates and, in fact, necessitated calculation of the four separate regressions for each ratio character. Fortunately some pooling of these regressions was possible since maternal effects, even though present, did not act to inflate the regressions on female parents. The general level of standard errors of the pooled estimates was higher than anticipated, in view of the sample sizes available for at least some of the characters, but admittedly was about the same as that previously achieved for the individual characters (Leamy, 1974). The differences in heritability estimates between the 5-month males and females are probably not so much inherent differences in the amount of genetic variance between the sexes as they are reflections of the degree of correspondence between each sex of parents and offspring in the developmental stage attained. The fact that the magnitude of these heritability differences is greater for the ratio characters, especially B and C, than that observed in the individual characters themselves tends to imply that the sexes differ more in relative size or "shape" than in absolute size.

Regressions of the 1-month and 3-month offspring on 5-month parents were calculated for the ratio traits primarily to assess the degree of repeatability of the heritability estimates over the three age groups. The means of the best estimates of heritability of the individual characters were 0.25, 0.47, and 0.41, respectively, for the 1-, 3-, and 5-month offspring (Leamy, 1974), the largest difference between the three estimates being 0.22. An improvement over this level of repeatability might be expected for the ratio characters only if reasonable isometry, i.e., little change in shape with growth, is obtained. Interestingly enough, the repeatability of the heritability estimates for the A and B ratios was somewhat better than that for the individual characters, the mean of the comparable largest differences among these heritabilities within each sex of offspring being 0.13. This probably explains why Underhill (1969) obtained an average heritability for five linear morphometric traits in frogs of -0.06 from regression of young, just metamorphosed, individuals on adults, but an average heritability of 0.19 for 10 ratios created from these traits. In fact this latter value may very well have increased had both parents and offspring been from the same ontogenetic stage. The repeatability of heritabilities across the age groups for the C ratios in this study was poorer (average difference = 0.31) than that for the A and B ratios and the individual characters, although most of the magnitude of this discrepancy is generated from the 1-month heritability estimates. However, the mean body weight (sexes averaged) of the 1-month mice is only about 67% of that of the 5-month parents, whereas the comparable figures for B_L and Sk_L are 82% and 90%, respectively (Table 1 in Leamy, 1974). Thus, the ontogenetic difference between parents and offspring obviously is more acute for body weight. The morphometric characters may grow more allometrically with respect to W than to either B_L or Sk_L; preliminary calculation of allometry coefficients indicates that a significant amount of allometry exists, but this matter will be more fully explored elsewhere.

To understand the general levels of heritability achieved for each of the three suites of ratios in the 5-month mice, it is useful to examine the formula for the theoretical heritability of a ratio (Sutherland, 1965):

Heritability of Morphometric Ratios

$$h^2_{x/y} = \frac{h^2_x V^2_x + h^2_y V^2_y - 2r_A h_x h_y V_x V_y}{V^2_x + V^2_y - 2r_P V_x V_y}$$

where V is the coefficient of variation, r_A is the genetic correlation, and r_P is the phenotypic correlation between x and y. Thus, from this it may be deduced that two characters which have relatively high heritabilities, and which are highly correlated phenotypically but not genetically, should tend to generate a high heritability if expressed as a ratio. As previously mentioned, the heritabilities of Sk_L, B_L, and W are high, low and moderate, respectively; the means of the phenotypic (Leamy, 1975) and genetic (Leamy, 1977) correlations of these three variables with the other variables (sexes averaged) are 0.55, 0.44, 0.32, and 0.43, 0.92, 0.40, respectively. From these values and especially the high heritability of Sk_L , it is more readily apparent why the A ratios exhibit the highest overall level of heritabilities. Since the level of heritability exhibited by the B and C ratios is quite similar, however, the high genetic correlation involving B_L seems more than compensated for by its low heritability, as apparently are the low correlations with body weight by its moderate heritability. Incidentally, heritabilities for the individual morphometric ratios were not calculated by this theoretical method primarily since sampling error would have produced inefficient estimates of r_P , and especially r_A , for many of the traits. In some animal breeding studies, however, heritabilities calculated via the theoretical formula generally have compared favorably with those calculated directly from the data (Eisen, 1966).

The overall level of heritability exhibited by the osteometric ratio characters ranges from moderate (B and C variables) to high (A variables), this being in essential agreement with previous results involving other anatomical ratios (Cock, 1966). It may well be that ample amounts of additive genetic variance are to be generally expected in these kinds of characters. Certainly if true, this implies that these traits would respond well to either natural or artificial selection, and also that they are not closely related to the overall fitness of the organism (Falconer, 1960). This is not to say that linear or stabilizing selection may not be operative on these characters, especially some of the B and C ratios, for they undoubtedly fit in some manner into the complex hierarchical fitness pattern of the organism.

The intraclass correlations (t) of full sibs were calculated primarily in order to obtain estimates of common environmental effects (V_{Ec}) for each of the morphometric ratios. Of course estimates of V_{Ec} obtained from t values, derived from offspring, plus heritabilities, derived from regression of offspring on parents, are at best approximate. This is especially true for those ratio characters where there are differences in relative growth between parents and offspring, even though both may be of the same age. As was true for the heritability estimates, the t values varied considerably among ages and sexes, and for the B characters, exhibited a pattern between the 5-month sexes exactly the reverse of that shown by the heritabilities. The t estimates from the male-female sib combinations were higher for the A but lower for the B and C characters over the ages; however, this may be more a consequence of their method of calculation (use of the sex-litter interaction as the error term in the analyses of variance) than any indication of a real difference between the characters. Actually the average of the means of the like-sexed t estimates decreased from 3 to 5 months for all three suites of characters, and the mean values for the A (0.27), B (0.27), and C (0.32) characters in the 5-month progeny were all less than the mean (0.41) for the individual characters themselves (Leamy, 1974).

Since these t estimates for the ratios were lower and the heritability estimates higher than, or nearly as high as, those for the individual characters, the estimates of common environmental effects (sexes averaged) for the A (0.05), B (0.13), and C (0.21) characters also are lower than the overall mean value (0.25) for the individual characters (Leamy, 1974). The very low estimates of maternal effects for the A ratios really are not unexpected if we assume that there has been linear (isometric) growth of most of the bone characters with respect to Sk ; in this case any size differences generally would be eliminated in the A ratios. Maternal effects, at least, seldom have been found to act independently of overall body size (Tenczar and Bader, 1966). This should be true for the B ratios as well since B_L also is a linear character, although perhaps to a lesser extent since it is an overall body rather than a bone dimension. Maternal effects were greater in the B compared with the A characters, although this general level is complicated by a considerable sexual dimorphism. The added dimensionality of body weight obviously was sufficient to generate considerable differences, and thus maternal effects, among the full-sib families for the C characters. Perhaps size differences could have been eliminated even for these C ratios had the cube root of body weight been used.

My results demonstrate that ratios are complicated characters whose properties depend upon those of the individual numerator and denominator variables. The use of assorted bone ratios, such as the A ratios of this study, by mammalian taxonomists would seem to be reasonable, at least as judged from the findings from the house mice used here. These kinds of characters had relatively high heritabilities, even over several age groups, and a reduced component of common environmental variance. Ratios of bone dimensions over gross body dimensions, e.g., body length or weight, however, are characters which actually may have quite low heritabilities and which may not eliminate size differences at all. Of course any ratio may exhibit allometric tendencies among age groups or sexes, and taxonomic inferences made from populations heterogeneous in either factor necessarily assume greater risk. The creation of morphometric ratios, especially of the type A, should produce the highly heritable kinds of characters useful in mammalian taxonomic, paleontological, and other studies.

Acknowledgments--I should like to thank Drs. Russell Lande, Robert Clover, and George Callison for helpful comments during the preparation of this manuscript. This work was supported in part by a faculty grant awarded by the Long Beach, California State University Foundation.

ANALYSIS OF GENIC HETEROGENEITY AMONG LOCAL POPULATIONS OF THE POCKET GOPHER, GEOMYS BURSARIUS

Earl G. Zimmerman and Neil A. Gayden

Abstract--The genetic structure of 21 populations along the Brazos River in Texas of the pocket gopher, Geomys bursarius, was analyzed electrophoretically for variation at 19 loci. The populations exhibited allozymic variation concordant with two established chromosomal races of G. bursarius. Populations of the two races had a mean genic identity of 0.696, while those within each chromosomal race had a mean genic identity of 0.860. Genic identities from populations occurring on the same or opposite sides of the Brazos River were essentially the same, indicating this barrier does little to restrict gene flow. Standardized variances of gene frequencies (F_{ST}) were high, averaging 0.52 for both races, while calculated migration rates were low, averaging 0.02 migrants per generation. Data from this and other studies support the hypothesis that restricted gene flow and high levels of inbreeding are major factors contributing to genic heterogeneity among geomyid rodent populations.

INTRODUCTION

A primary goal in recent investigations of genetic variation in natural populations has been to characterize the amount of genic differentiation among populations of a species to ascertain the evolutionary mechanisms involved in such differentiation. Fossorial rodents have received much attention in this regard because of their specialized habits for a subterranean existence. Nevo et al. (1974), Patton and Yang (1977), and Penney and Zimmerman (1976) summarized certain properties that appear to be common to the biology of all fossorial rodents studied thus far: (1) a low individual vagility, (2) the occupation of a rather uniform habitat; (3) the establishment of disjunct populations based on soil type; and (4) the development of a high degree of territoriality. All of these contribute to a distribution which fits the island-type model; as a result, there is often extreme differentiation in both chromosomal and allozymic systems. Therefore, fossorial rodents offer an excellent opportunity for elucidating mechanisms contributing to local population differentiation with isolation.

Most studies of genic variation among fossorial rodents have established levels of genetic differentiation over a wide geographic region with little emphasis on genic divergence among local demes. Nevertheless, genic variation in species such as the plains pocket gopher, *Geomys bursarius*, is typically extreme and permits the prediction that microgeographic differentiation might also be evident. Much is known about chromosomal variation in *G. bursarius* (Hart, 1971; Kim, 1972; Baker et al., 1973; Honeycutt, 1978). Three chromosomal groups have been characterized with the following karyotypic features: (1) *lutescens* group with diploid and autosomal arm numbers ranging from 69-72 and 68-72, respectively, X and Y both acrocentric; (2) *attwateri* group with diploid and autosomal arm numbers of 70 or 72(2N) and 74(AN), a submetacentric X and acrocentric Y; (3) *breviceps* group with diploid and autosomal arm numbers of 74(2N) and 72 or 74(AN), a submetacentric X and acrocentric Y. This variation is indicative of extensive genomic restructuring of *G. bursarius* populations.

This investigation was designed to interpret genic variation among and between local populations (demes) of two of the chromosomal groups, chromosomal race D of the *lutescens* group and chromosomal race E of the *breviceps* group (Honeycutt, 1978). Our specific aims were to establish the effects of inbreeding, migration, and barriers on genomic structuring in *G. bursarius*. Analysis of such phenomena should provide a basis for recognizing factors contributing to microgeographic differentiation among these rodents.

MATERIALS AND METHODS

Pocket gophers (N = 163) were sampled from 21 local populations along the Brazos River in McLennan and Falls Counties, Texas. Sample localities (Fig. 1A) are given with sample sizes in parentheses as follows (even-numbered localities were on the east side of the Brazos

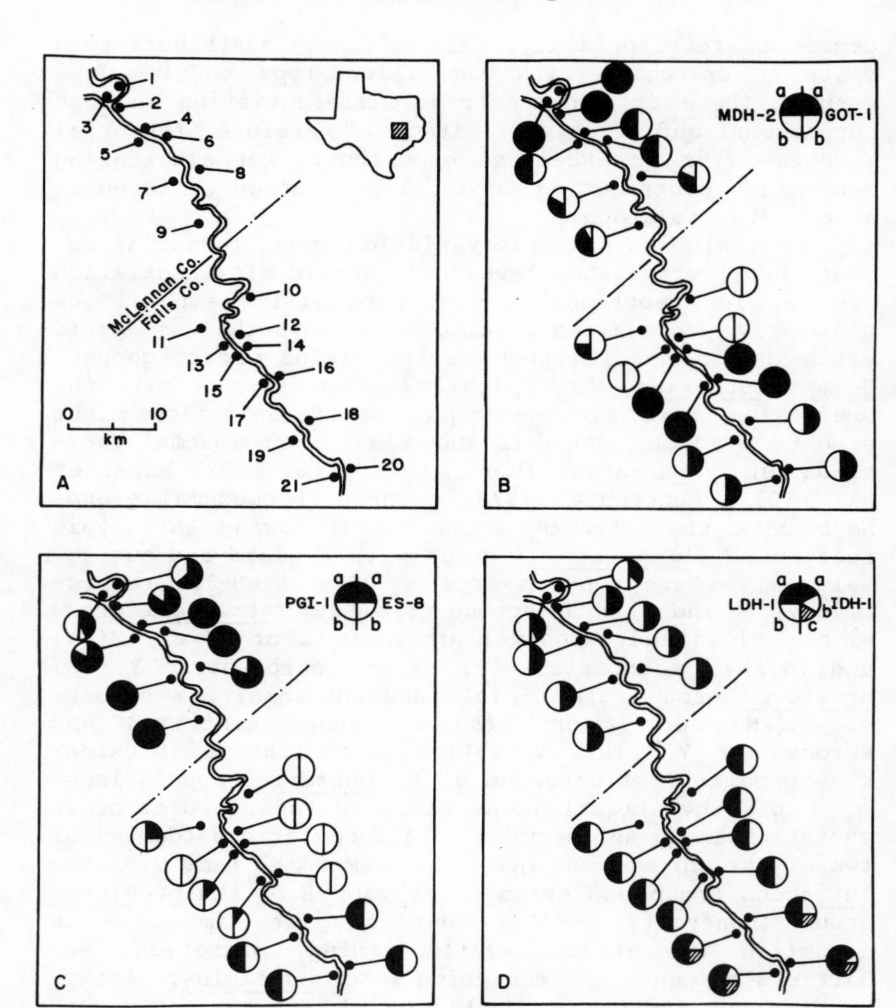

Fig. 1. Sample localities for 21 local populations of *Geomys bursarius* along the Brazos River in Texas and genic variation at the malate dehydrogenase (MDH)-2, glutamate oxaloacetate transaminase (GOT)-1, phosphoglucose isomerase (PGI)-1, esterase (ES)-8, lactate dehydrogenase (LDH)-1, and isocitrate dehydrogenase (IDH)-1 loci. Allele frequencies at each locus are indicated by the proportion of the left or right halves of the circles that are shaded according to their identification code.

River, odd-numbered localities were on the west side of the Brazos River): McLennan Co.: (1) Waco, Jct. Steinbeck Bend Rd. and W Brazos River (8); (2) Waco, Jct. Lake Brazos Drive and E Brazos River (24); (3) Waco, Jct. Lake Brazos Drive and W Brazos River (5); (4) Waco, Jct. Hwy. 81 and E Brazos River (3); (5) Waco, 8.1 km SE Jct. Hwy. 81 and FM 434 (16); (6) Waco, 1.1 km ENE Jct. Hwy. 81 and FM 434 (10); (7) 8.5 km WNW Riesel (5); (8) 4.8 km N Downsville (4); (9) 5.0 km NE Asa (5). Falls Co.: (10) 7.3 km NW Marlin (2); (11) 7.3 km W Marlin (2); (12) 3.4 km S Satin (2); (13) 10.1 km WSW Marlin (3); (14) 11.3 km E Chilton (9); (15) 11.3 km E Chilton (7); (16) 4.8 km SSW Marlin (8); (17) 4.8 km SSW Marlin (11); (18) 7.6 km W Reagan (7); (19) 2.6 km E Cedar Springs (8); (20) 9.7 km SW Reagan (11); and, (21) 9.7 km SW Reagan (13).

Blood samples and homogenates of kidney and liver were prepared and electrophoresed according to the methods described by Selander et al. (1971) and Kilpatrick and Zimmerman (1975). A 12% concentration of hydrolyzed starch (Sigma Chemical Co., St. Louis, Mo.) was used. Alleles were designated alphabetically in order of decreasing mobility.

Proteins encoded by 19 loci were examined as follows: lactate dehydrogenase (LDH-1, LDH-2), 6-phosphogluconate dehydrogenase (6-PGD-1), isocitrate dehydrogenase (IDH-1, IDH-2), glutamate oxaloacetate transaminase (GOT-1, GOT-2), phosphoglucose isomerase (PGI-1), malic enzyme (MOD-1), leucine aminopeptidase (LAP-1), indophenol oxidase (IPO-1), malate dehydrogenase (MDH-1, MDH-2), α-glycerophosphate dehydrogenase (α-GPD-1), three esterases (ES-3, ES-4, ES-8), and hemoglobin (HB-α, HB-β).

The normalized identity (I) between populations at the j locus was calculated as

$$I_j = \frac{x_i y_i}{(\Sigma x_i^2 \Sigma y_i^2)}$$

where x_i and y_i are the frequencies of the ith allele in populations x and y (Nei, 1972). The mean genic identity over all loci is

$$I = \frac{J_{xy}}{(J_x J_y)^{1/2}}$$

where J_x and J_y are means of x_i^2 and y_i^2, respectively. The average genetic distance per locus was estimated as $D = -\log_e I$. D is useful in that it is a measure of the average number of electrophoretically detectable allelic substitutions that have occurred since the populations in question diverged from a common ancestor.

The standardized variance of frequencies at polymorphic loci across populations was calculated as

$$F_{ST} = S_p^2/\bar{p}(1-\bar{p})$$

where p is the frequency of the allele at each sample site, and S_p^2 is the unstandardized variance of p. Effort was made to sample all populations completely to establish effective population sizes (uneven sex ratios considered) to characterize effective migration by solving for m in the following equation:

$$F_{ST} = 1/(4N_e m + 1)$$

where N_e is the effective population size and F_{ST} is the previously determined standardized variance of frequencies (Nei, 1975).

Multivariate biometric routines were used to analyze genic data among populations using NT-SYS programs (Rohlf et al., 1969). Matrices of Pearson's product-moment correlations were computed and phenetic distance coefficients were derived from character values (gene frequencies). Populations were clustered using unweighted pair-group method on the correlation matrix. We extracted the first three principal components, and projections of localities were prepared from these data. A shortest minimally connected network was computed in the original character space and was utilized to connect most similar localities in the three-dimensional plot.

RESULTS AND DISCUSSION

Correlation of Chromosomal and Allozymic Data

Significant allozymic differences existed between chromosomal races D and E discussed by Honeycutt (1978). Race D has a diploid number of 72 and an autosomal arm number of 70, with 35 pairs of large to small acrocentric autosomes. The X chromosome is a large acrocentric, and the Y chromosome is a small acrocentric. Race E has a diploid number of 72 and an autosomal arm number of 74. The karyotype is characterized by one pair of large submetacentric, one pair of medium submetacentric, and one pair of small metacentric autosomes. The X chromosome is a medium submetacentric, and the Y chromosome is a small acrocentric. Honeycutt (1978) reported that race D reaches its southern limits along the Brazos River in southern McLennan County, Texas. Race E occurs along both banks of the Brazos River in Falls County, Texas and southeast Texas. A hiatus of 12.8 km exists between the two chromosomal races along the McLennan-Falls County line. This corresponds to the hiatus of 14 km separating populations we examined from this same area. No clear-cut habitat differences occur in this area, and the work by Honeycutt (1978) and our subsequent collecting did not establish any causal factors for this differentiation. We have separated our assessment of allozymic variation for the two chromosomal races in the remainder of our analysis.

Seven loci were monomorphic in all populations: MDH-1, IDH-2, LDH-2, LAP-1, IPO-1, and the two hemoglobin loci. Variation at the 12 remaining loci can be assigned to one of two classes of allozymic variation among the populations. The first class includes variable loci which are polymorphic in most or all populations but do not serve to differentiate between the two chromosomal forms; these loci included PGI-1, MOD-1, MDH-2, α-GPD-1, GOT-1, 6-PGD-1, ES-3, and ES-4. This is not to say that these loci do not show a great deal of heterogeneity among and between the chromosomal forms. In fact, loci such as PGI-1, MDH-2, and GOT-1 demonstrated the highest degree of heterogeneity among loci in this class with fixation of alternate alleles being common (Figs. 1B and 1C). The second class of variable loci included those which were significantly heterogeneous between the two chromosomal races (Figs. 1C and 1D). The most striking variation occurred at the LDH-1 locus, where the $Ldh\text{-}1^{b}$ allele was fixed in all populations of chromosomal race D, while the $Ldh\text{-}1^{a}$ allele was fixed in those of chromosomal race E. Although the pattern of variation did not involve complete fixation of alternate alleles, loci such as IDH-1 and ES-8 showed distributions of predominant alleles more or less characteristic of one or the other chromosomal race. Therefore, an abrupt geographic shift in allelic frequencies existed for three loci, and this shift occurred at the zone of contact of the two chromosomal races.

Nei's (1972) genic identity (I) over all loci calculated for paired combinations of the 21 populations further exemplified the allozymic differences between the two chromosomal races (Table 1). Mean genic identity between the two races was 0.696, with $\bar{D}$ = 0.362, while mean I values among populations of the chromosomal races were 0.891 for race D and 0.828 for race E. Clearly, the differences in chromosomes varied concomitantly with allozyme frequencies.

Additional analysis of this differentiation is represented by projecting the 21 populations with respect to the first three principal components extracted from a matrix of correlations among allelic frequencies (Fig. 2). Loadings, indicating the correlations of characters with the first three principal components, are given in Table 2. The first three components explained 65.7% of the total interlocality variance. The cophenetic correlation of the model was 0.95, indicating that reducing the 32-dimensional character space to three dimensions resulted in some distortion of the original distances.

Component I was influenced by the expression of alternate alleles at the LDH-1, ES-8, and IDH-1 loci and effectively separated populations of chromosomal race D from those of chromosomal race E. Interestingly, populations 10 through 13 of chromosomal race E projected the farthest from populations 1 through 9 to which they were the closest geographically. Component II, influenced by the GOT-2 locus and frequencies of the 6-Pgd^{c} and Es-3^{a} alleles, further separated populations 10 through 13 from populations 17 through 21 of chromosomal race E; however, this separation may be only superficial since Component II accounted for only 21.1% of the total interlocality variance. Component III was influenced by alternate alleles expressed by the 6-PGD, PGI-1, and ES-3 loci, and extensive polymorphism at these loci was reflected by projections of this principal component.

The $\bar{D}$ of 0.362 between the two chromosomal races of *G. bursarius* permits the interpretation of the degree of genomic modification since these forms diverged, and approximately 36.2 allelic substitutions per 100 loci have occurred. This value is not appreciably higher than a D of 0.345 reported between five species of *Geomys* by Penney and Zimmerman (1976). The concordance

Table 1. Genic identity among 21 local populations of two chromosomal races of the plains pocket gopher, *Geomys bursarius*.

	Race D								Race E											
	2	3	4	5	6	7	8	9	10	11	12	13	14	15	16	17	18	19	20	21
1	.955	.988	.780	.836	.780	.815	.868	.831	.553	.648	.616	.597	.792	.793	.801	.808	.739	.747	.697	.697
2	---	.941	.857	.880	.857	.867	.880	.890	.555	.634	.599	.580	.750	.777	.778	.832	.786	.793	.757	.754
3		---	.750	.817	.750	.780	.850	.800	.557	.643	.619	.600	.788	.786	.804	.774	.716	.726	.676	.676
4			---	.987	1.00	.914	.920	.945	.684	.669	.640	.621	.676	.709	.669	.697	.662	.641	.705	.694
5				---	.987	.937	.962	.966	.697	.719	.688	.668	.730	.756	.719	.712	.663	.652	.683	.671
6					---	.914	.920	.945	.684	.669	.640	.621	.676	.709	.669	.697	.662	.641	.705	.694
7						---	.962	.973	.624	.757	.688	.668	.693	.723	.691	.725	.680	.699	.660	.634
8							---	.981	.679	.753	.743	.724	.748	.767	.732	.718	.689	.701	.659	.646
9								---	.648	.730	.710	.691	.724	.758	.718	.747	.701	.712	.666	.656
10									---	.888	.933	.941	.785	.784	.777	.694	.743	.709	.757	.770
11										---	.928	.936	.829	.839	.831	.753	.699	.705	.647	.646
12											---	.981	.852	.849	.842	.760	.776	.774	.712	.726
13												---	.831	.829	.823	.739	.756	.754	.692	.707
14													---	.993	.989	.905	.812	.810	.753	.764
15														---	.989	.914	.824	.822	.761	.774
16															---	.933	.848	.847	.783	.797
17																---	.922	.921	.852	.870
18																	---	.993	.957	.971
19																		---	.937	.940
20																			---	.986

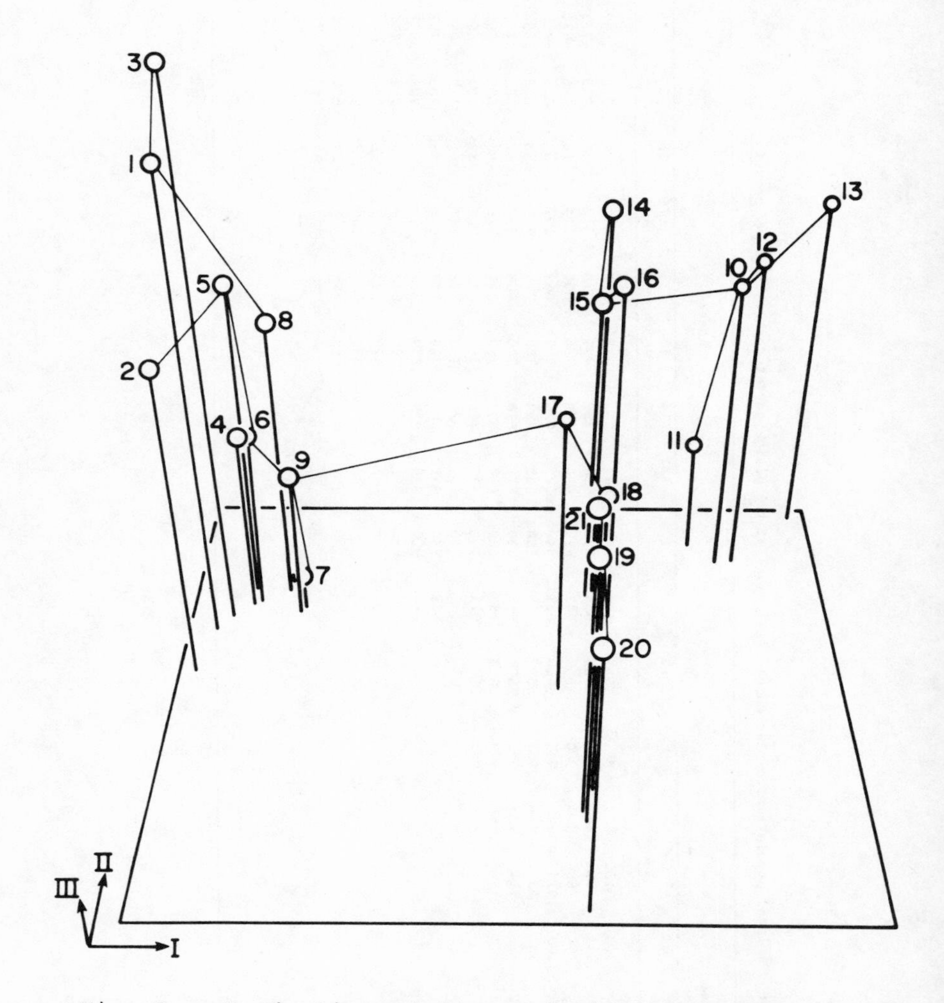

Fig. 2. Projection of 21 populations onto the axes of the first three principal components of variation in the matrix of correlations of 32 allelic frequencies for *Geomys bursarius*.

Genic Heterogeneity Among Pocket Gophers

Table 2. Character loadings on the first three principal components of interlocality phenetic variation in *Geomys bursarius*.

Characters*	Principal Components		
	I	II	III
Ldh-1[a]	.948	-.233	.068
Ldh-1[b]	-.948	.233	-.068
Es-8[a]	-.915	.306	-.161
Es-8[b]	.915	-.306	.161
Idh-1[a]	-.937	.145	-.174
Idh-1[b]	.873	.248	.203
Got-2[a]	.097	.769	-.457
Got-2[b]	-.091	-.786	.449
Es-3[a]	.134	-.739	-.015
Es-3[b]	-.117	.129	.523
Es-3[c]	.057	.219	-.536
Pgi-1[a]	-.441	-.651	-.495
Pgi-1[b]	.441	.651	.495
6-Pgd-1[a]	-.530	.117	.696
6-Pgd-1[b]	.513	.009	-.698
6-Pgd-1[c]	.123	-.694	-.024

*Abbreviations are identified in text.

of chromosomal and allozymic data for these chromosomal races is indiative of a greater amount of genomic modification than expected for conspecific populations. Additional investigation is necessary to clarify the systematic status of these forms.

Until the systematic status of these chromosomal races is established, it seems appropriate to provide a suitable designation to such genetically fragmented populations. Levins (1970) introduced the term "metapopulation" to describe a cluster of populations belonging to the same species spread over a fixed number of patches. Extinction of populations in old patches and colonization of empty patches by immigrants results in a pattern of occupancy which is constantly changing. This is confirmed by Kennerly (1954) who described the distribution of *G. personatus* populations as an intricate "mosaic pattern," with populations displaying an island-model pattern of distribution (Wright, 1943). Under Wright's (1943) model, this pattern of distribution with limited migration between "islands" should result in genic divergence, as has been demonstrated in *Geomys* populations by Selander et al. (1974), Penney and Zimmerman (1976), and by our data (Fig. 1). Hence, *Geomys* populations can be discussed in terms of a group of genically fragmented islands or metapopulations. Interpretation of phenomena such as inbreeding, migration, and effect of barriers on genic structuring within these two metapopulations will have more biological meaning in light of this designation.

Genomic Structure in Geomys Metapopulations

To further elucidate factors involved in genomic structure and genic heterogeneity within individual metapopulations, we calculated the standardized variance of gene frequencies (F_{ST}) according to Wright (1965). These values were calculated for demes having a population size of 12 or greater for six loci demonstrating heterogeneity within each metapopulation (Table 3). Estimates of this measure had a mean of 0.520 for both metapopulations. Values of F_{ST} ranged from 0.030 at the 6-PGD-1 locus to 1.00 at the GOT-2 locus in race D, while F_{ST} ranged from 0.147 at the 6-PGD-1 locus to 1.00 at the PGI-1 locus in race E. Similarity of F_{ST} values among demes of both metapopulations is suggestive of a great deal of similarity in demographic patterns and population structure.

Unfortunately, there are few reports of F_{ST} values for other organisms. However, for the few standardized variances available, the values for the geomyids are among the highest. An F_{ST} of 0.412 was reported by Patton and Yang (1977) for polymorphic loci in *Thomomys bottae* populations over a wide geographic area. F_{ST} values were also high on a regional basis in *T. bottae* from California (0.321), Arizona (0.198), and New Mexico (0.361). These values can be compared to those of 0.03 for *Drosophila pseudoobscura* (Lewontin, 1974), 0.148 for humans (Cavalli-Sforza, 1966), 0.043 and 0.291 for two species of minnows, genus *Campostoma* (Zimmerman et al., 1980), and 0.421 for darters, genus *Etheostoma* (Echelle et al., 1976).

High F_{ST} values can be explained by the effects of inbreeding (Wright, 1965; Lewontin and Krakauer, 1973;

Table 3. Mean and variance of allele frequencies and estimates of inbreeding (F_{ST}) for populations of *Geomys bursarius*.

Locus and Allele	Mean Gene Frequency Chromosomal Race		Variance Chromosomal Race		F_{ST} Chromosomal Race	
	D	E	D	E	D	E
6-PGD-1						
a	.288	.040	.114	.007	.558	.174
b	.709	.935	.114	.010	.551	.160
c	.003	.025	.000	.004	.030	.147
GOT-1						
a	.689	.582	.219	.217	1.00	.893
b	.311	.418				
ES-3						
a	.009	.061	.000	.017	.047	.304
b	.908	.852	.048	.053	.576	.418
c	.083	.087	.049	.047	.643	.589
ES-8						
a	.989	.937	.001	.023	.101	.381
b	.011	.063				
IDH-1						
a	.954	.036		.007		.193
b	.046	.797	.011	.093	.251	.578
c	.000	.167		.060		.436
PGI-1						
a	.716	.424	.121	.259	.595	1.00
b	.284	.576				
MDH-2						
a	.897	.375	.027	.233	.287	.994
b	.103	.625				
α-GPD-1						
a	.684	.726	.198	.198	.915	.997
b	.316	.274				
			Weighted Mean F_{ST}		.465	.575

Selander and Kaufman, 1975). Since F_{ST} is an index of probability that allelic frequencies at a given locus in populations are not drawn from one panmictic population, the higher the F_{ST}, the greater the effect inbreeding may have on increasing interdemic heterogeneity. The high F_{ST} values established for pocket gophers suggest that inbreeding in small, isolated demes is appreciable and contributes significantly to genic heterogeneity among demes.

Heterogeneity is a function of effective population size and migration rate. Examination of the equation

$$F_{ST} = 1/(4N_e m + 1)$$

reveals that the degree of differentiation of gene frequencies increases as the product of effective population size (N_e) and migration rate (m) decreases. We will attempt to establish that the product of these two parameters is indeed small and contributes to the high levels of heterogeneity. Since our sampling was designed to collect all individuals of a local population when possible, we were able to estimate the effective population size (with unequal sex ratios) of local demes to be 12.01 (Wright, 1931). Using this value of N_e, the F_{ST} values for the polymorphic loci, and solving for m, we obtained mean migration rates in metapopulations D and E of 0.024 and 0.015, respectively. In terms of a migration rate, an average of about 0.02 migrants occurred per generation, a level which is indicative of a restriction of gene flow among local demes.

A final analysis of the effect of a barrier, the Brazos River, on genic differentiation in *Geomys* can be made from our design of a double-dendritic distribution pattern. Demes within their respective metapopulations and occurring on the same side of the Brazos River had a $\bar{I} = 0.928$, while those from opposite banks of the river had a $\bar{I} = 0.962$. A comparison of deme location (Fig. 1A) and pair-wise combinations represented in the dendrogram in Fig. 3 demonstrated demes from opposite sides of the river occurred in more closely paired combinations than did adjacent demes on the same side. Kennerly (1963), commenting on the swimming ability of pocket gophers, depicted this double-dendritic distribution as a ladder with the rungs formed at sites where occasional crossings occur. Kennerly also established that pocket gophers are "rather vigorous swimmers." Our findings

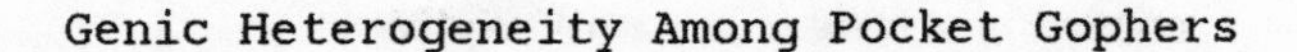

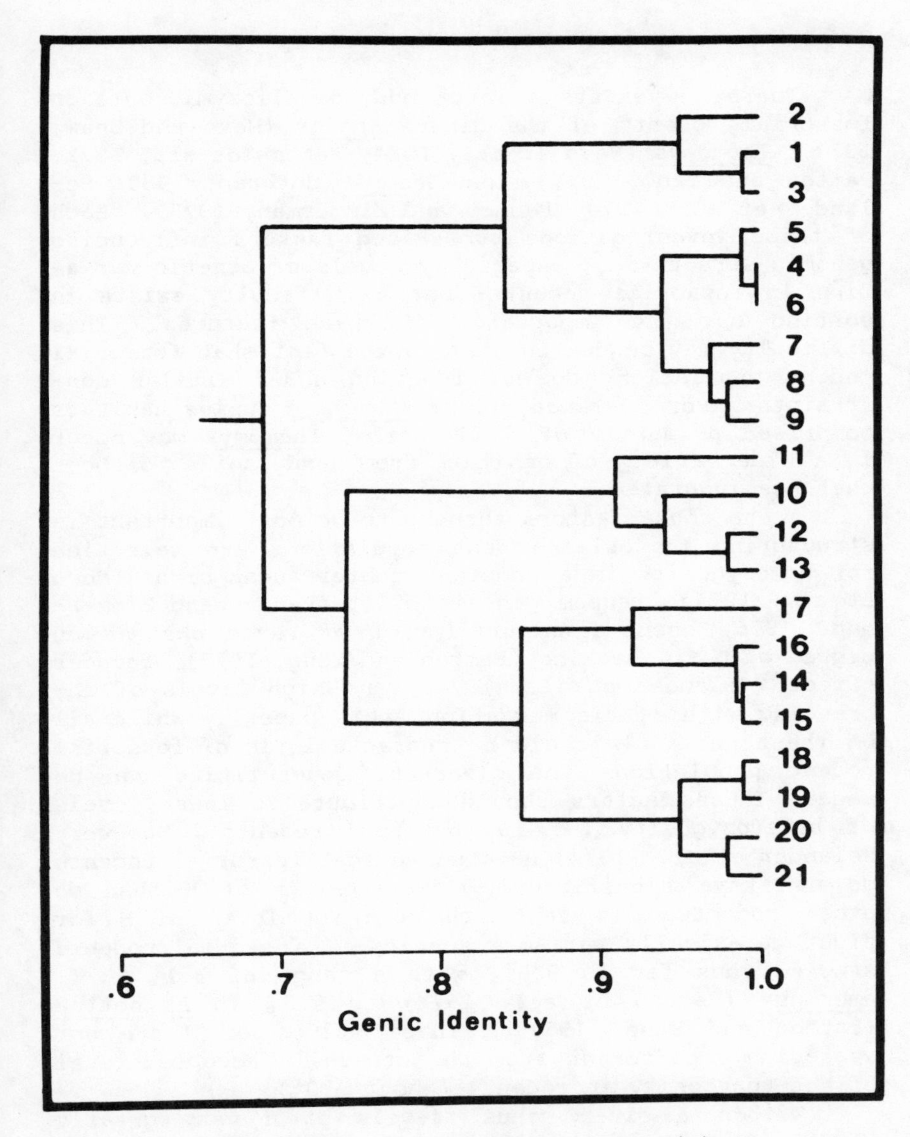

Fig. 3. Dendrogram with genic identities (I) for 21 local populations of *Geomys bursarius*.

support Kennerly's model but are contrary to previous reports that waterways act as effective barriers to migration in geomyids (Davis, 1940).

Genic Evolution in Fossorial Rodents

There now exists a large body of allozymic data on fossorial rodents of the genera Spalax (Nevo and Shaw, 1972), Thomomys (Nevo et al., 1974; Patton et al., 1972; Patton and Yang, 1977), and Geomys (Cothran, 1980; Selander et al., 1974; Penney and Zimmerman, 1976). Each of these investigations summarized factors influencing genomic structuring, especially levels of genetic variation in fossorial rodents; but a difficulty exists in sorting out the importance of these features. This difficulty may be due in part to the fact that fossorial rodent populations do not function under similar constraints. For instance, while Geomys occupies habitats comprised primarily of sandy soils, Thomomys may occur in a wide variety of habitats from sandy soils to those that are indurate.

Among those factors thought to be most important in structuring fossorial rodent populations are selection for homozygosity in a constant subterranean niche (Nevo et al., 1974), random genetic drift (Penney and Zimmerman, 1976), gene flow and impact of range changes on degree of bottlenecking (Patton and Yang, 1977), founder effect (Selander et al., 1974), and high levels of inbreeding with little migration (this paper). While all of these certainly could be characteristic of fossorial rodent populations, no clear-cut generalities can be made. These factors should contribute to lower levels of heterozygosity (H) in fossorial rodents. However, Selander et al. (1974) have shown that fossorial rodents do not have significantly lower levels of H than do other rodents. In fact, the overall level of $\bar{H}$ for eight genetically variable species of fossorial rodents studied thus far is 5.6%, with a range of 3.1% in T. umbrinus (Patton et al., 1972) to 9.3% in T. bottae (Patton and Yang, 1977). These values of $\bar{H}$ are not appreciably different from the generally accepted level of heterozygosity in rodents (about 6.0%).

Within a given genus, levels of $\bar{H}$ vary greatly also. $\bar{H}$ in the genus Thomomys ranges from 3.1% in T. umbrinus (Patton et al., 1972) to 9.3% in T. bottae (Patton and Yang, 1977), with T. talpoides having an intermediate $\bar{H}$ of 4.7%. Within G. bursarius, levels of H were rather uniform over a broad geographic area, and $\bar{H}$ varied from 2.3% (this paper) to 5.6% (Cothran, 1980), with a mean of 3.9%.

Genic Heterogeneity Among Pocket Gophers

There appears to be little support for the contention that selection in a uniform habitat plays an important role in genomic structuring among fossorial rodents. The high levels of genic heterogeneity characteristic of many fossorial rodent populations occupying similar habitats seem to contradict this as a mechanism where homogeneity would be expected. The effect of random genetic drift and degree of bottlenecking would appear to play some role in genomic structuring and are in need of additional study.

Most of the evidence accumulated thus far suggests that restricted gene flow and high levels of inbreeding may be major factors contributing to heterogeneity among populations of geomyids. Patton and Yang (1977) contend that gene flow in *Thomomys* is rather extensive only over areas of continuous suitable habitat and is "minimal at present." If heterogeneous standardized variances of gene frequencies (F_{ST}) are a reflection of high levels of inbreeding, there is a good indication that inbreeding may be an important force structuring geomyid populations. Values of F_{ST} for *T. bottae* (Patton and Yang, 1977) and *G. bursarius* are among the highest reported for any organism. The combination of small effective population size and low rate of migration undoubtedly contribute to the high level of inbreeding, a phenomenon that should be investigated in other fossorial rodents.

Acknowledgments--Dr. Michael Kennedy kindly performed the multivariate analysis and provided insight into its application to the genic data. We express our appreciation to Stuart Calhoun for aiding in the collection of specimens. This research was supported by Faculty Research Grant 34888 from North Texas State University.

PHENETIC AND CLADISTIC ANALYSES OF BIOCHEMICAL EVOLUTION IN PEROMYSCINE RODENTS

John C. Patton, Robert J. Baker and John C. Avise

Abstract--An approach to applying principles of Hennigian cladistics to qualitative, uncoded electrophoretic data is suggested. This approach is contrasted with more conventional phenetic analyses, using a data base of newly collected biochemical information on nine genera and 14 species of New World rodents, Cricetidae. Phenetic dendrograms and cladistic trees based on protein electrophoretic data were evaluated against a model phylogeny for these taxa derived from nonprotein information. In two cases, the cladistic analysis of our electrophoretic data gives a somewhat better fit to the model phylogeny than does the phenetic analysis: (1) In suggesting only a distant affiliation between *Ochrotomys* and *Neotoma*, and (2) in defining *Peromyscus*-*Neotomodon* as a distinct clade. In most other respects, the phenetic and cladistic interpretations fit the model phylogeny equally well. Whether or not this cladistic method of analysis will ultimately prove more precise than standard phenetic approaches, it does offer advantages of greater testability since individual character states are defined along all branches of a tree. Our results indicate that electrophoretic data provide useful systematic information even at intergeneric taxonomic levels.

INTRODUCTION

In the past decade numerous authors have documented the power and efficiency of protein electrophoresis to examine genetic relatedness of populations (Avise, 1974). While the general utility of the electrophoretic approach is no longer in serious question, one remaining area of contention concerns how best to analyze and interpret data derived from these techniques. At least two basic philosophies are prevalent: phenetics and cladistics.

Phenetic and Cladistic Analyses

Pheneticists conclude that it is difficult or impossible to be certain whether a phylogenetic reconstruction is correct and that classifications can only be based on how similar (genetically or otherwise) organisms are to one another (Sokal and Sneath, 1963; Sneath and Sokal, 1973); however, cladists argue that classifications should be based on genealogical affinities which can in fact be determined (Farris et al., 1970; Nelson, 1972, 1973; Cracraft, 1974). Proponents of both philosophies have suggested algorithms for data analysis which can be applied to electrophoretic information. Often, Q-type phenetic analyses involve manipulation of quantitative values in similarity or distance matrices which index overall phenetic resemblance. Output in the form of dendrograms summarizies the phenetic relationships among organisms. On the other hand, cladistic analyses may employ original or coded qualitative characters to define presumed branch points of phylogenetic trees (Hennig, 1965, 1966). Surprisingly, Henningian principles have seldom been applied to analysis of the distribution of protein electromorphs (Mickevich and Johnson, 1976).

In this paper we introduce a new approach in applying principles of Hennigian cladistics to qualitative electrophoretic information. Data from peromyscine rodents were chosen for this purpose because this assemblage has been well studied and a model phylogeny based on nonelectrophoretic data can be proposed. Here we will compare phenetic dendrograms and cladistic trees based on protein electrophoretic data against the model phylogeny. The results should indicate whether phenetic or cladistic analyses better summarize the genealogical relationships among these peromyscine species.

Peromyscine Rodents

Hesperomyine rodents of the subfamily Cricetinae, family Cricetidae comprise an assemblage of approximately 50 genera and 350 species (Simpson, 1945; Arata, 1967). The peromyscine rodent complex consists of seven genera and over 80 species of hesperomyine rodents (Arata, 1967) found principally in the southern half of North America. These rodents appear to form a natural assemblage separate from the closely related neotomine complex. Hooper (1968) stated that sufficient evidence exists to recognize three distinctive phyletic groups of about equal taxonomic rank (the implication being the subfamilial level) in the New World: the neotomine-peromyscines, the South American cricetines, and the microtines.

This study was designed to test several hypotheses about relationships within the peromyscine complex and the ability of electrophoresis to elucidate intergeneric relationships. The neotomine-peromyscine taxa have been studied by numerous researchers employing a variety of techniques. A resulting model phylogeny derived largely from chromosomal data is given in Fig. 1. Data on G-banded chromosomes (Yates et al., 1979 and Baker et al., 1979) indicate the 2n = 48 genera (Baiomys, Onychomys, Neotomodon, and Peromyscus) form a natural assemblage. The basal diploid number appears to be 2n = 52 for Neotoma, Ochrotomys, and Reithrodontomys (Mascarello and Hsu, 1976; Patton and Hsu, 1967; Carleton and Myers, 1979). The cladogram shown in Fig. 1 does not resolve the relationship between Reithrodontomys, Ochrotomys and Neotoma. However, several lines of evidence support Reithrodontomys and Ochrotomys as peromyscine forms (Rinker, 1963; Hooper and Musser, 1964; Arata, 1964).

The South American cricetines Sigmodon and Oryzomys were added to this study to assess possible conservation of alleles (electromorphs) across higher taxonomic levels. If such alleles are observed, this will allow the proposal of plesiomorphic characters from which to develop a phylogenetic tree (Hennig, 1966). They may also be used to gain insight into the relative time of divergence of the neotomine-peromyscine forms from one another with respect to time of divergence from the South American cricetines.

Thus, the main purposes of this study are (1) to further assess the value of electrophoretic data as a taxonomic tool, particularly at higher levels of evolutionary divergence; (2) to introduce a Hennigian cladistic procedure to analyze electrophoretic data; (3) to contrast phenetic and cladistic approaches to analyses of those data; and, (4) to provide new information on evolutionary relationships among peromyscine rodents.

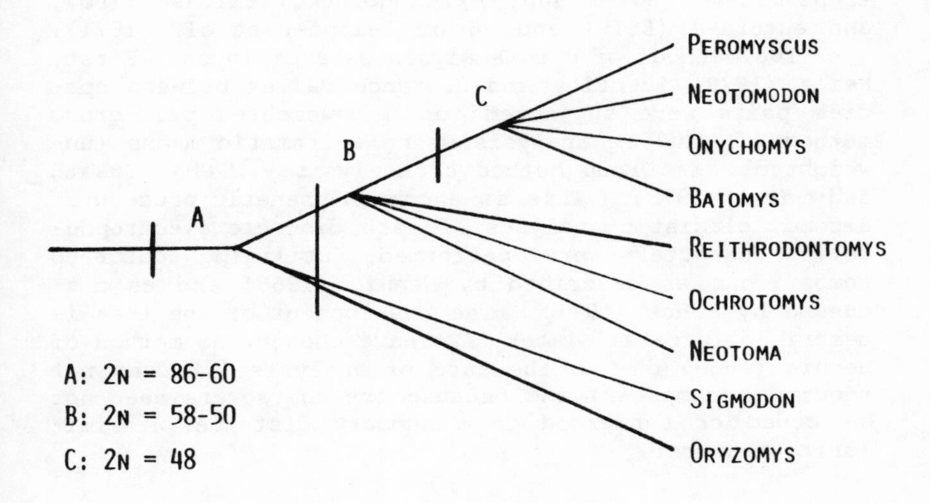

Fig. 1. Cladogram representing the model phylogeny for cricetine rodents examined, determined largely from chromosomal data. Level A represents clades which likely diverged at 2 n = 86-60. Level B represents clades which appear to have diverged between 2n = 58-50. Level C represents 2n = 48 genera.

MATERIALS AND METHODS

Specimens were live trapped from natural populations (see SPECIMENS EXAMINED for locality data). Tissue preparation, electrophoretic conditions, and staining procedures were basically as described by Selander et al. (1971), Ayala et al. (1972), Ridgeway et al. (1970), Shaw and Prasad (1970) and Smith et al. (1973). Proteins were subjected to an electric current for longer times than originally suggested and were scored on as many buffer types as possible. Allelic designations basically followed those of Smith et al. (1973) and Avise et al. (1974). The 10 proteins (15 loci) examined were lactate dehydrogenase (LDH-1 and -2), malate dehydrogenase (MDH-1 and -2), malic enzyme (ME), isocitrate dehydrogenase (IDH-1 and -2), 6-phosphogluconate dehydrogenase (6-PGD), sorbitol dehydrogenase (SDH), creatine kinase (CK-2), glutamate oxalate transaminase (GOT-1 and -2), indophenol oxidase (IPO), and esterase (EST-1 and -6 of Selander et al., 1971).

Two methods of data analysis were employed. First, Nei's (1972) identity and distance values between species pairs were subjected to an unweighted pair-group method of cluster analysis using arithmetic means (Unweighted Pair Group Method of Analysis - UPGMA; Sneath and Sokal, 1973). This is a common phenetic procedure. Second, cladistic analyses of each discrete electrophoretic character were performed, utilizing outgroup comparisons as described by Hennig (1966) and as discussed by Bonde (1977). The development of the tree is described in detail later. We have chosen the method of Hennig (1966) due to the ease of analysis (it does not require a computer) and because the characters need not be coded or submerged in a summary distance or similarity measure.

RESULTS

Electromorph mobilities and electrophoretic conditions for 15 loci are given in Table 1. Banding patterns and tissue specificities were similar to gene products previously reported in _Peromyscus_ (Selander et al., 1971; Mascarello and Shaw, 1973). Gels stained for creatine kinase (CK) showed two areas of activity in samples from heart and kidney. The most anodal band (CK-1) was indistinct and could not be scored reliably. The more cathodal zone of activity (CK-2) banded sharply but no variation was found within or between the taxa analyzed. All other loci demonstrated interspecific variation.

Phenetic Analysis

Table 2 lists genetic identities ($\bar{I}$) and distances ($\bar{D}$) between species, computed according to Nei's (1972) formulae. These data were summarized in a dendrogram which provides a two-dimensional representation of phenetic relationships among the taxa (Fig. 2).

As noted in Table 1, only 13 loci were analyzed in _N. alstoni_; hence, three different estimates of $\bar{I}$ and $\bar{D}$ were obtained. First, the 13 loci analyzed for _N. alstoni_ were examined over all taxa. The second esti-

Table 1. Designations of electromorphs of loci examined using <u>Peromyscus polionotus</u> as the standard. Electromorphs taken from Avise et al. (1979a) are enclosed in parentheses. Species abbreviations are as given in Table 2 and locus abbreviations in text.

Locus	Species														Buffers*	Tissue**
	P.m.	P.p.	P.l.	P.f.	N.a.	O.l.	O.t.	B.t.	R.f.	R.m.	O.n.	N.m.	S.h.	O.p.		
CK-2	100	100	100	100	100	100	100	100	100	100	100	100	100	100	7,5	H&K
MDH-2	-100	-100	-100	-100	-100	-100	-100	-100	-100	-100	-80	-80	-80	-80	1,6,4	H&K
LDH-2	100	100	100	100 30	100	20	20	20	20	20 105	20	20	20	120	1,6,2,4,5	H&K,He
MDH-1	100	100	100	100	100	100	100	100	70	70	100	100	90	100	1,6,4	H&K,He
IDH-2	-100	-100	-100	-100	-100	-50	-50 20	-50	-50	-50	-50	-120	-120	-120	1,6,4	H&K
GOT-2	-100	-100	-100	-64	-64	-100	-100	-75	-64	-64	-100	-100	-64	-70	1,2,6	H&K,L
EST-6	100 102	100	100	100		100	100	100	95	65	100	102	100	94	4	He
IDH-1	100	100	100	127	127	100	100	60	100	100	110	100	135	120	1,6,4	H&K
GOT-1	100 130	100 (130)	100	170 (130)	170	100	100	170	100	100	170	170	175	180	1,2,6	H&K,L
SDH	100	100	5	100	100	100	100	350	100	100	-10	-50	140	120	1,2,6	H&K
ME	100	100	120	100	200 175	160	180	40	165	105	90	168	170	210	1,4	H&K
IPO	100	100	100 60	130	100	100	100	-20	10	10	40	105	95	45	5,7,2,6	H&K,L
6-PGD	100 (117)	100 (117)	117	139	139	132	150	95 85	90	141 135	125	160	104 82	134	6,1	H&K,He
EST-1	100 90	100	90 120	98		50 40 30 10	50 40	70	55	65 72	102	75	25	60	3	He
LDH-1	98	100	98	95	99	98.5	98.5	99	99	99	85	110	99	99	4,1,6,2,5	H&K

*1 = T.C. 6.7/T.C. 6.3[1] 2 = T.C. 8.0[1] 3 = T HCl[1] 4 = T MAL[1] 5 = Poulik[1] 6 = JRP[2] 7 = RSL[3] with recipes given in references below.

**H&K = heart and kidney, L = liver, and He = hemolysate

[1]Selander et al., 1971. [2]Ayala et al., 1972, (modified as described in text). [3]Ridgeway et al., 1970.

Table 2. Genic similarities reported for taxa examined. Figures in upper right diagonal are Nei's $\bar{I}$ and those in lower left are Nei's $\bar{D}$.

Species*	P.m.	P.p.	P.l.	P.f.	N.a.	O.l.	O.t.	B.t.	R.f.	R.m.	O.n.	N.m.	S.h.	O.p.
P.m.		.892	.739	.516	.498	.591	.591	.248	.332	.347	.245	.280	.107	.140
P.p.	.114		.670	.526	.475	.615	.614	.270	.333	.348	.267	.267	.136	.133
P.l.	.303	.401		.403	.402	.545	.545	.278	.275	.287	.275	.275	.140	.137
P.f.	.661	.643	.909		.672	.345	.345	.341	.270	.282	.270	.202	.206	.135
N.a.	.691	.745	.918	.398		.347	.347	.343	.339	.354	.203	.203	.207	.203
O.l.	.526	.487	.606	1.064	1.058		.847	.415	.478	.481	.410	.341	.208	.137
O.t.	.526	.487	.606	1.064	1.058	.166		.396	.461	.464	.393	.341	.208	.137
B.t.	1.395	1.309	1.279	1.075	1.069	.880	.923		.338	.335	.405	.270	.275	.203
R.f.	1.102	1.099	1.292	1.311	1.082	.738	.775	1.086		.749	.267	.200	.271	.133
R.m.	1.058	1.055	1.248	1.268	1.038	.731	.769	1.094	.290		.192	.192	.266	.139
O.n.	1.407	1.322	1.292	1.311	1.593	.892	.935	.904	1.322	1.653		.467	.271	.200
N.m.	1.274	1.322	1.292	1.599	1.593	1.075	1.075	1.309	1.609	1.653	.762		.271	.267
S.h.	2.238	1.998	1.968	1.582	1.576	1.569	1.569	1.292	1.305	1.326	1.305	1.305		.271
O.p.	1.967	2.015	1.985	2.004	1.593	1.991	1.991	1.597	2.015	1.971	1.609	1.322	1.305	

* P.m. = *Peromyscus maniculatus*; P.p. = *Peromyscus polionotus*; P.l. = *Peromyscus leucopus*; P.f. = *Peromyscus floridanus*; N.a. = *Neotomodon alstoni*; O.l. = *Onychomys leucogaster*; O.t. = *Onychomys torridus*; B.t. = *Baiomys taylori*; R.f. = *Reithrodontomys fulvescens*; R.m. = *Reithrodontomys megalotis*; O.n. = *Ochrotomys nuttalli*; N.m. = *Neotoma micropus*; S.h. = *Sigmodon hispidus*; O.p. = *Oryzomys palustris*.

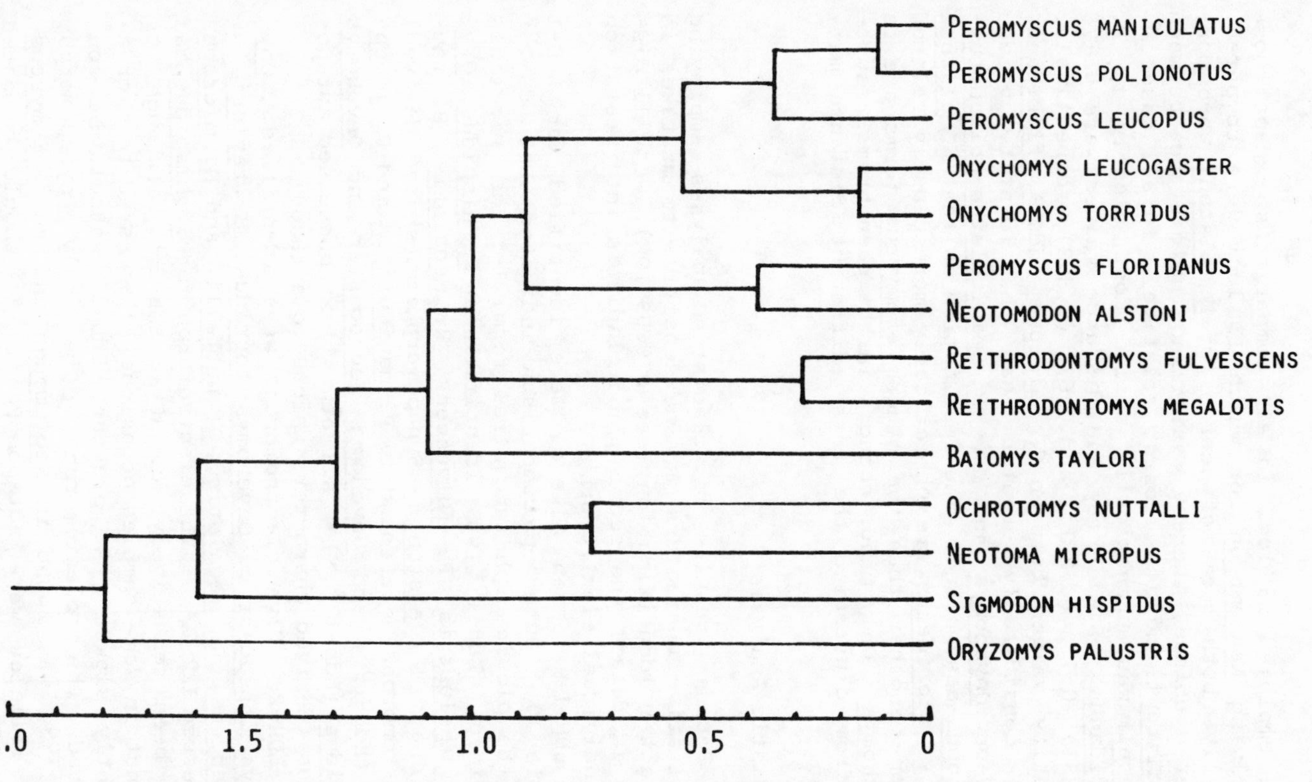

Fig. 2. Unweighted Pair Group Method of Analysis (UPGMA) dendrogram depicting possible relationships within the cricetine rodents based on genetic distances at 13-15 protein loci.

mate employed 15 loci in all comparisons except those involving _Neotomodon_ for which 13 loci were compared. Finally, both loci not scored for _N. alstoni_ were considered unique, thereby maximizing the distance between _N. alstoni_ and its closest relative. In all cases _N. alstoni_ was phenetically allied most closely to _P. floridanus_. Similarity and distance values were 0.778 (0.251), 0.778 (0.251), and 0.672 (0.398), respectively. Identity values between _N. alstoni_ and _P. floridanus_ were considerably greater than the identity values between _Onychomys_ and the _P. maniculatus-polionotus-leucopus_ assemblage. The implication is that _N. alstoni_ and _P. floridanus_ are phenetically more closely related to one another than the other _Peromyscus_ forms are to _Onychomys_. Only minor changes in the phenograms resulted from clustering the three different distance matrices.

Cladistic Analysis

Because results of cladistic analyses employing qualitative characters may be sensitive to mistakes in character identification (see DISCUSSION), we will discuss in detail the scoring and cladistic interpretation of individual electromorphs.

LDH-1: Seven alleles were identified for LDH-1 (Table 1). Outgroup comparison indicated that Ldh-1^{99} was the plesiomorph (ancestral form) for the cricetines examined. Ldh-1^{99} was identified in _S. hispidus_, _Oryzomys palustris_, _R. fulvescens_, _R. megalotis_, _B. taylori_, and _N. alstoni_. Synapomorphic (shared derived) electromorphs were found in _P. maniculatus_ and _P. leucopus_ (Ldh-1^{98}) and _Onychomys leucogaster_ and _Onychomys torridus_ (Ldh-$1^{98.5}$). All other taxa possessed autapomorphs (derived characters unique to a taxon).

LDH-2: Five electromorphs were identified. _Onychomys leucogaster_, _Onychomys torridus_, _B. taylori_, _R. fulvescens_, _R. megalotis_, _O. nuttalli_, and _N. micropus_ possessed Ldh-2^{20}. The electromorph for _S. hispidus_ has also been tentatively scored as Ldh-2^{20}, although its product in some samples occasionally appeared to migrate slightly slower. At present, Ldh-2^{20} should be considered the plesiomorph. The only synapomorph, Ldh-2^{100}, was found in _Neotomodon_ and all _Peromyscus_. Other alleles were autapomorphs. It should be noted that this scoring does not agree with that presented by

Smith et al. (1973) where Ldh-2^{95} was described in *P. floridanus*. The causes of the discrepancy are unknown.

MDH-1: Mdh-1^{100} is proposed as the plesiomorph because it was found both in the neotomine-peromyscine line and in *Oryzomys palustris*. A synapomorph (Mdh-1^{70}) appeared in the *Reithrodontomys* species. *Sigmodon hispidus* possessed an autapomorph (Mdh-1^{90}).

MDH-2: Since Mdh-2^{-80} was common to *S. hispidus*, *Oryzomys palustris*, *N. micropus*, and *O. nuttalli*, it has been designated as the plesiomorph. The synapomorph is Mdh-2^{-100}, common to all *Reithrodontomys*, *Baiomys*, *Onychomys*, *Neotomodon*, and *Peromyscus* taxa examined.

ME: Due to the rapid rate of evolution in this protein, it was impossible to designate a plesiomorph for all taxa examined. Thirteen electromorphs were identified and each was considered unique (autapomorphic) for its respective taxon. The exception involves the *Peromyscus*-*Neotomodon* complex (cladistically defined by LDH-2), within which three taxa shared Me-1^{100} (*P. maniculatus*, *P. polionotus*, and *P. floridanus*).

IDH-1: Six electromorphs have been identified for IDH-1. No electromorphs appeared to be shared between the South American cricetines and the neotomine-peromyscine taxa. Within the neotomine-peromyscine line 8 of the 12 taxa shared Idh-1^{100}: *N. micropus*, *R. fulvescens*, *R. megalotis*, *Onychomys leucogaster*, *Onychomys torridus*, *P. leucopus*, *P. polionotus*, and *P. maniculatus*. Whether Idh-1^{100} represented an apomorph or plesiomorph for the South American, neotomine-peromyscine forms cannot be resolved at present. However, Idh-1^{100} must be considered plesiomorphic for the neotomine-peromyscine complex. Idh-1^{127} appeared to be a synapomorph for *P. floridanus* and *N. alstoni*. All other derived electromorphs appeared to be restricted to a single assayed taxon.

IDH-2: Four IDH-2 electromorphs were observed. Idh-2^{-120} was shared by *N. micropus*, *S. hispidus*, and *Oryzomys palustris* and was therefore plesiomorphic. Idh-2^{-50} was shared by the *Ochrotomys*, *Reithrodontomys*, *Baiomys*, and *Onychomys* taxa, thus defining Idh-2^{50} as a synapomorph which was also the apparent plesiomorph for the peromyscine forms. Idh-2^{-100} was synapomorphic for *Neotomodon* and *Peromyscus* in the peromyscine lineage. Idh-2^{20} was unique to *Onychomys torridus*.

6-PGD: Fifteen electromorphs were identified for the taxa examined. No allele could be proposed as plesiomorphic except within the Neotomodon-Peromyscus complex. Data presented here and in Avise et al. (1974a) indicated that 6-Pgd^{117} appears synapomorphic for P. maniculatus, P. polionotus, and P. leucopus. Peromyscus maniculatus and P. polionotus shared 6-Pgd^{100}, while P. floridanus and N. alstoni shared 6-Pgd^{139}. All other alleles have been considered autapomorphic.

SDH: No electromorph was shared between the South American cricetines and the neotomine-peromyscines. Reithrodontomys megalotis, R. fulvescens, Onychomys torridus, Onychomys leucogaster, N. alstoni, P. floridanus, P. polionotus, and P. maniculatus possessed Sdh^{100}. Six other electromorphs were presumed autapomorphic.

GOT-1: Five electromorphs have been identified at this locus. Since none were shared by the neotomine-peromyscine forms and the South American cricetines, it is impossible at this time to propose a plesiomorph for the entire assemblage. Got-1^{170} was found in N. micropus, O. nuttalli, B. taylori, N. alstoni and P. floridanus, and is apparently plesiomorphic for the neotomine-peromyscine forms. Got-1^{100} was found in R. megalotis, R. fulvescens, Onychomys torridus, Onychomys leucogaster, P. leucopus, P. polionotus, and P. maniculatus and is apomorphic. Got-1^{130} was found only in P. maniculatus in the specimens examined here, but Smith et al. (1973) reported the electromorph in both P. polionotus and P. floridanus. Got-1^{130} must also be considered apomorphic. Sigmodon hispidus and Oryzomys palustris possessed autapomorphs.

GOT-2: Got-2^{-64} was present in both the neotomine-peromyscine forms and S. hispidus and is therefore considered plesiomorphic. Got-2^{-100} was found in N. micropus, O. nuttali, Onychomys torridus, Onychomys leucogaster, P. leucopus, P. polionotus, and P. maniculatus. Autapomorphic alleles were found in B. taylori and Oryzomys palustris.

IPO: No electromorph was shared between any neotomine-peromyscine form and any South American cricetine form. Onychomys torridus, Onychomys leucogaster, N. alstoni, P. leucopus, P. polionotus, and P. maniculatus all shared Ipo^{100}. Reithrodontomys fulvescens and R. megalotis shared Ipo^{70}. Seven other electromorphs were considered autapomorphs.

EST-1: This highly variable esterase was represented by 16 electromorphs in the specimens examined.

Phenetic and Cladistic Analyses

No electromorph could be designated as a plesiomorph for all taxa examined. Synapomorphs were found only for members of *Peromyscus* and *Onychomys*. All other forms possessed autapomorphic alleles.

EST-6: This erythrocytic esterase was represented by four electromorphs. Est-6^{100} was present in both major lineages examined and is therefore ancestral. All other observed alleles were autapomorphic.

DISCUSSION

Phenetic Analysis

The UPGMA dendrogram revealed a fair concordance with current ideas regarding systematic relationships of peromyscine rodents. However, at least four questionable results were apparent. First, the phenetic linking of *P. maniculatus*, *P. polionotus*, and *P. leucopus* to *Onychomys* is surprising since their supposed congener, *P. floridanus*, is phenetically much more distinct. Morphological and chromosomal data do not support these phenetic results (Hooper, 1968; Baker et al., 1979). A second possible problem is the apparent phenetic similarity between *N. alstoni* and *P. floridanus*. In this case, recent morphological and chromosomal data support such a relationship. Morphological similarities were noted between the external genitalia of *N. alstoni* and *P. floridanus* by Hooper (1959). Also, studies of G- and C-banded chromosomes of *Peromyscus*, *Neotomodon*, and *Baiomys* recently led to the conclusion that *N. alstoni* is more closely related to the *P. floridanus*, *P. maniculatus*, and *P. leucopus* groups, respectively, than to the *eremicus* group (Yates et al., 1979).

A third area of apparent disagreement is the placement of *B. taylori* outside the 2n = 48 group of rodents. If the assumptions of Yates et al. (1979) are correct that the 2n = 48 peromyscines are a monophyletic assemblage, *Baiomys* is the more distantly related (basal) form of the clade. This basal relationship is also supported by Hooper's (1959) work with external reproductive structures. If *Reithrodontomys* is the taxon most closely related to the 2n = 48 forms, a slight difference in the rates of biochemical evolution between *Baiomys* and *Reithrodontomys* (as reflected by sampling biases in loci examined) could cause the supposed reversal seen in the phenetic dendrogram.

The final concern about accuracy of the phenetic groupings is the alignment of *O. nuttalli* with *N. micropus*. Although the systematic status of *Ochrotomys* is poorly understood, evidence from other sources does not suggest placing *Ochrotomys* with neotomines.

Cladistic Analysis

Five stages in the development of the cladistic tree for peromyscine rodents are presented in Figs. 3 and 4. For reasons discussed later, data for loci were added to the analysis in the order listed in Table 1. Figure 3A shows the result of joint consideration of CK-2 and MDH-2; Fig. 3B shows the effect of the further addition of LDH-2 and MDH-1; Fig. 3C shows the effect of the addition of IDH-2, GOT-2, and EST-6 to Fig. 3B; and, Fig. 3D shows the effects of the addition of IDH-1, GOT-1, and SDH to Fig. 3C. Figure 4 shows the final cladistic tree for the 103 character states identified for the 15 loci examined.

The composite cladistic tree is composed of a stem and 24 branches (Fig. 4). The distributions of electromorphs along all stems and branches are presented in Table 3. For loci 1-7 the plesiomorph for all taxa could be designated. These loci were used to form the skeleton of the cladistic tree. Plesiomorphs at loci 8-14 could not be designated for all taxa and, therefore, characters were initially assessed as apomorphic. Data for locus 15 (LDH-1) were added after those for EST-1 because only Est-1^{100} defines the *P. maniculatus-polionotus* complex. Adding the data for LDH-1 prior to adding those for EST-1 and 6-PGD would result in a phylogenetic tree linking *P. maniculatus* to *P. leucopus* rather than to *P. polionotus*. Such a relationship is refuted by evidence from morphology (Hooper, 1968), hybridization potential (Dice, 1968), chromosome banding (Greenbaum et al., 1978), protein electrophoresis (Avise et al., 1979a; phenetic analysis in this paper), and mitochondrial DNA sequence relatedness (Avise et al., 1979b).

The distributions of electromorphs along all stems and branches are presented in Table 3. A major advantage of this cladistic analysis is that discrete character states, such as those listed in Table 3, define the

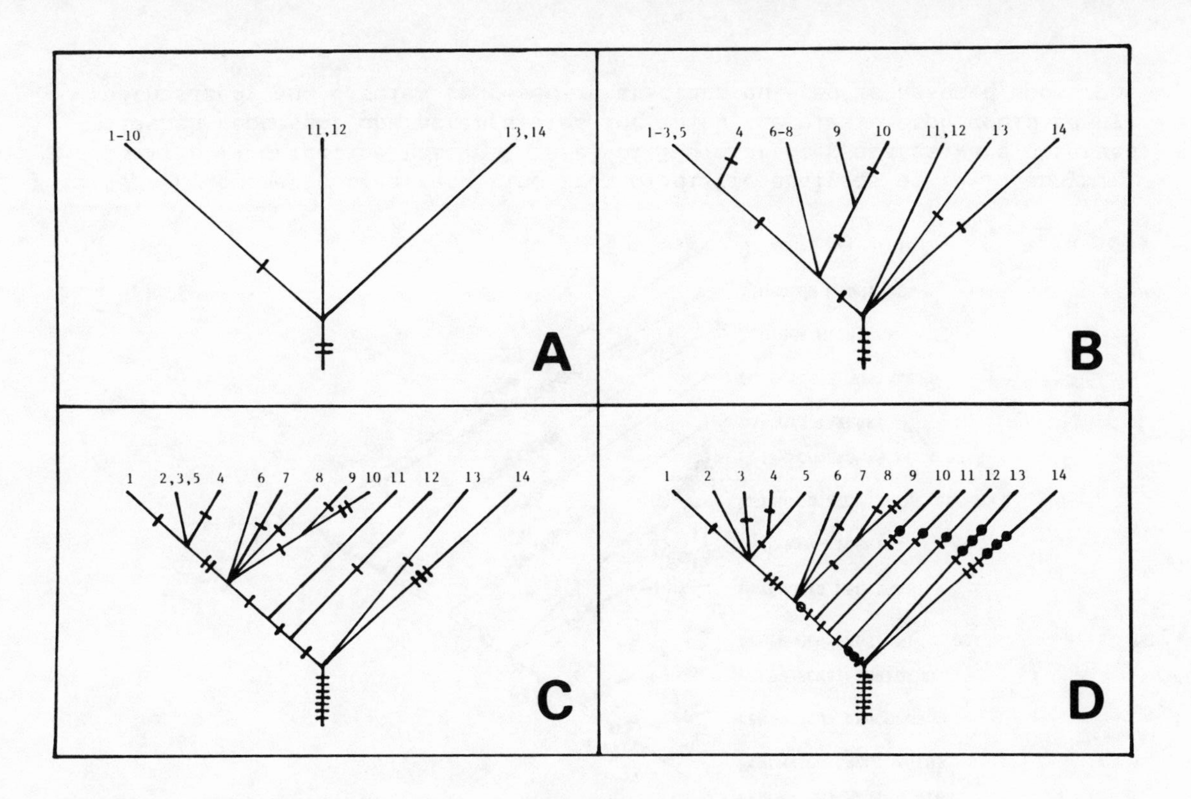

Fig. 3. Various stages in the development of the phylogenetic tree as discussed in text.

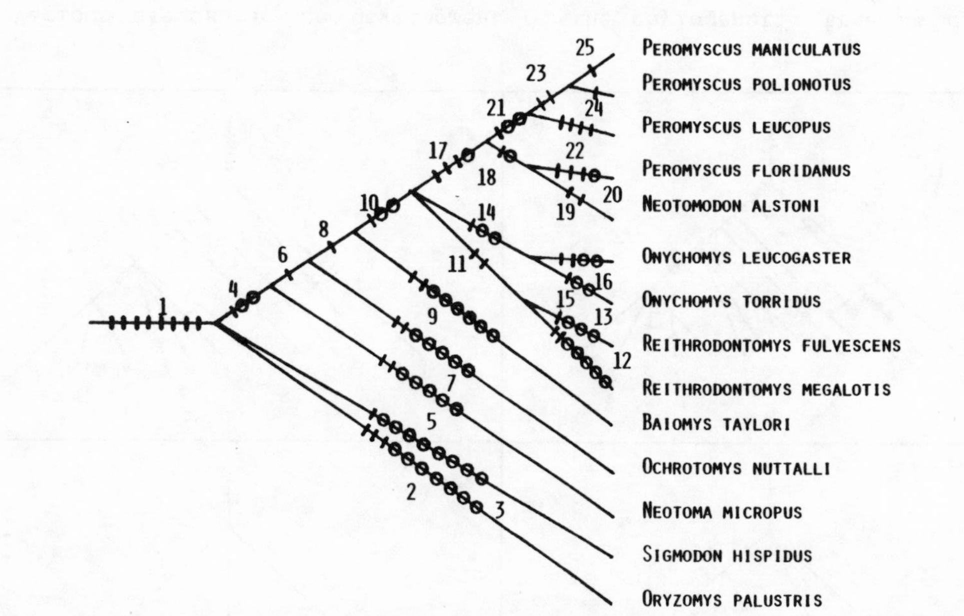

Fig. 4. Phylogenetic tree resulting from cladistic analysis of 13-15 enzyme loci in 14 cricetine taxa. A total of 103 character states were observed. Slashes represent character states for which the plesiomorph could be hypothesized, and circles represent characters entered as assumed apomorphs.

Table 3. Discrete character states identified in rodent taxa examined. Numbers correspond to stem and branches shown in Fig. 4. Item 1 lists characters considered plesiomorphic (ancestral) for all taxa examined. Items 2 through 25 list apomorphic (derived) characters which appear to define the respective clades (Fig. 4A). Characters placed within parentheses are characters considered polymorphic in stem taxa but subsequently lost in the respective clades. Characters marked with asterisks are apomorphic electromorphs found or determined to be polymorphic. Locus abbreviations are given in text.

1) Ck-2^{100}, Mdh-^{100}, Mdh-2^{-80}, Ldh-1^{99}, Ldh-2^{20}, Idh-2^{-120}, Got-2^{-64}, Est-1^{100}
2) Ldh-2^{120}, Me^{210}, Idh-1^{120}, 6-Pgd^{134}, Sdh^{120}, Got-1^{180}, Got-2^{-70}, Ipo^{45}, Est-6^{94}
3) Mdh-1^{90}, Me^{170}, Idh-1^{135}, 6-Pgd^{104}, 6-Pgd^{82}, Sdh^{140}, Got-1^{175}, Ipo^{95}, Est-1^{25}
4) Idh-1^{100}, Got-1^{170}, Got-2^{-100*}
5) Ldh-1^{110}, Me^{168}, 6-Pgd^{160}, Sdh^{-50}, Ipo^{105}, Est-1^{75}, Est-6^{102}, (Got-2^{-64})
6) Idh-2^{-50}
7) Ldh-1^{85}, Me^{90}, Idh-1^{110}, 6-Pgd^{125}, Sdh^{-10}, Ipo^{40}, Est-1^{102}, (Got-2^{-64})
8) Mdh-2^{-100}
9) Me^{40}, Idh-1^{60}, 6-Pgd^{95}, 6-Pgd^{85}, Sdh^{350}, Got-2^{-75}, Ipo^{-20}, Est-1^{70}
10) Sdh^{100}, Got-1^{100*}, Ipo^{100}
11) Mdh-1^{70}, Ipo^{10}, (Got-1^{170}, Got-2^{-100})
12) Ldh-2^{105*}, Me^{105}, 6-Pgd^{141}, 6-Pgd^{135}, Est-1^{65}, Est-1^{72}, Est-6^{65}
13) Me^{165}, 6-Pgd^{90}, Est-1^{55}, Est-6^{95}
14) Ldh-1$^{98.5}$, Est-1^{50}, Est-1^{40} (Got-1^{170}, Got-2^{-64})
15) Me^{180}, Idh-2^{20*}, 6-Pgd^{150}
16) Me^{160}, 6-Pgd^{132}, Est-1^{30*}, Est-1^{10*}
17) Ldh-2^{100}, Me^{100}, Idh-2^{-100}, Got-1^{130*}
18) Idh-1^{127}, 6-Pgd^{139}, (Got-1^{100}, Got-2^{-100})
19) Me^{200}, Me^{175}, (Got-1^{130})
20) Ldh-1^{95}, Ldh-2^{30*}, Ipo^{130}, Est-1^{98}
21) Ldh-1^{98}, Est-1^{90*}, (Got-1^{170}, Got-2^{-64})
22) Me^{120}, Sdh^{5}, Ipo^{60*}, Est-1^{120*}, (Got-1^{130})
23) Est-1^{100*}, 6-Pgd^{100}
24) Ldh-1^{100}, (Est-1^{90})
25) Est-6^{102*}

respective branches. This facilitates hypothesis testing. For example, Got-1^{100} is presently considered a synapomorph for the Peromyscus-Neotomodon-Onychomys-Reithrodontomys (PNOR) clade. The sister clade contains only B. taylori which possesses Got-1^{170}. Since Got-1^{170} is also found in O. nuttalli and N. micropus, it must be considered the plesiomorph for the neotomine-peromyscine clade. Got-1^{170} has also been identified in taxa within the PNOR clade, so this electromorph seems to have been retained in a polymorphic condition after incorporation of Got-1^{100}. If the 2n = 48 peromyscine forms are monophyletic, it is possible that the B. taylori population examined is monomorphic, or nearly so, for the ancestral allele. According to this procedure of cladistic analysis, the demonstration of Got-1^{100} in even a single individual of either B. taylori or B. musculus would define Baiomys as a member of the PNOR clade.

Both Ipo^{100} and Sdh^{100}, which also appear to distinguish the PNOR group from Baiomys, are assumed apomorphs which do not define the PNOR clade in a cladistic sense. No plesimorph can be designated below the PNOR level from which these respective alleles could have arisen.

Figure 4 represents relationships within the peromyscine rodent complex as defined by the derived character states observed. The cladistic analysis supports not only the validity of the inclusion of Neotomodon within Peromyscus, as now defined by Yates et al. (1979), but also indicates that P. floridanus and N. alstoni may have been derived from the same basal stock. As previously discussed, Baiomys is currently placed outside the PNOR clade. Cladistically peromyscine rodents form a clade distinct from neotomines. Sigmodon hispidus and Oryzomys palustris also form clades as distinct from one another as either is from the neotomine-peromyscines.

Comparison of Analyses

One important characteristic of the cladistic approach is that it attempts to trace the evolutionary history of each individual, qualitative character. Ancestral character states are distinguished from derived character states on which phylogenies must be based. Individual character states are defined along the stem and all branches of the tree. Thus, whether or not this cladistic method of analysis ultimately prove to be more precise than standard phenetic approaches, it does offer advantages of greater testability. As already illustrated, points of ambiguity in cladograms can often be recognized and the source of the dilemma specifically identified for further study.

Electrophoretic information can be cladistically analyzed because the data are expressed as discrete characters. One potential drawback is the possibility of obtaining several alternative cladistic trees from a common data base. This may result from any process leading to inconsistencies in information content of different distributions of electromorphs within a tree. For example, distinct allelic products could be scored incorrectly as synapormorphs or symplesiomorphs.

It is unclear whether sampling errors due to small sample size will be more severe for phenetic or cladistic analyses. The effect of small sample size in phenetics has recently been addressed from both theoretical and empirical standpoints (Nei, 1978; Gorman and Renzi, 1979). The bias introduced through sampling error appears not to be a serious problem, even for sample sizes as small as one or a few individuals, as long as the genetic distance ($\bar{D}$) between the OTU's (operational taxonomic units) and branch points under study is ≥ 0.15 and heterozygosity ($\bar{H}$) is ≤ 0.05. Both of these conditions held for the great majority of taxa examined in this study.

Because the present cladistic analysis utilizes a qualitative data base, it might appear to be more prone to sampling error due to small sample size. Several kinds of information could be introduced into a cladistic analysis with the discovery of additional electromorphs at the loci assayed. First, additional autapomorphic alleles might be identified. These would have no effect on the branching sequence of the cladistic tree. Second, symplesiomorphs not observed in the original sample might be found. An example from the present study would be the discovery of Got-1^{170} in *Onychomys*. This again would have no effect on the branching sequence of the cladistic tree. Third, an electromorph could be identified in clade B which was formerly considered synapomorphic or autapomorphic to clade A. This latter case could indeed cause a possible reassessment of phylogenetic relatedness of species. An example of this situation has been discussed for the Got-1 locus of Baiomys. Fourth, entirely new sets of synapomorphs could be detected, which could also cause a reassessment of the branching framework. Since the current study is exploratory in design, we suggest that future work empirically address the issue of sampling error in cladistic analyses.

Cladistic analysis of our electrophoretic data gives a somewhat better fit to the model phylogeny for peromyscine rodents than does the phenetic analysis. First, there are no cladistic characters suggesting the close affiliation between Ochrotomys and Neotoma implied by the phenetic analysis. The synapomorph Idh-2^{-50} demonstrates Ochrotomys to have cladistic affinities with the other peromyscine forms. Second, Ldh-2^{100} and Idh-2^{-100} define Peromyscus-Neotomodon as a distinct clade, an association not apparent in the phenetic groupings. In both instances, the cladistic analysis may resolve apparent discrepancies between results of the phenetic analysis and the model phylogeny. In most other respects, the phenetic and cladistic interpretations fit the model phylogeny for peromyscine rodents equally well.

Systematic Implications

The cladistic analyses of the biochemical data yielded inferences relevant to the systematics of cricetine rodents. The classically recognized split between peromyscine and neotomine rodents is supported. Ochrotomys seems to represent a basal divergence from peromyscine stock. Beyond this level, at least four distinctive clades can be recognized from the biochemical data: Baiomys, Reithrodontomys, Onychomys, and Peromyscus (including Neotomodon). The retention of Neotomodon as a genus would require the recognition of Neotomodon, Podomys (P. floridanus), and a redefined Peromyscus (including P. maniculatus, P. polionotus, and P. leucopus) as genera, or the inclusion of P. floridanus within a redefined Neotomodon. The splitting of Peromyscus would not, however, seem warranted from these data, since none of the proposed genera would be as distinct genetically as are Onychomys, Reithrodontomys, and Baiomys from one another or from Peromyscus. Thus we agree with the suggestion of Yates et al. (1979) which includes Neotomodon within Peromyscus.

Acknowledgments--We wish to express our gratitude to a number of people who aided in the acquisition of various specimens: C. F. Aquadro, J. W. Bickham, J. R. Choate, J. Laerm, J. Layne, B. S. McDonald, G. McDonald, F. J. Patton, S. J. Patton, and V. L. Patton. We especially would like to thank H. H. Genoways, M. J. Smolen, and S. L. Williams for acquiring specimens of Neotomodon through field work supported by the M. Graham Netting Research Fund through a grant from the Cordelia Scaife May Charitable Trust. We thank I. F. Greenbaum for both his review of the manuscript and technical assistance. We also thank C. W. Kilpatrick, J. Porter, D. J. Schmidly and three anonymous reviewers for their constructive criticisms of the manuscript. This work was supported by a USPHS training grant to J. C. P., by a grant from the American Philosophical Society, and by NSF grant DEB7814195.

SPECIMENS EXAMINED

Peromyscus maniculatus - Texas (Brewster Co.): 10.5 km N Alpine, N = 2. North Carolina (Highlands Co.): 3.2 km W Norton, Coweeta Hydrologic Laboratory, N = 2.

Peromyscus polionotus - Georgia (Barrow Co.): 8.1 km NE Winder, N = 2.

Peromyscus leucopus - Georgia (Clarke Co.): 4.8 km S Athens, N = 3.

Peromyscus floridanus - Florida (Highlands Co.): 8.1 km S, 3.2 km W Lake Placid, N = 3.

Neotomodon alstoni - Mexico (Districto Federal): 1 km N Morelos, N = 5.

Onychomys leucogaster - Texas (Winkler Co): 3.2 km W Wink, N = 1; 10.5 km E Wink, N = 1.

Onychomys torridus - Arizona (Cochise Co.): 5.6 km SE Portal, N = 1; 22.5 km SE Portal, N = 1.

Baiomys taylori - Texas (Galveston Co.): 6.4 km N Texas City, N = 2; (Brazos Co.): Bryan, N = 1.

Reithrodontomys fulvescens - Texas (Galveston Co.): 6.4 km N Texas City, N = 3.

Reithrodontomys megalotis - Kansas (Ellis Co.): 3.2 km W Hays (Relict Area), N = 2.

Ochrotomys nuttalli - Georgia (Clarke Co.): 4.8 km S Athens, N = 2. North Carolina (Highlands Co.): 3.2 km W Norton, Coweeta Hydrologic Laboratory, N = 2.

Neotoma micropus - Texas (Brewster Co.): Alpine, City Dump, N = 1.

Sigmodon hispidus - Georgia (Barrow Co.): 8.1 km NE Winder, N = 4.

Oryzomys palustris - Texas (Galveston Co.): 6.4 km N Texas City, N = 3.

ORGANIZATION AND CHAOS IN POPULATION STRUCTURE: SOME THOUGHTS ON FUTURE DIRECTIONS FOR MAMMALIAN POPULATION GENETICS

William Z. Lidicker, Jr.

The organizers of this symposium have asked me to perform an unrealistic task. To briefly summarize and criticize the four major symposium contributions, all of which are interesting, complex, and scholarly, is formidable enough, but in addition they have requested some thoughts on the future of this field of investigation. Population genetics is not a compact, coherent, and contented field of intellectual activity. It is burgeoning, provocative, and interdisciplinary in the extreme. I hope readers will therefore be tolerant if these remarks are perceived to lack the comprehensiveness that the subject deserves. I will attempt to briefly summarize the important issues addressed in this symposium, offer some specific comments on the major papers, and then suggest some future directions for the field.

The objectives of mammalian population genetics can be summarized as follows: (1) to achieve a description of the genetic structure of populations, recognizing that this will likely vary both among species and within species over time and space; it is thus a dynamic rather than a static description that we seek; (2) to explain the origin of this structure in both proximate and ultimate terms; and, (3) to understand its consequences. Consistent with the interdisciplinary nature of the field, we can point to a variety of kinds of data which can now be utilized in pursuing these difficult objectives. The traditional sources of morphology, physiology, and behavior continue to be major contributors, and many novel kinds of such data are being evaluated. These kinds of characters generally suffer from a lack of understanding of their genetic bases and heritabilities. It is important that much more be learned about their inheritance, especially for clearly relevant traits such as litter size, aspects of growth dynamics, and behavior (dispersal, aggressive, and maternal).

This is also true for the nonmetric morphological traits (epigenetic) which are proving to be surprisingly useful measures of population relatedness (Berry et al., 1967; Berry, 1969b; Berry, 1974). In recent years, karyotypic and biochemical traits have been increasingly used by the mammalian population geneticist. In fact the major thrust of this symposium has been an examination and evaluation of the role of biochemical variation in this field, generally as analyzed by electrophoresis. By itself, this speaks to its phenomenal success. Nevertheless, perhaps it will be appropriate for future symposia in this field to examine the integration of a broader data base.

CURRENT ISSUES

The first subject to be mentioned is that of the role of biochemical variation in reconstructing phylogenies above the species level. I raise it first because I will largely ignore it on the grounds that it falls mostly outside the realm of population genetics. The topic is nevertheless developed in this symposium in an interesting and innovative manner by Straney (this volume) and touched on as well by Schnell and Selander (this volume) and by Kilpatrick (this volume). Moreover, it has important connections to the analysis of population structure in that some of the same data are used. Where the data base overlaps, as for example with morphological or biochemical traits, we need to know how much historical (cladistic) information these characters contain; and population genetics will clearly be useful in making such assessments. Of course phylogenetic analysis can and does utilize additional sources of information not especially relevant to the population geneticist, such as the fossil record, biogeography, DNA hybridization, mitochondrial DNA characterizations, etc. A second area of overlap between population genetics and phylogenesis is in the identification of species boundaries (Patton and Yang, 1977; Sage, 1978).

In the case of biochemical traits, the question of the extent to which they are influenced by natural selection versus other processes is extremely important and is being addressed by numerous investigators (including all four major symposium papers in this volume). On the intraspecific level, this issue has implications for our understanding of almost all the other major current questions in population genetics. For phylogenetic analysis, it is also clearly important to know what kinds of evolutionary processes are implied by biochemical variations. If structural genes turn out generally to be subject to natural selection (Nevo, 1978, but see Schnell and Selander for contrary arguments), their patterns of change should then correlate with other kinds of adaptive characters and thereby contribute to reconstruction of the adaptive histories of lineages. However, we might expect that since some of these genes evolve relatively rapidly, the record of history which they reveal will be short. This is because of the rapid loss in our ability to recognize homologous allozymes with increasing genetic distance between groups. On the other hand, if biochemical traits are neutral, or nearly so, they will not provide much information on the adaptive history of lineages. However, they may then usefully serve to fuel a molecular clock and hence help us to time cladistic branch points (Sarich and Cronin, 1976; Wilson, D. S. et al., 1977).

Three other precautions need to be mentioned with regard to the question of selection versus neutrality in structural genes. The first is that we should not expect the answer to be the same for all loci or even for all populations of a particular species. Moreover, the situation may turn out to be quite complex where various genomic interactions may be involved. The second point is that it is important to be explicit about the scale on which we are asking the question. The selective importance of, say, a hemoglobin variant in *Mus musculus* will likely have one answer on the scale of the social group or local deme and perhaps another on the scale of macrogeographic variation such as across Texas or North America. Lastly, we must keep in mind that some biochemical traits seem to evolve at a much faster rate than others (Sarich, 1977), and this may affect our interpretation of data. Intertwined with this major issue is the matter of correlation and functional relations between different data sources. A good review of this for morphological versus biochemical data is presented by Schnell and Selander with emphasis being placed on the intraspecific level.

The review by Gaines (this volume) expertly summarizes the controversial and fascinating area of the relation between genetic structure of populations and demographic processes. This subject has been explored by biologists at least as far back as 1949 (Franz, 1949). Tremendous impetus has been provided, however, by the advent of electrophoretic techniques which permit the genetic analysis of populations in parallel with demographic analysis. Considerable data have now accumulated which establish that changes in genetic composition can often be correlated with demographic events. The major question, as pointed out by Gaines, is to determine the causative relationships among these correlations. Whatever the answers, their impact on population genetic theory will be momentous.

One demographic aspect not sufficiently emphasized by Gaines is the importance of density bottlenecks to future structure and the temporal extent of these presumed effects. Related to this is the determination of the geneticist's effective population size (N_e). This parameter is critical to understanding the genetic and evolutionary consequences of population structure. I would suggest that it will generally be futile to search for <u>the</u> N_e of a given species. Rather we can expect this value to vary over time and space. Where temporal variation is found, we will have to discover which times are the most influential in determining the genetic structure of those populations and their evolutionary responses. Relevant to a search for N_e will be the need to increase our understanding of the social systems of most species of mammals.

Another major question which has been extensively addressed (especially by the ancillary papers in this symposium) is that of the relationship between genetic variability in populations and environmental grain. A common assumption has been that increased variability will be associated with coarse-grain environments. This idea is also expressed as the niche-width variation hypothesis, the idea being that species or populations exhibiting broad niches will also carry more extensive genetic variation. A search for support of this correlation has generally been disappointing, but then only a few of the important niche dimensions are usually measured. Nevertheless, Nevo (1978) concludes that the theory is supportable by available evidence and speculates further that generalists are probably more heterozygous and polymorphic than specialists. Moreover, there are species such as <u>Peromyscus</u> <u>maniculatus</u> which seem definitely to support this hypothesis. It is a niche generalist on the species level, and several studies have shown it to have relatively high levels of variability (Smith, 1978; Smith et al., 1978; Kilpatrick, this volume). On the other hand, pocket gophers (Thomomys) which are generally acknowledged to have relatively narrow niches also have high levels of biochemical variability (Patton and Yang, 1977). Still another possibility is that some generalist species will possess a monomorphic, but phenotypically plastic, genotype; Rattus rattus may be an example of this phenomenon (Patton et al., 1975). Thus, no simple solution to this intriguing question seems imminent.

Related to the niche-width issue is the question of whether heterozygosity itself represents some kind of an adaptation on the individual level. This is not unreasonable considering the wealth of data on heterosis in a large variety of organisms and clearly warrants further investigation (see also Gaines, this volume). Several workers have reported changes in heterozygosity levels with density fluctuations (Smith et al., 1975; Gaines et al., 1978; Massey and Joule, this volume). On the population level, we need to explore the hypothesis that genetic variation is correlated with general adaptations (Haldane, 1937; Brown, 1958) which imply improved capabilities of coping with future environmental change, that is, change not previously or at least recently experienced by the lineage.

The final subject I would like to mention is the complex one of assessing the levels of gene flow between groups. Electrophoretic data are becoming increasingly useful in this regard, and the extent to which various allozymes turn out to be selectively neutral, or nearly so, will enhance their utility for these purposes. On the other hand, caution must be exercized in interpreting situations on a scale where gene frequencies are strongly influenced by stochastic buffeting. Attempts to assess gene flow have been reported for levels ranging from the local population (Bowen, 1978) to biogeographic phenomena (Kilpatrick). Their usefulness in discovering species boundaries in cryptic or sibling species has already been noted. As is well known, karyotypic characters have already been widely used for this purpose. An area that has been barely explored is that of using biochemical data for defining subspecific boundaries (e.g., for Mus, Selander et al., 1969a; Hunt and Selander, 1973). Such an application is particularly appropriate if one adopts an evolutionary view of the subspecies concept (Lidicker, 1962b), because it focuses attention on boundaries of putative subspecies rather than on a comparison of topotypes. Biochemical characters are also proving useful in the study of dispersal (Myers and Krebs, 1971a; Pickering et al., 1974), a process which potentially has both genetic and demographic consequences (Lidicker, 1975). We need to know much more about dispersal, and the studies of islands (Kilpatrick) and semi-isolated populations should make significant contributions. Two other biogeographic subjects of relevance here are characterizing and quantifying long-distance dispersal and analyzing peripheral versus central populations in species distribution.

SPECIFIC COMMENTS

In this section I will make a few critical, but I hope constructive, comments on the primary symposium contributions. Of course some of my views on these topics have already been revealed in the preceding section, and still others will become apparent in the final part. What remains are either a few points of clarification or suggestions on how the authors could have made their contributions even longer.

In his paper on the importance of genetics to population dynamics, Gaines makes the prediction that if in fact genetic changes can influence demographic parameters, changes in gene frequency will be due to natural selection. While this is likely to be the case, it should not be interpreted so as to exclude genetic changes not due to selection that may be influencing demographic behavior. Gaines points out that age structure changes alone can influence gene frequencies, and these can feed back on demography. Moreover, we now know that on the local level there may be profound changes in gene frequencies through stochastic processes as long as population sizes are small. A second point concerns the use of the term "natural selection" as a factor postulated to drive cyclic changes in microtines (also used by Tamarin, 1978a and others) or as in the expression "natural selection model" of density regulation. This seems to be a device for intimidating holders of other views by implying that their models are non-Darwinian and hence inferior. Actually, of course, many other models of regulation are posed in an evolutionary context, and this particular one should be more accurately called the "density dependent selection model."

Three additional comments on Gaines' generally excellent review are: (1) The work on plasma esterase of Semeonoff and Robertson (1968) cited as an example of gene frequency changes correlated with density offers weaker support than originally supposed. This particular plasma esterase system does not behave in a simple Mendelian fashion (Bowen and Yang, 1978); (2) The statistic F_{ST} gives a measure of heterogeneity among populations and not of inbreeding (although this is of course one possible contribution to heterogeneity) and tests using F_{ST} such as the Lewontin-Krakauer test for selective neutrality are in fact properly applied to populations arrayed through time; it is the use of tests through space that requires the precautions noted by Gaines (Nei and Maruyama, 1975; Lewontin and Krakauer, 1975); and, (3) The reported correlation between population density and average heterozygosity for the pocket gophers *Thomomys bottae* and *T. umbrinus* (Patton and Yang, 1977; Patton and Feder, 1978) is based on an understandable misinterpretation of the term "population size"; these authors meant gene pool size which is of course a function of both density and neighborhood area (Patton, pers. comm.).

Schnell and Selander have given us a rigorously developed treatise which concludes that there exist no convincing and adaptively meaningful correlations between variability in structural genes and either environmental heterogeneity or morphological variability. It is wise to counsel caution in interpreting correlations between data sets, but it is unwise to expect that all structural gene variation will have a common explanation. In their enthusiasm for generality, Schnell and Selander weaken their case by sometimes attempting to refute data which objectively can support other hypotheses as well as their own. For example, they suggest that the correlation between genic heterozygosity and morphological variability reported by Patton et al. (1975) for *R. rattus* on the Galapagos Islands is "biologically spurious" because of ambiguities in their sampling procedures. In fact their procedures are given explicitly and the correlation is statistically significant ($P < 0.05$) whether the measurements are based on a per island estimate (Patton et al., 1975) or per population (Patton, pers. comm.). This of course does not provide an explanation for the correlation, but at least it cannot simply be dismissed as invalid.

Finally, it would be well to remind ourselves that their carefully documented conclusions apply most specifically to the population level of organization as revealed by structural genes. Extending these conclusions above the population level will, I suspect, lead to increasingly shaky results. I have already pointed out the potential utility of biochemical data in assessing species and subspecies boundaries, and Straney outlines their usefulness in phyletic reasoning above the species level. Moreover, note how even the subspecies boundary for M. musculus described in detail by Schnell and Selander is clearly identified by "genic" as well as morphological characters. These authors do give some attention in their paper to karyotypic variation, but conclusions are based primarily on "genic" data. I certainly agree with their conclusion "that evolution proceeds more or less independently at the genic, karyotypic and morphological levels..."

Kilpatrick has given us a useful summary of the evidence for genetic structure of insular populations. I agree that islands have the potential for providing us with a treasure of insights into evolutionary and ecological processes. Unfortunately, the paper is limited by (1) being largely restricted to biochemical data and (2) ignoring habitat islands in terrestrial situations. In view of the reduced variability that Kilpatrick has documented for islands, it would seem worth pursuing how this relates to the low variabilities in morphometric traits sometimes also reported for islands (Lidicker, 1960:140). If a common cause were implicated, this would be extremely interesting in view of the presumed differences in rates of evolution and genetic bases for these two data sets.

Straney has written a thoughtful and provocative analysis of the application of biochemical data to phyletic reasoning above the species level. It is thus largely outside the province of population genetics per se, but as already pointed out much of common interest remains. The advantages of being able to use structural genes, which generally are inherited in a Mendelian fashion, are balanced against the disadvantages of using characters that evolve rapidly and often vary within populations or species. I find it particularly cogent to be reminded that in our search for genetic relatedness there are really three varieties of this concept. How we view information gained from largely neutral "genic" characters will thus depend on whether we are searching for genealogies or evolutionary relatedness.

FUTURE DIRECTIONS

Any thoughts on future directions for such a multifaceted field as mammalian population genetics must bear the aura of personal provincialism. Besides, future research will be strongly influenced and encompassed by the current major issues mentioned above. What I will attempt in this section, therefore, is to point to five issues which I see emerging from our current efforts and which would I believe justify our serious attention. Some, of course, are merely explicit subsets of the more general current issues.

Body Size Versus Genetic Variability

While the extent to which biochemical variability may be correlated meaningfully with morphological variation remains controversial, there is some support for the notion that body size is inversely correlated with "genic" variability. The problem applies only to the species level and above and has been discussed by several speakers in this symposium (Lidicker et al., this volume). Such a relationship has also been reported for nonmammals (Merkle et al., 1977). However, just as with humans, there are at least three species of cervids which are both large and exhibit high levels of biochemical variability. Whether these exceptions weaken the generalization or actually enhance it by demonstrating alternate and competing influences on variability remains to be determined.

If this currently weak generalization should gain further support, it raises the spectre of Cope's Law which states that in mammalian evolution species tend to get progressively larger and then become extinct (Stanley, 1973). So, if we ultimately should find that large size is associated with low genetic variability, it follows plausibly that this will lead to a reduced capability for coping with changing conditions, and that in turn implies an increased probability of extinction. Of course there are other explanations for Cope's Law which have nothing to do with genetic variability, but this presumed loss of "general adaptiveness" can conceivably be a contributing factor to the extinction of large-sized species. Conservation and wildlife management considerations alone demand that we know more about this relationship.

Commonness Versus Rarity and Population Structure

Although most species appear to be rare, relatively little is known about them, and this is particularly true for their population genetics. The ubiquitous and successful strategy of being rare needs to be explored on all fronts, including an analysis of the genetic structure of such populations. We may be in for some surprises.

Heterozygosity as an Adaptation

Is there something significant in general heterosis? Do high levels of heterozygosity actually increase viability and lead to increased levels of aggression, greater exploratory behavior, etc., as has been claimed by some (Smith et al., 1975; Berger, 1976; Garten, 1976, 1977). Aside from possible direct effects on dispersal behavior, the existence of such a phenomenon would also influence the selective basis for the evolution of dispersal behavior by improving genetic payoffs to successful individuals (Lidicker, 1962a). The matter needs to be explored not only with structural gene traits but with other kinds of characters as well.

Levels of Selection

One of the most fundamental questions in evolutionary biology today concerns the levels at which natural selection operates and their relative importance to the evolutionary process. I see definite possibilities for contributions of fundamental importance to this area by mammalogists. At the genomic level of organization we need to have greater understanding about interactions among loci, effects of chromosomal rearrangements, behavior of supergene complexes, etc. Another intriguing question at this level is the determination of genetic influences, other than those of the sex chromosomes, on the sex ratio of an individual's offspring and their interactions with environmental, ontogenetic, and parental condition factors. At the group level of organization, we can point first to the considerable contributions already being made by mammalogists to kin selection theory. With regard to the more controversial interdemic selection, I suggest that the "trait group" model of Wilson (D. S., 1975, 1977) will likely have applicability to mammals. The basic ingredients for this model, namely alternating demographic phases in which individuals are first restricted to small interactive groups and then the groups are conjoined into much larger panmictic populations, seem to be widespread among mammals. Critical to this and other interdemic selection models is vastly improved information on demic survival statistics and particularly extinction rates.

Levels of Chaos in Population Processes

Rapidly accumulating information is forcing us to admit, at the point of despair for some, that stochastic processes can be critically important in explaining the biology of populations. We can be reassured, however, that the biosphere is indeed organized; there is negentropy, or life would not be possible. The problem is finding, documenting and understanding this organization in spite of a great deal of stochastic "noise." I would venture to guess that this random factor accounts for 50 to 90% of what we observe, depending of course to a large extent on the level and scope of the question being asked (an inverse relationship is expected between the size of the random component and the level of inquiry). Although this makes our job more difficult, we should not despair as stochastic processes are a normal part of life. We need to determine not only how much randomness there is, but how organisms cope with it. How fine tuned can adaptations be? This problem is, of course, not unique to population biology. Biologists working at the organismal and other levels have faced this problem, perhaps less intensively, for a long time; yet they have preservered and prospered. Population biologists will succeed as well, and work on mammals will surely be an important part of this effort.

Acknowledgments--I appreciate the confidence placed in me by the organizers of this symposium in offering me this intriguing assignment. Ramone Baccus and Drs. B. S. Bowen and J. L. Patton kindly read and criticized an early draft of this manuscript.

LITERATURE CITED

Adamczyk, K., and W. Walkowa. 1971. Compensation of numbers and production in a Mus musculus population as a result of partial removal. Ann. Zool. Fenn. 8:145-53.

Anderson, J. L. 1975. Phenotypic correlations among relatives and variability in reproductive performance in populations of the vole Microtus townsendii. Ph.D. diss. Univ. British Columbia, Vancouver.

Anderson, P. K. 1965. The role of breeding structure in evolutionary processes of Mus musculus populations. Proc. Symp. Mutational Processes, Prague Academia: 17-21.

Anderson, P. K. 1970. Ecological structure and gene flow in small mammals. Symp. Zool. Soc. Lond. 26:299-325.

Andrewartha, H. G., and L. C. Birch. 1954. The Distribution and Abundance of Animals. Univ. Chicago Press, Chicago. 782 p.

Aquadro, C. F. 1978. Evolutionary genetics of insular Peromyscus: electrophoretic, morphological, and chromosomal variation. M. S. thesis, Univ. Vermont, Burlington.

Arata, A. A. 1964. The anatomys and taxonomic significance of the male accessory reproductive glands of muroid rodents. Bull. Fla. State Mus. 9:1-42.

Arata, A. A. 1967. Muroid, gliroid, and dipodoid rodents. pp. 226-53. In Recent Mammals of the World. S. Anderson, and J. K. Jones, Jr. (eds.). Ronald Press Co., New York.

Armitage, K. B. 1975. Social behavior and population dynamics of marmots. Oikos 26:341-54.

Armitage, K. B. 1977. Social variety in the yellow-bellied marmot: a population-behavioural system. Anim. Behav. 25:585-93.

LITERATURE CITED

Armitage, K. B., and J. F. Downhower. 1974. Demography of yellow-bellied marmot populations. *Ecology* 55:1233-45.

Ashton, E. H., R. M. Flinn, C. E. Oxnard, and T. F. Spence. 1976. The adaptive and classificatory significance of certain quantitative features of the forelimb in primates. *J. Zool.* 179:515-56.

Atchley, W. R., C. T. Gaskins, and D. Anderson. 1976. Statistical properties of ratios. I. Empirical results. *Syst. Zool.* 25:137-48.

Avise, J. C. 1974. Systematic value of electrophoretic data. *Syst. Zool.* 23:465-81.

Avise, J. C. 1976. Genetic differentiation during speciation. pp. 106-22. *In Molecular Evolution.* F. J. Ayala (ed.). Sinauer Assoc., Inc., Sunderland, Mass.

Avise, J. C., and R. K. Selander. 1972. Evolutionary genetics of cave-dwelling fishes of the genus *Astyanax*. *Evolution* 26:1-19.

Avise, J. C., and M. H. Smith. 1977. Gene frequency comparisons between sunfish (Centrarchidae) populations at various stages of evolutionary divergence. *Syst. Zool.* 26:319-35.

Avise, J. C., M. H. Smith, and R. K. Selander. 1974b. Biochemical polymorphism and systematics in the genus *Peromyscus*. VI. The *boylii* species group. *J. Mammal.* 55:751-63.

Avise, J. C., M. H. Smith, and R. K. Selander. 1979a. Biochemical polymorphism and systematics in the genus *Peromyscus*. VII. Geographic differentiation in members of the *truei* and *maniculatus* species groups. *J. Mammal.* 60:177-92.

Avise, J. C., M. H. Smith, R. K. Selander, T. E. Lawlor, and P. R. Ramsey. 1974a. Biochemical polymorphism and systematics in the genus *Peromyscus*. V. Insular and mainland species of the subgenus *Haplomylomys*. *Syst. Zool.* 23:226-38.

LITERATURE CITED

Avise, J. C., R. A. Lansman, and R. O. Shade. 1979b. The use of restriction endonucleases to measure mitochondrial DNA sequence relatedness in natural populations. I. Population structure and evolution in the genus *Peromyscus*. *Genetics*, 92:279-295.

Ayala, F. J. 1974. Biological evolution: natural selection or random walk? *Am. Sci.* 62:692-701.

Ayala, F. J. 1975. Genetic differentiation during the speciation process. pp. 1-78. *In Evolutionary Biology*, Vol. 8. T. Dobzhansky, M. K. Hecht, and W. C. Steere (eds.). Plenum Press, New York.

Ayala, F. J. (ed.). 1976. *Molecular Evolution*. Sinauer Assoc., Inc., Sunderland, Mass. 277 p.

Ayala, F. J. 1977. Protein evolution: nonrandom patterns in related species. pp. 177-205. *In Measuring Selection in Natural Populations*. F. B. Christiansen and T. M. Fenchel, (eds.). Springer-Verlag, New York, New York.

Ayala, F. J., and W. W. Anderson. 1973. Evidence of natural selection in molecular evolution. *Nature* 241:274-76.

Ayala, F. J., and M. L. Tracey. 1974. Genetic differentiation within and between species of the *Drosophila willistoni* group. *Proc. Natl. Acad. Sci. USA* 71:999-1003.

Ayala, F. J., J. R. Powell, and T. Dobzhansky. 1971. Polymorphisms in continental and island populations of *Drosophila willistoni*. *Proc. Natl. Acad. Sci. USA* 68:2480-83.

Ayala, F. J., J. R. Powell, M. L. Tracey, C. A. Mourao, and S. Perez-Salas. 1972. Enzyme variability in the *Drosophila willistoni* group. IV. Genic variation in natural populations of *Drosophila willistoni*. *Genetics* 70:113-39.

Ayala, F. J., M. L. Tracey, D. Hedgecock, and R. C. Richmond. 1974. Genetic differentiation during

LITERATURE CITED

the speciation process in *Drosophila*. Evolution 28:576-92.

Baker, R. H. 1968. Habitats and distribution. pp. 98-126. *In Biology of Peromyscus* (Rodentia). J. A. King (ed.). Spec. Publ. Am. Soc. Mammal. No. 2.

Baker, R. J. 1979. Karyology. pp. 107-156. *In Biology of Bats of the New World Family Phyllostomatidae*. Part III. R. J. Baker, J. K. Jones, Jr., and D. C. Carter (eds.). Spec. Publ. Mus., Texas Tech. Univ. 16.

Baker, R. J., R. K. Barnett, and I. F. Greenbaum. 1979. Chromosomal evolution in grasshopper mice (*Onychomys*: Cricetidae). *J. Mammal*. 60:297-306.

Baker, R. J., S. L. Williams, and J. C. Patton. 1973. Chromosomal variation in the plains pocket gopher, *Geomys bursarius major*. *J. Mammal*. 54:765-69.

Baker, W. K. 1975. Linkage disequilibrium over space and time in natural populations of *Drosophila montana*. *Proc. Natl. Acad. Sci. USA* 72:4095-99.

Bekoff, M. 1977. Mammalian dispersal and the ontogeny of individual behavioral phenotypes. *Am. Nat.* 111:715-32.

Bellamy, D., R. J. Berry, M. E. Jakobson, W. Z. Lidicker, J. Morgan, and H. M. Murphy. 1973. Ageing in an island population of the house mouse. *Age and Ageing* 2:235-50.

Berger, E. 1976. Heterosis and the maintenance of enzyme polymorphism. *Am. Nat*. 110:823-39.

Berry, A. C. 1974. The use of non-metrical variations of the cranium in the study of Scandanavian population movements. *Am. J. Phys. Anthro*. 40:345-58.

Berry, A. C., and R. J. Berry. 1972. Origins and relationships of the ancient Egyptians. *J. Hum. Evolution* 1:199-208.

LITERATURE CITED

Berry, D. L., and R. J. Baker. 1971. Apparent convergence of karyotypes in two species of pocket gogophers of the genus Thomomys (Mammalia, Rodentia). Cytogenetics 10:1-9.

Berry, R. J. 1963. Epigenetic polymorphism in wild populations of Mus musculus. Genet. Res. 4:193-220.

Berry, R. J. 1964. The evolution of an island population of the house mouse. Evolution 18:468-83.

Berry, R. J. 1967. Genetical changes in mice and men. Eugen. Rev. 59:78-96.

Berry, R. J. 1969a. History in the evolution of Apodemus sylvaticus (Mammalia) at one edge of its range. J. Zool. 159:311-28.

Berry, R. J. 1969b. Non-metrical skull variation in two Scottish colonies of the Grey seal. J. Zool. 157:11-18.

Berry, R. J. 1970. Covert and overt variation, as exemplified by British mouse populations. Symp. Zool. Soc. Lond. 26:3-26.

Berry, R. J. 1975. On the nature of genetical distance and island races of Apodemus sylvaticus. J. Zool. 176:292-96.

Berry, R. J. 1977a. Inheritance and Natural History. London: Collins New Naturalist. 350 p.

Berry, R. J. 1977b. The population genetics of the house mouse. Sci. Prog. 61:341-70.

Berry, R. J. 1977c. Variability in mammals - concepts and complication. pp. 5-25. In Advances in Modern Theriology. Sokolov, V. E. (ed.). Moscow: Nauka (in Russian, with English summary).

Berry, R. J. 1978. Genetic variation in wild house mice: where natural selection and history meet. Am. Sci. 66:52-60.

LITERATURE CITED

Berry, R. J. 1979. Genetical factors in animal population dynamics. pp. 53-80. In Population Dynamics. R. M. Anderson, L. R. Taylor, and B. D. Turner (eds.). Oxford: Blackwell.

Berry, R. J., and M. E. Jakobson. 1974. Vagility in an island population of the House mouse. J. Zool. 173:341-54.

Berry, R. J., and M. E. Jakobson. 1975a. Adaptation and adaptability in wild-living House mice. J. Zool. 176:391-402.

Berry, R. J., and M. E. Jakobson. 1975b. Ecological genetics of an island population of the House mouse (Mus musculus). J. Zool. 175:523-40.

Berry, R. J., and H. M. Murphy. 1970. The biochemical genetics of an island population of House mouse. Proc. R. Soc. Lond. B. Biol. Sci. 176:87-103.

Berry, R. J., and J. Peters. 1975. Macquarie Island House mice: a genetical isolate on a sub-Antarctic island. J. Zool. 176:375-89.

Berry, R. J., and J. Peters. 1976. Genes, survival and adjustment in an island population of the house mouse. pp. 23-48. In Population Genetics and Ecology. S. Karlin and E. Nevo (eds.). Academic Press, New York.

Berry, R. J., and J. Peters. 1977. Heterogeneous heterozygosities in Mus musculus populations. Proc. R. Soc. Lond. B. Biol. Sci. 197:485-503.

Berry, R. J., and H. N. Southern (eds.). 1970. Variation in Mammalian Populations. Symp. Zool. Soc. Lond. No. 26. Academic Press, London.

Berry, R. J., and T. Warwick. 1974. Field mice (Apodemus sylvaticus) on the Castle Rock, Edinburgh: an isolated population. J. Zool. 174:325-31.

LITERATURE CITED

Berry, R. J., W. N. Bonner, and J. Peters. 1979. Natural selection in mice from South Georgia (South Atlantic Ocean). J. Zool. 189:385-398.

Berry, R. J., I. M. Evans, and B. F. C. Sennitt. 1967. The relationships and ecology of Apodemus sylvaticus from the Small Isles of the Inner Hebrides, Scotland. J. Zool. 152:333-46.

Berry, R. J., M. E. Jakobson, and J. Peters. 1978a. The House mice of the Faroe Islands: a study in microdifferentiation. J. Zool. 185:73-92.

Berry, R. J., J. Peters, and R. J. Van Aarde. 1978b. Sub-antarctic House mice: colonization, survival and selection. J. Zool. 184:127-41.

Bertram, B. C. R. 1977. Kin selection in lions and in evolution. pp. 281-301. In Growing Points In Ethology. R. A. Hinde and P. P. G. Bateson (eds.). Cambridge University Press, Cambridge.

Best, T. L., and G. D. Schnell. 1974. Bacular variation in kangaroo rats (genus Dipodomys). Am. Midl. Nat. 91:257-70.

Birch, L. C. 1960. The genetic factor in population ecology. Am. Nat. 94:5-24.

Blackwell, T. L., and P. R. Ramsey. 1972. Exploratory activity and lack of genotypic correlates in Peromyscus polionotus. J. Mammal. 53:401-03.

Blaffer Hrdy, S. 1977. The Langurs of Abu. Female and Male Strategies of Reproduction. Harvard University Press, Cambridge. 361 p.

Blair, W. F. 1947. Estimated frequencies of the buff and gray genes (G, g) in adjacent populations of deer-mice (Peromyscus maniculatus blandus) living on soils of different colors. Contrib. Lab. Vert. Biol. Univ. Mich. 36:1-16.

Blair, W. F. 1950. Ecological factors in speciation of Peromyscus. Evolution 4:253-75.

LITERATURE CITED

Blair, W. F. 1951. Population structure, social behavior, and environmental relations in a natural population of the beach mouse (*Peromyscus polionotus leucocephalus*). *Contrib. Lab. Vert. Biol. Univ. Mich*. 48:1-47.

Blake, E. T. 1977. Genetic markers in human semen. Unpub. Ph.D. diss., Univ. Calif. Berkeley. 277 p.

Blumberg, B. S., A. C. Allison, and B. Garry. 1960. The haptoglobins, hemoglobins and serum proteins of the Alaskan fur seal, ground squirrel and marmot. *J. Cell Comp. Physiol*. 55:61-71.

Bodmer, W., and L. L. Cavalli-Sforza. 1974. The analysis of genetic variation using migration matrices. pp. 45-61. *In Genetic Distance*. J. F. Crow and C. Denniston (eds.). Plenum Press, New York.

Bonde, N. 1977. Cladistic classification as applied to vertebrates. pp. 741-804. *In Major Patterns in Vertebrate Evolution*. M. K. Hecht, P. C. Goody, and B. M. Hecht (eds.). Plenum Press, New York.

Bonhomme, F., and R. K. Selander. 1978. Estimating total genic diversity in the house mouse. *Biochem. Genet*. 16:287-97.

Bonnell, M. L., and R. K. Selander. 1974. Elephant seals: genetic variation and near extinction. *Science* 184:908-09.

Boonstra, R. 1978. Effect of adult Townsend voles (*Microtus townsendii*) on survival of young. *Ecology* 59:242-48.

Boonstra, R., and C. J. Krebs. 1977. A fencing experiment on a high-density population of *Microtus townsendii*. *Can. J. Zool*. 55:1166-75.

Bowen, B. S. 1978. Spatial and temporal patterns of genetic variation in a local population of the California vole, *Microtus californicus* (Rodentia: Cricetidae). Ph.D. diss., Univ. of Calif. at Berkeley.

LITERATURE CITED

Bowen, B. S., and S. Y. Yang. 1978. Genetic control of enzyme polymorphisms in the California vole, *Microtus californicus*. *Biochem. Genet.* 16:455-67.

Bowen, W. W. 1968. Variation and evolution of Gulf Coast populations of beach mice, *Peromyscus polionotus*. *Bull. Fla. State Mus.* 12:1-91.

Bowen, W. W., and W. D. Dawson. 1977. Genetic analysis of coat pattern variation in oldfield mice (*Peromyscus polionotus*) of western Florida. *J. Mammal.* 58:521-30.

Bowers, J. H. 1974. Genetic compatibility of *Peromyscus maniculatus* and *Peromyscus melanotis*, as indicated by breeding studies and morphometrics. *J. Mammal.* 55:720-37.

Bowers, J. H., R. J. Baker, and M. H. Smith. 1973. Chromosomal, electrophoretic, and breeding studies of selected populations of deer mice (*Peromyscus maniculatus*) and black-eared mice (*P. melanotis*). *Evolution* 27:378-86.

Breen, G. A. M., A. J. Lusis, and K. Paigen. 1977. Linkage of genetic determinants for mouse β-galactosidase electrophoresis and activity. *Genetics* 85:73-84.

Britten, R. J., and E. H. Davidson. 1969. Gene regulation for higher cells: a theory. *Science* 164:349-57.

Britten, R. J., and E. H. Davidson. 1971. Repetitive and non-repetitive DNA sequences and a speculation on the origins of evolutionary novelty. *Quart. Rev. Biol.* 46:111-33.

Britton, J., and L. Thaler. 1978. Evidence for the presence of two sympatric species of mice (genus *Mus* L.) in southern France based on biochemical genetics. *Biochem. Genet.* 16:213-25.

Brown, A. H. D. 1975a. Samples sizes required to detect linkage disequilibrium between two or three loci. *Theor. Pop. Biol.* 8:184-201.

LITERATURE CITED

Brown, A. J. L. 1977. Genetic changes in a population of field mice (Apodemus sylvaticus) during one winter. J. Zool. 182:281-89.

Brown, H. L. 1945. Evidence of winter breeding of Peromyscus. Ecology 26:308-9.

Brown, J. H. 1971. Mammals on mountaintops: non-equilibrium insular biogeography. Am. Nat. 105: 467-78.

Brown, J. H. 1975b. Geographical ecology of desert rodents. pp. 315-41. In Ecology and Evolution of Communities. M. L. Cody and J. M. Diamond (eds.). Harvard University Press, Cambridge.

Brown, W. L., Jr. 1958. General adaptation and evolution. Syst. Zool. 7:157-68.

Browne, R. A. 1977. Genetic variation in island and mainland populations of Peromyscus leucopus. Am. Midl. Nat. 97:1-9.

Bryant, E. H. 1974a. On the adaptive significance of enzyme polymorphisms in relation to environmental variability. Am. Nat. 108:1-19.

Bryant, E. H. 1974b. An addendum on the statistical relationship between enzyme polymorphisms and environmental variability. Am. Nat. 108:698-701.

Bulfield, G., E. A. Moore, and H. Kacser. 1978. Genetic variation in activity of the enzymes of glycolysis and gluconeogenesis between inbred strains of mice. Genetics 89:551-61.

Bush, G. L., S. M. Case, A. C. Wilson, and J. L. Patton. 1977. Rapid speciation and chromosomal evolution in mammals. Proc. Natl. Acad. Sci. USA 74:3942-46.

Caire, W., E. G. Zimmerman. 1975. Chromosomal and morphological variation and circular overlap in the deer mouse, Peromyscus maniculatus, in Texas and Oklahoma. Syst. Zool. 24:89-95.

LITERATURE CITED

Cameron, A. W. 1958. Mammals of the islands in the Gulf of St. Lawrence. Bull. Nat. Mus. Canada 154:1-165.

Camin, J. H., and P. R. Ehrlich. 1958. Natural selection in water snakes (Natrix sipedon L.) on islands in Lake Erie. Evolution 12:504-11.

Canham, R. P. 1969. Serum protein variation and selection in fluctuating populations of cricetid rodents. Ph.D. diss., Univ. Alberta, Alberta.

Capanna, E., M. Cristaldi, P. Perticone, and M. Rizzoni. 1975. Identification of chromosomes involved in the 9 Robertsonian fusions of the Apennine mouse with a 22-chromosome karyotype. Experientia 31:294-96.

Capanna, E., A. Gropp, H. Winking, G. Noack, and M. V. Civitelli. 1976. Robertsonian metacentrics in the mouse. Chromosoma 58:341-53.

Capanna, E., and M. Valle. 1977. A Robertsonian population of Mus musculus L. in the Orobian Alps. Acc. Naz. Licei, Rend. Cl. Sci. Fis. Mat. Nat. Ser. 8, in press.

Carleton, M. D. 1977. Interrelationships of populations of the Peromyscus boylii species group (Rodentia, Muridae) in western Mexico. Occ. Pap. Mus. Zool. Univ. Mich. No. 675:1-47.

Carleton, M. D., and P. Myers. 1979. Karyotypes of some harvest mice, genus Reithrodontomys. J. Mammal. 60:307-13.

Carson, H. L. 1955. The genetic characteristics of marginal populations of Drosophila. Cold Spring Harbor Symp. Quant. Biol. 20:276-87.

Carson, H. L. 1959. Genetic conditions which promote or retard the formation of species. Cold Spring Harbor Symp. Quant. Biol. 24:87-105.

LITERATURE CITED

Cavalli-Sforza, L. L. 1966. Population structure and human evolution. *Proc. R. Soc. Lond. B. Biol. Sci.* 164:362-79.

Chakraborty, R., and M. Nei. 1977. Bottleneck effects on average heterozygosity and genetic distance with the stepwise mutation model. *Evolution* 31:347-56.

Chapman, V. M., E. Z. Nichols, and H. Ruddle. 1974. Esterase-8 (*Es-8*): characterization, polymorphism and linkage of an erythrocyte esterase locus on chromosome 7 of *Mus musculus*. *Biochem. Genet.* 11:347-58.

Charlesworth, B., and J. T. Giesel. 1972. Selection in populations with overlapping generations. II. Relations between gene frequency and demographic variables. *Am. Nat.* 106:388-401.

Cherry, L. M., S. M. Case, and A. C. Wilson. 1978. Frog perspective on the morphological difference between humans and chimpanzees. *Science* 200:209-11.

Chipman, R. K. 1965. Age determination of the cotton rat (*Sigmodon hispidus*). *Tulane Stud. Zool.* 12:19-38.

Chitty, D. 1960. Population processes in the vole and their relevance to general theory. *Can. J. Zool.* 38:99-113.

Chitty, D. 1967. The natural selection of self-regulatory behavior in animal populations. *Proc. Ecol. Soc. Australia* 2:51-78.

Choate, J. R. 1973. Identification and recent distribution of white-footed mice (*Peromyscus*) in New England. *J. Mammal.* 54:41-49.

Clarke, B. C. 1968. Balanced polymorphism and regional differentiation in land snails. pp. 351-68. *In Evolution and Environment*. E. T. Drake (ed.). Yale Univ. Press, New Haven.

LITERATURE CITED

Clarke, B. 1970a. Darwinian evolution of proteins. *Science* 168:1009-11.

Clarke, B. 1970b. Selective constraints on animo-acid substitutions during the evolution of proteins. *Nature* 228:159-60.

Clarke, B. 1974. Causes of genetic variation. *Science* 186:524-25.

Clarke, J. R. 1956. The aggressive behaviour of the vole. *Behaviour* 9:1-23.

Cock, A. G. 1966. Genetical aspects of metrical growth and form in animals. *Quart. Rev. Biol.* 41:131-90.

Cockrum, E. L. 1948. Distribution of the cotton rat in Kansas. *Trans. Kansas Acad. Sci.* 51:306-12.

Coombs, M. C. 1975. Sexual dimorphism in chalieotheres (Mammalia, Perissodactyla). *Syst. Zool.* 24:55-62.

Copi, I. M. 1972. *Introduction to logic*. Macmillan Co., New York. 540 p.

Corbet, G. B. 1961. Origin of the British insular races of small mammals and of the "Lusitanian" fauna. *Nature* 191:1037-40.

Corbet, G. B. 1964. Regional variation in the bank-vole *Clethrionomys glareolus* in the British Isles. *Proc. Zool. Soc. Lond.* 143:191-219.

Corruccini, R. S. 1975. Multivariate analysis in biological anthropology: some considerations. *J. Hum. Evol.* 4:1-19.

Cothran, E. G. 1980. Patterns of electromorphic variation between populations of the fossorial rodent, *Geomys bursarius* (Rodentia: Geomyidae). Submitted *S. W. Natur.*

Cothran, E. G., E. G. Zimmerman, and C. F. Nadler. 1977. Genic differentiation and evolution in the ground squirrel subgenus *Ictidomys* (genus *Spermophilus*). *J. Mammal.* 58:610-22.

LITERATURE CITED

Coyne, J. A. 1976. Lack of genic similarity between two sibling species of Drosophila as revealed by varied techniques. Genetics 84:593-607.

Cracraft, J. 1974. Phylogenetic models and classification. Syst. Zool. 23:71-90.

Creagan, R. P., and F. H. Ruddle. 1977. New approaches to human gene mapping by somatic cell genetics. pp. 89-142. In Molecular Structure of Human Chromosomes. J. J. Yunis (ed.). Academic Press, New York.

Croizat, L. 1958. Panbiogeography. Published by the author, Caracas.

Crow, J. F., and M. Kimura. 1970. An Introduction to Population Genetics Theory. Harper and Row, New York. 591 p.

Crowell, K. L. 1973. Experimental zoogeography: introductions of mice to small islands. Am. Nat. 107:535-58.

Danske Meteorologiske Institut. 1933. Danmarks Klima. Copenhagen.

Darwin, C. 1859. On the Origin of Species by Means of Natural Selection. John Murray Publishers, London.

Davies, E. (Ed.). 1944. Denmark: Geographic Handbook Series. [British] Navel Intelligence Division, London.

Davis, B. L., S. L. Williams, and G. Lopez. 1971. Chromosomal studies of Geomys. J. Mammal. 52: 617-20.

Davis, W. B. 1940. Distribution and variation of pocket gophers (genus Geomys) in the southwestern United States. Tex. Agri. Exp. St. Bull. 590:1-38.

DeFries, J. C., and G. E. McClearn. 1972. Behavioral genetics and the fine structure of mouse populations: a study in microevolution. pp. 279-91. In Evolutionary Biology, Vol. 5. T. Dobzhansky, M. K.

LITERATURE CITED

Hecht, and W. C. Steere (eds.). Appleton-Century-Crofts, New York.

Dessauer, H. C., and E. Nevo. 1969. Geographic variation of blood and liver proteins in cricket frogs. Biochem. Genet. 3:171-88.

Dice, L. R. 1940. Ecologic and genetic variability within species of Peromyscus. Am. Nat. 74:212-21.

Dice, L. R. 1968. Speciation. pp. 75-97. In Biology of Peromyscus (Rodentia). J. A. King (ed.). Spec. Publ. Am. Soc. Mammal. No. 2.

Dice, L. R., and W. E. Howard. 1951. Distance of dispersal by prairie deer-mice from birthplaces to breeding sites. Contrib. Lab. Vertebr. Biol. Univ. Mich. 50:1-15.

Dickerson, R. E. 1971. Structure of cytochrome c and the rates of molecular evolution. J. Molec. Evol. 1:26-45.

Dickinson, H., and J. Antonovics. 1973. The effects of environmental heterogeneity on the genetics of finite populations. Genetics 73:713-35.

Dixon, W. J. 1975. BMDP: Biomedical Computer Programs. Univ. California Press, Los Angeles. 792 p.

Dobzhansky, T. 1951. Genetics and the Origin of Species. Third ed. Columbia Univ. Press, New York. 364 p.

Dobzhansky, T. 1970. Genetics of the Evolutionary Process. Columbia Univ. Press, New York. 505 p.

Dodson, P. 1978. On the use of ratios in growth studies. Syst. Zool. 27:62-67.

Downhower, J. F., and K. B. Armitage. 1971. The yellow-bellied marmot and the evolution of polygamy. Am. Nat. 105:355-70.

LITERATURE CITED

Dubach, J. M. 1975. Biochemical analysis of Peromyscus maniculatus over an altitudinal gradient. M. S. thesis, Univ. Colorado, Denver.

Duffey, P. A. 1972. Chromosome variation in Peromyscus a new mechanism. Science 176:1333-34.

Dunaway, P. B., and S. V. Kaye. 1961. Cotton rat mortality during severe winter. J. Mammal. 42: 265-68.

Dunn, L. C., D. Bennett, and A. B. Beasley. 1962. Mutation and recombination in the vicinity of a complex gene. Genetics 47:285-303.

Eanes, W. F. 1978. Morphological variance and enzyme heterozygosity in the monarch butterfly. Nature 276:263-64.

Echelle, A. A., A. F. Echelle, and B. A. Taber. 1976. Biochemical evidence for congeneric competition as a factor restricting gene flow between populations of a darter (Percidae: Etheostoma). Syst. Zool. 25:228-35.

Ehrlich, P. R., and P. H. Raven. 1969. Differentiation of populations. Science 165:1228-32.

Eicher, E. M., R. H. Stern, J. E., Womack, M. T. Davisson, T. H. Roderick, and S. C. Reynolds. 1976. Evolution of mammalian carbonic anhydrase loci by tandem duplication: close linkage of Car-1 and Car-2 to the centromere region of chromosome 3 of the mouse. Biochem. Genet. 14:651-60.

Eisen, E. J. 1966. Effect of the biometrical relationship among total milk yield, milk constituent yield, and percent of milk constituent on response to selection. J. Dairy Sci. 49:1230-34.

Eldredge, N., and S. J. Gould. 1977. Punctuated equilibria: the tempo and mode of evolution reconsidered. Paleobiology 3:115-51.

Elton, C. 1927. Animal Ecology. Sidgwich and Jackson, London. 207 p.

LITERATURE CITED

Endler, J. A. 1977. *Geographic Variation, Speciation, and Clines*. Princeton Univ. Press, Princeton. 246 p.

Engel, W., K. Bender, S. Kadir, J. O. Hof, and U. Wolf. 1970. Zur Genetic der 6-Phosphogluconatdehydrogenase (EC: 1.1.1.44) bei Saugern. I. Untersuchungen and 6 Species der Familie *Microtinae*, *Rodentia*. Isoenzympolymorphismen und Familienbefunde bei *Microtus oeconomus* und *Microtus ochrogaster*. *Humangenetik* 10:151-57.

Fairbairn, D. J. 1977a. The spring decline in deer mice: death or dispersal? *Can. J. Zool*. 55:84-92.

Fairbairn, D. J. 1977b. Why breed early? A study of reproductive tactics in *Peromyscus*. *Can. J. Zool*. 55:862-71.

Fairbairn, D. J. 1978a. Behaviour of dispersing deer mice (*Peromyscus maniculatus*). *Behav. Ecol. Sociobiol*. 3:265-82.

Fairbairn, D. J. 1978b. Dispersal of deer mice, *Peromyscus maniculatus*: proximal causes and effects on fitness. *Oecologia* 32:171-93.

Falconer, D. S. 1960. *Introduction to Quantitative Genetics*. Ronald Press Co., New York. 365 p.

Falconer, D. S. 1963. Quantitative inheritance. pp. 193-216. *In Methodology in Mammalian Genetics*. W. J. Burdette (ed.). Holden-Day Inc., San Francisco.

Farris, J. S. 1970. Methods for computing Wagner trees. *Syst. Zool*. 19:83-92.

Farris, J. S. 1972. Estimating phylogenetic trees from distance matrices. *Am. Nat*. 106:645-68.

Farris, J. S., A. G. Kluge, and M. J. Eckardt. 1970. A numerical approach to phylogenetic systematics. *Syst. Zool*. 19:172-89.

LITERATURE CITED

Felsenstein, J. 1978. Cases in which parsimony or compatibility methods will be positively misleading. Syst. Zool. 27:401-10.

Fincham, J. R. S. 1972. Heterozygous advantage as a likely general basis for enzyme polymorphisms. Heredity 28:387-91.

Finnegan, M., and M. A. Faust. 1974. Bibliography of human and nonhuman non-metric variation. Res. Rep. No. 14, Dept. Anthropol., Univ. Mass., Amherst.

Fitch, W. M. 1976. Molecular evolutionary clocks. pp. 160-78. In Molecular Evolution. F. J. Ayala (ed.). Sinauer Assoc., Inc., Sunderland, Mass.

Fitch, W. M., and E. Margoliash. 1967. Construction of phylogenetic trees. Science 155:279-84.

Fleharty, E. D., J. R. Choate, and M. A. Mares. 1972. Fluctuations in population density of the hispid cotton rat: factors influencing a "crash". Bull. So. Calif. Acad. Sci. 71:132-38.

Forejt, J. 1972. Chiasmata and crossing-over in the male mouse (Mus musculus): suppression of recombination and chiasma frequencies in the ninth linkage group. Folia Biol. 18:161-70.

Foster, J. B. 1965. The evolution of mammals of the Queen Charlotte Islands, British Columbia. Occas. Pap. British Columbia Prov. Mus. 14:1-130.

Franklin, I., and R. C. Lewontin. 1970. Is the gene the unit of selection? Genetics 65:707-34.

Franz, J. 1949. Über die genetischen Grundlagen das Zusammenbrucks einer Massenvermchrung aus inneren ursachen. Zeit. Angew. Ent. 31:228-60.

Gaines, M. S., and C. J. Krebs. 1971. Genetic changes in fluctuating vole populations. Evolution 25:702-23.

LITERATURE CITED

Gaines, M. S., and R. K. Rose. 1976. Population dynamics of *Microtus ochrogaster* in eastern Kansas. *Ecology* 57:1145-61.

Gaines, M. S., L. R. McClenaghan, Jr., and R. K. Rose. 1978. Temporal patterns of allozymic variation in fluctuating populations of *Microtus ochrogaster*. *Evolution* 32:723-39.

Gaines, M. S., J. H. Myers, and C. J. Krebs. 1971. Experimental analysis of relative fitness in transferrin genotypes of *Microtus ochrogaster*. *Evolution* 25:443-50.

Garnett, I. 1976. The genetic relationship between the *Hbb* locus and body size in a population of mice divergently selected for six-week body weight. *Can. J. Genet. Cytol.* 18:519-23.

Garnett, I., and D. S. Falconer. 1975. Protein variation in strains of mice differing in body size. *Genet. Res.* 25:45-57.

Garten, C. T., Jr. 1974. Relationships between behavior, genetic heterozygosity, and population dynamics in the oldfield mouse, *Peromyscus polionotus*. M. S. thesis, Univ. Georgia, Athens.

Garten, C. T., Jr. 1976. Relationships between aggressive behavior and genic heterozygosity in the oldfield mouse, *Peromyscus polionotus*. *Evolution* 30:59-72.

Garten, C. T., Jr. 1977. Relationships between exploratory behaviour and genic heterozygosity in the oldfield mouse. *Anim. Behav.* 25:328-32.

Garten, C. T., Jr., and M. H. Smith. 1974. Movement by oldfield mice and population regulation. *Acta Theriol.* 19:513-14.

Getz, L. L. 1978. Speculation on social structure and population cycles of microtine rodents. *Biologist* 60:134-47.

LITERATURE CITED

Gier, H. T. 1967. The Kansas small mammal census: Terminal report. Trans. Kansas Acad. Sci. 70: 505-18.

Gill, A. E. 1977. Maintenance of polymorphism in an island population of the California vole, Microtus californicus. Evolution 31:512-25.

Gillespie, J. H. 1975. The role of migration in the genetic structure of populations in temporally (sic) and spatially varying environments. I. Conditions for polymorphism. Am. Nat. 109:127-35.

Gillespie, J. H., and K. Kojima. 1968. The degree of polymorphisms in enzymes involved in energy production compared to that in nonspecific enzymes in two Drosophila ananassae populations. Proc. Natl. Acad. Sci. USA 61:582-85.

Gipson, P. S., J. A. Selander, and J. E. Dunn. 1974. The taxonomic status of wild Canis in Arkansas. Syst. Zool. 23:1-11.

Glover, D. G., M. H. Smith, L. Ames, J. Joule, and J. M. Dubach. 1977. Genetic variation in pika populations. Can. J. Zool. 55:1841-45.

Goertz, J. W. 1964. The influence of habitat quality upon density of cotton rat populations. Ecol. Monogr. 34:359-81.

Goodman, M., and R. E. Tashian. 1976. Molecular Anthropology. Plenum Press, New York. 466 p.

Gorman, G. C., and J. R. Renzi, Jr. 1979. Genetic distance and heterozygosity estimates in electrophoretic studies: effects of samples size. Copeia 1979:242-49.

Gorman, G. C., M. Soule, S. Y. Yang, and E. Nevo. 1975. Evolutionary genetics of insular Adriatic lizards. Evolution 29:52-71.

Gould, S. J. 1966. Allometry and size in ontogeny and phylogeny. Biol. Rev. 41:587-640.

LITERATURE CITED

Gould, S. J. 1977. *Ontogeny and Phylogeny*. Harvard Univ. Press, Cambridge. 501 p.

Grant, P. R. 1965. The adaptive significance of some size trends in island birds. *Evolution* 19:355-67.

Grant, V. 1971. *Plant Speciation*. Columbia Univ. Press, New York. 435 p.

Greenbaum, I. F., and R. J. Baker. 1976. Evolutionary relationships in *Macrotus* (Mammalia: Chiroptera): biochemical variation and karyology. *Syst. Zool.* 25:15-25.

Greenbaum, I. F., R. J. Baker, and P. R. Ramsey. 1978. Chromosomal evolution and the mode of speciation in three species of *Peromyscus*. *Evolution* 32:646-54.

Grinnell, J. 1921. Revised list of the species in the genus *Dipodomys*. *J. Mammal.* 2:94-7.

Guthrie, R. D. 1965. Variability in characters undergoing rapid evolution, an analysis of *Microtus* molars. *Evolution* 19:214-33.

Haldane, J. B. S. 1937. The effect of variation of fitness. *Am. Nat.* 71:337-49.

Hall, E. R., and K. R. Kelson. 1959. *The Mammals of North America*. Vol. II. Ronald Press Co., New York. 547-1083 p.

Handford, P. T., and J. C. Pernetta. 1974. The origin of island races of *Apodemys sylvaticus*: an alternative hypothesis. *J. Zool.* 174:534-37.

Harris, H. 1966. Enzyme polymorphisms in man. *Proc. R. Soc. Lond. B. Biol. Sci.* 164:298-310.

Harris, H., and D. A. Hopkinson. 1972. Average heterozygosity per locus in man: an estimate based on the incidence of enzyme polymorphisms. *Ann. Hum. Genet.* 36:9-20.

Harris, H., and D. A. Hopkinson. 1976. *Handbook of Enzyme Electrophoresis*. Amsterdam: North-Holland.

LITERATURE CITED

Harris, H., D. A. Hopkinson, and E. B. Robson. 1974. The incidence of rare alleles determining electrophoretic variants: data on 43 enzyme loci in man. Ann. Hum. Genet. 37:237-53.

Harris, V. T. 1954. Experimental evidence of reproductive isolation between two subspecies of Peromyscus maniculatus. Contrib. Lab. Vertebr. Biol. Univ. Mich. 70:1-13.

Hart, E. B. 1971. Karyology and evolution of the plains pocket gopher, Geomys bursarius. Ph.D. diss., Univ. Oklahoma, Norman.

Healey, M. C. 1967. Aggression and self-regulation of population size in deer-mice. Ecology 48:377-92.

Heaney, L. R. 1978. Island area and body size of insular mammals: evidence from the tri-colored squirrel (Callosciurus prevosti) of South-east Asia. Evolution 32:29-44.

Heath, C. E. 1978. Serum protein analysis of grey seals. Mammal Rev. 8:47-51.

Hecht, M. K., and J. L. Edwards. 1977. The methodology of phylogenetic inference above the species level. pp. 3-51. In Major Patterns in Vertebrate Evolution. M. K. Hecht, P. C. Goody, and B. M. Hecht (eds.). Plenum Press, New York.

Hedrick, P., S. Jain, and L. Holden. 1978. Multilocus systems in evolution. Evol. Biol. 11:101-84.

Hedrick, P. W. 1971. A new approach to measuring genetic similarity. Evolution 25:276-80.

Hedrick, P. W. 1975. Genetic similarity and distance: comments and comparisons. Evolution 29:362-66.

Hedrick, P. W., M. E. Ginevan, and E. P. Ewing. 1976. Genetic polymorphism in heterogeneous environments. Ann. Rev. Ecol. Syst. 7:1-32.

Hennig, W. 1965. Phylogenetic systematics. Ann. Rev. Entomol. 10:97-116.

LITERATURE CITED

Hennig, W. 1966. Phylogenetic Systematics. Univ. Illinois Press, Urbana. 263 p.

Hibbard, C. W., D. E. Ray, D. E. Savage, D. W. Taylor, and J. E. Guilday. 1965. Quaternary mammals of North America. pp. 509-25. In The Quaternary of the United States. H. E. Wright, Jr., and D. G. Frey (eds.). Princeton Univ. Press, Princeton.

Highton, R., and T. P. Webster. 1976. Geographic protein variation and divergence in populations of the salamander Plethodon cinereus. Evolution 30:33-45.

Hill, W. G. 1974. Estimation of linkage disequilibrium in randomly mating populations. Heredity 33:229-39.

Hill, W. G., and A. Robertson. 1968. Linkage disequilibrium in finite populations. Theor. Appl. Genet. 38:226-31.

Holdgate, M. W. and N. M. Wace. 1961. The influence of man on the floras and faunas of southern islands. Polar Rec. 10:465-93.

Honeycutt, R. L. 1978. Chromosomal and morphological variation in Geomys bursarius (Shaw) from Texas and adjacent states with comments on factors influencing distribution. M.S. thesis, Texas A & M Univ., College Station.

Hooper, E. T. 1949. Faunal relationships of recent North American rodents. Univ. Mich. Misc. Publ. Mus. Zool. 72:1-28.

Hooper, E. T. 1959. The glans penis in five genera of cricetid rodents. Occas. Pap. Mus. Zool. Univ. Mich. 613:1-11.

Hooper, E. T. 1968. Classification. pp. 27-74. In Biology of Peromyscus (Rodentia). J. A. King, (ed.). Spec. Publ. Am. Soc. Mamm. No. 2.

Hooper, E. T., and G. G. Musser. 1964. The glans penis in neotropical cricetines (Family Muridae) with

LITERATURE CITED

comments on classification of muroid rodents. *Misc. Publ. Mus. Zool. Univ. Mich.* 123:1-57.

Howard, W. E. 1960. Innate and environmental dispersal of individual vertebrates. *Am. Midl. Nat.* 63:152-61.

Hsu, T. C., and F. E. Arrighi. 1968. Chromosomes of *Peromyscus* (Rodentia, Cricetidae). I. Evolutionary trends in 20 species. *Cytogenetics* 7:417-46.

Hubby, J. L., and R. C. Lewontin. 1966. A molecular approach to the study of genic heterozygosity in natural populations. I. The number of alleles at different loci in *Drosophila pseudoobscura*. *Genetics* 54:577-94.

Hunt, W. G., and R. K. Selander. 1973. Biochemical genetics of hybridisation in European house mice. *Heredity* 31:11-33.

Jaenike, J., E. D. Parker, Jr., and R. K. Selander. 1979. Clonal niche structure in the parthenogenetic earthworm *Octolasion tyrtaeum*. *Am. Nat.*, in press.

Jaenike, J. R. 1973. A steady state model of genetic polymorphism on islands. *Am. Nat.* 107:793-95.

Jameson, E. W., Jr. 1950. Determining fecundity in male small mammals. *J. Mammal.* 31:433-36.

Jameson, E. W., Jr. 1953. Reproduction of deer mice (*Peromyscus maniculatus* and *P. boylei*) in the Sierra Nevada, California. *J. Mammal.* 34:44-58.

Jensen, J. N., and D. I. Rasmussen. 1971. Serum albumins in natural populations of *Peromyscus*. *J. Mammal.* 52:508-14.

Jensen, J. N., and D. I. Rasmussen. 1972. Erratum in serum albumins in natural populations of *Peromyscus*. *J. Mammal.* 53:413.

LITERATURE CITED

Johns, D., and K. B. Armitage. 1979. Behavioral ecology of alpine yellow-bellied marmots. Behav. Ecol. Sociobiol., 5:133-157.

Johnson, F. M. 1971. Isozyme polymorphisms in Drosophila annanassae: genetic diversity among island populations in the South Pacific. Genetics 68: 77-95.

Johnson, G. B. 1974. Enzyme polymorphism and metabolism. Science 184:28-37.

Johnson, G. B. 1977. Assessing electrophoretic similarity: the problem of hidden heterogeneity. Ann. Rev. Ecol. Syst. 8:309-28.

Johnson, W. E., and R. K. Selander. 1971. Protein variation and systematics in kangaroo rats (genus Dipodomys). Syst. Zool. 20:377-405.

Johnson, W. E., R. K. Selander, M. H. Smith, and Y. J. Kim. 1972. Biochemical genetics of sibling species of the cotton rat (Sigmodon). Univ. Texas Publ. 7213, Stud. Genet. 7:297-305.

Johnston, C. H., and G. R. Carmody. 1980. Allozyme variation within and between populations of the cave spider Meta menardi (Latreille). Submitted to Can. J. Zool.

Johnston, R. F. 1974. Studies in phenetic and genetic covariation. pp. 333-53. In Proc. 8th Int. Conf. Num. Taxonomy. G. F. Estabrook, (ed.). W. H. Freeman, San Francisco.

Jolly, G. M. 1965. Explicit estimates from capture-recapture data with both death and immigration-stochastic model. Biometrika 52:225-47.

Joule, J., and G. N. Cameron. 1974. Field estimation of demographic parameters: influence of Sigmodon hispidus population structure. J. Mammal. 55:309-18.

Justice, K. E. 1962. Ecological and genetical studies of evolutionary forces acting on desert populations

LITERATURE CITED

of Mus musculus. Tucson, Arizona: Arizona-Sonora Desert Museum Inc.

Kahler, A. L., and R. W. Allard. 1970. Genetics of isozyme variants in barley. I. Esterases. Crop Sci. 10:44-448.

Karlin, S., and B. Levikson. 1974. Temporal fluctuations in selection intensities: case of small population size. Theor. Pop. Biol. 6:383-412.

Kennerly, T. E., Jr. 1954. Local differentiation in the pocket gopher (Geomys personatus) in southern Texas. Tex. J. Sci. 6:297-329.

Kennerly, T. E., Jr. 1963. Gene flow pattern and swimming ability of the pocket gopher. S. W. Natur. 8: 85-88.

Kenyon, K. W., and D. W. Rice. 1961. Abundance and distribution of the Steller sea lion. J. Mammal. 42:223-34.

Kidd, K. K., and L. L. Cavalli-Sforza. 1974. The role of genetic drift in the differentiation of Icelandic and Norwegian cattle. Evolution 28:381-95.

Kilpatrick, C. W., and K. L. Crowell. 1980a. Genetic variation in insular and mainland populations of Microtus pennsylvanicus. Submitted to Evolution.

Kilpatrick, C. W., and K. L. Crowell. 1980b. Genetic variation of the rock vole, Microtus chrotorrhinus. Submitted to J. Mammal.

Kilpatrick, C. W., and E. G. Zimmerman. 1975. Genetic variation and systematics of four species of mice of the Peromyscus boylii species group. Syst. Zool. 24:143-62.

Kim, Y. J. 1972. Studies of biochemical genetics and karyotypes in pocket gophers (family Geomyidae). Ph.D. diss., Univ. Texas, Austin.

Kimura, M. 1977. Causes of evolution and polymorphism at the molecular level. pp. 1-28. In Molecular

LITERATURE CITED

Evolution and Polymorphism. M. Kimura (ed.). National Institute of Genetics, Mishima, Japan.

Kimura, M., and T. Ohta. 1971. *Theoretical Aspects of Population Genetics*. Princeton Univ. Press, Princeton. 219 p.

Kimura, M., and T. Ohta. 1972. Population genetics, molecular biometry, and evolution. pp. 43-68. *In Darwinian, Neo-Darwinian, and Non-Darwinian Evolution*, Proc. 6th Berkely Symp. Math. Stat. Prob., Vol. 5. L. M. Le Cam, J. Neyman, and E. L. Scott (eds.). Univ. California Press, Berkely.

Kimura, M., and T. Ohta. 1974. On some principles governing molecular evolution. *Proc. Natl. Acad. Sci. USA* 71:2848-52.

King, J. L., and T. H. Jukes. 1969. Non-Darwinian evolution. *Science* 164:788-98.

King, J. L., and T. Ohta. 1975. Polyallelic mutational equilibria. *Genetics* 79:681-91.

King, M. C., and A. C. Wilson. 1975. Evolution at two levels in humans and chimpanzees. *Science* 188: 107-16.

Kohn, P. H., and R. H. Tamarin. 1978. Selection of electrophoretic loci for reproductive parameters in island and mainland voles. *Evolution* 32:15-28.

Kojima, K., J. Gillespie, and Y. N. Tobari. 1970. A profile of *Drosophila* species' enzymes assayed by electrophoresis. I. Number of alleles, heterozygosities, and linkage disequilibrium in glucose-metabolizing systems and some other enzymes. *Biochem. Genet*. 4:627-37.

Krebs, C. J. 1970. *Microtus* population biology: behavioral changes associated with the population cycle in *M. ochrogaster* and *M. pennsylvanicus*. *Ecology* 51:34-52.

LITERATURE CITED

Krebs, C. J. 1978a. A review of the Chitty hypothesis of population regulation. Can. J. Zool. 56:2463-80.

Krebs, C. J. 1978b. Ecology: The Experimental Analysis of Distribution and Abundance. Second. Ed. Harper and Row, N. Y.

Krebs, C. J., and J. H. Myers. 1974. Population cycles in small mammals. Adv. Ecol. Res. 8:267-399.

Krebs, C. J., M. S. Gaines, B. L. Keller, J. H. Myers, and R. H. Tamarin. 1973. Population cycles in small rodents. Science 179:35-41.

Krebs, C. J., B. L. Keller, and R. H. Tamarin. 1969. Microtus population biology: demographic changes in fluctuating populations of M. ochrogaster and M. pennsylvanicus in southern Indiana. Ecology 50: 587-607.

Krebs, C. J., I. Wingate, J. LeDuc, J. A. Redfield, M. Taitt, and R. Hilborn. 1976. Microtus population biology: dispersal in fluctuating populations of M. townsendii. Can. J. Zool. 54:79-95.

Kreizinger, J. D., and M. W. Shaw. 1970. Chromosomes of Peromyscus (Rodentia, Cricetidae). II. The Y chromosome of Peromyscus maniculatus. Cytogenetics 9:52-70.

Kurten, B. 1972. The Age of Mammals. Columbia Univ. Press, New York. 250 p.

Laing, C., G. R. Carmody, and S. B. Peck. 1976. Population genetics and evolutionary biology of the cave beetle Ptomaphagus hirtus. Evolution 30: 484-98.

Lande, R. 1976. Natural selection and random genetic drift in phenotypic evolution. Evolution 30:314-34.

Lande, R. 1977. On comparing coefficients of variation. Syst. Zool. 26:214-17.

LITERATURE CITED

Lande, R. 1979. Effective deme sizes during long-term evolution estimated form rates of chromosomal rearrangement. Evolution 33:234-51.

Lawlor, T. E. 1974. Chromosomal evolution in Peromyscus. Evolution 28:689-92.

Leamy, L. 1974. Heritability of osteometric traits in a randombred population of mice. J. Hered. 65:109-20.

Leamy, L. 1975. Component analysis of osteometric traits in randombred house mice. Syst. Zool. 24:176-90.

Leamy, L. 1977. Genetic and environmental correlations of morphometric traits in randombred house mice. Evolution 31:357-69.

LeDuc, J., and C. J. Krebs. 1975. Demographic consequences of artificial selection at the LAP locus in voles (Microtus townsendii). Can. J. Zool. 53:1825-40.

Lee, M. R., D. J. Schmidly, and C. C. Huheey. 1972. Chromosomal variation in certain populations of Peromyscus boylii and its systematic implications. J. Mammal. 53:697-707.

Lerner, I. M., 1954. Genetic Homeostatis. John Wiley and Sons, New York. 134 p.

Levene, H. 1949. On a matching problem arising in genetics. Ann. Math. Stat. 20:91-99.

Levene, H. 1953. Genetic equilibrium when more than one ecological niche is available. Am. Nat. 87:331-33.

Levin, B. R., M. L. Petras, and D. I. Rasmussen. 1969. The effect of migration on the maintenance of a lethal polymorphism in the house mouse. Am. Nat. 103:647-61.

Levins, R. 1968. Evolution in Changing Environments. Princeton Univ. Press, Princeton. 120 p.

LITERATURE CITED

Levins, R. 1970. Extinction. pp. 77-107. In Some Mathematical Questions In Biology. Lectures on Mathematics in the Life Sciences. M. Gerstenhaber (ed.). Amer. Math. Soc., Vol. 2, Providence, R.I.

Lewontin, R. C. 1958. Studies on heterozygosity and homeostatis. II: Loss of heterosis in a constant environment. Evolution 12:494-503.

Lewontin, R. C. 1967. An estimate of average heterozygosity in man. Am. J. Hum. Genet. 19:681-85.

Lewontin, R. C. 1970. The units of selection. Ann. Rev. Ecol. Syst. 1:1-18.

Lewontin, R. C. 1972. The apportionment of human diversity. Evol. Biol. 6:381-98.

Lewontin, R. C. 1974. The Genetic Basis of Evolutionary Change. Columbia Univ. Press, New York. 346 p.

Lewontin, R. C., and L. C. Dunn. 1960. The evolutionary dynamics of a polymorphism in the house mouse. Genetics 45:705-22.

Lewontin, R. C., and J. L. Hubby. 1966. A molecular approach to the study of genic heterozygosity in natural populations. II. Amount of variation and degree of heterozygosity in natural populations of Drosophila pseudoobscura. Genetics 54:595-609.

Lewontin, R. C., and J. Krakauer. 1973. Distribution of gene frequency as a test of the theory of the selective neutrality of polymorphisms. Genetics 74:175-95.

Lewontin, R. C., and J. Krakauer. 1975. Testing the heterogeneity of F values. Genetics 80:397-98.

Lewontin, R. C., L. R. Ginzburg, and S. D. Tuljapurkar. 1978. Heterosis as an explanation for large amounts of genic polymorphism. Genetics 88:149-70.

LITERATURE CITED

Li, C. C. 1955. The stability of an equilibrium and the average fitness of a population. Am. Nat. 89:281-95.

Lidicker, W. Z., Jr. 1960. An analysis of intraspecific variation in the kangaroo rat *Dipodomys merriami*. Univ. Calif. Publ. Zool. 67:125-218.

Lidicker, W. Z., Jr. 1962a. Emigration as a possible mechanism permitting the regulation of population density below carrying capacity. Am. Nat. 96:29-33.

Lidicker, W. Z., Jr. 1962b. The nature of subspecies boundaries in a desert rodent and its implications for subspecies taxonomy. Syst. Zool. 11:160-71.

Lidicker, W. Z., Jr. 1973. Regulation of numbers in an island population of the California vole, a problem in community dynamics. Ecol. Monogr. 43:271-302.

Lidicker, W. Z., Jr. 1975. The role of dispersal in the demography of small mammals. pp. 103-28. In *Small Mammals: Their Production and Population Dynamics*. F. B. Golley, K. Petrusewicz, and L. Ryszkowski (eds.). Cambridge Univ. Press, London.

Lidicker, W. Z., Jr. 1976. Social behavior and density regulation in house mice living in large enclosures. J. Anim. Ecol. 45:677-97.

Lomnicki, A. 1974. Evolution of the herbivore-plant, predator-prey and parasite-host systems: a theoretical model. Am. Nat. 108:167-80.

Lomnicki, A. 1977. Evolution of plant resistance and herbivore population cycles. Am. Nat. 111:198-200.

Long, C. A. 1968. An analysis of patterns of variation in some representative Mammalia. I. A review of estimates of variability in selected measurements. Trans. Kansas Acad. Sci. 71:201-77.

Long, C. A. 1969. An analysis of patterns of variation in some representative Mammalia. II. Studies on the nature and correlation of measurements of

LITERATURE CITED

variation. pp. 289-302. In Contributions in Mammalogy. J. K. Jones (ed.). Misc. Publ. Mus. Nat. Hist. Univ. Kansas 51.

Ludwig, W. 1950. Zur theorie der konkurrenz. Die annidation (Einnischung) als funther Evolutionsfaktor. Neue. Ergeb. Probleme Zool. Klatt-Festschrift 1950:516-37.

McCabe, T. T., and I. McT. Cowan. 1945. Peromyscus maniculatus macroninus and the problem of insularity. Roy. Can. Inst. Trans. 172-215.

McClenaghan, L. R., Jr. 1977. Genic variability, morphological variation and reproduction in central and marginal populations of Sigmodon hispidus. Ph.D. diss., Univ. Kansas, Lawrence.

McClenaghan, L. R., Jr., and M. S. Gaines. 1978. Reproduction in marginal populations of the hispid cotton rat (Sigmodon hispidus) in northeastern Kansas. Occas. Pap. Univ. Mus. Nat. Hist. 74:1-16.

McCracken, G. F., and J. W. Bradbury. 1978. Paternity and genetic heterogeneity in the polygynous bat, Phyllostomus hastatus. Science 198:303-06.

McDermid, E. M., R. Ananthakrishnan, and N. S. Agar. 1972. Electrophoretic investigation of plasma and red cell proteins and enzymes of Macquarie Island elephant seals. Anim. Blood Groups Biochem. Genet. 3:85-94.

McKinney, C. O., R. K. Selander, W. E. Johnson, and S. Y. Yang. 1972. Genetic variation in the side-blotched lizard (Uta stansburiana). Univ. Texas Publ. 7213, Stud. Genet. 7:307-18.

McNab, B. K. 1971. On the ecological significance of Bergmann's rule. Ecology 52:845-54.

MacArthur, R. H., and E. O. Wilson. 1967. The Theory of Island Biogeography. Princeton Univ. Press, Princeton. 203 p.

LITERATURE CITED

Manwell, C., and C. M. A. Baker. 1970. Molecular Biology and the Origin of Species: Heterosis, Protein Polymorphism, and Animal Breeding. Univ. of Washington Press, Seattle. 394 p.

Mares, M. A. 1976. Convergent evolution of desert rodents: multivariate analysis and zoogeographic implications. Paleobiology 2:39-63.

Marquardt, R. B. 1976. Linkage and selection component analysis of biochemical variants in the deer mouse Peromyscus maniculatus. M. S. thesis, Univ. of Colorado, Denver.

Mascarello, J. T., and T. C. Hsu. 1976. Chromosome evolution in woodrats, genus Neotoma (Rodentia: Cricetidae). Evolution 30:152-69.

Mascarello, J. T., and C. R. Shaw. 1973. Analysis of multiple zymograms in certain species of Peromyscus, Rodentia: Cricetidae. Tex. Rep. Biol. Med. 31:507-18.

Massey, D. R. 1977. Spatial-temporal changes in the genetic composition of deer mouse populations (Peromyscus maniculatus). M. S. thesis, Univ. Colorado, Denver.

Matthew, W. D. 1915. Climate and evolution. Ann. N. Y. Acad. Sci. 24:171-318.

Matthey, R., and J. van Brink. 1960. Nouvelle contribution à la cytologie comparée des Chamaeleontidae (Reptilia, Lacertilia). Bull. Soc. Vaud. Sci. Nat. 67:333-48.

Mayr, E. 1954. Change of genetic environment and evolution. pp. 157-80. In Evolution as a Process. J. Huxley, A. C. Hardy, and E. B. Ford (eds.). Allen and Unwin, London.

Mayr, E. 1963. Animal Species and Evolution. Harvard Univ. Press, Cambridge. 797 p.

Mayr, E. 1969. Principles of Systematic Zoology. McGraw-Hill, New York. 428 p.

LITERATURE CITED

Merkle, D. A., S. I. Guttman, and M. A. Nickerson. 1977. Genetic uniformity throughout the range of the hellbender, Cryptobranchus alleganiensis. Copeia 1977:549-53.

Michie, D. 1953. Affinity: a new genetic phenomenon in the house mouse: evidence from distance crosses. Nature 171:26-27.

Mickevich, M. F. 1978. Taxonomic congruence. Syst. Zool. 27:143-58.

Mickevich, M. F. and M. S. Johnson. 1976. Congruence between morphological and allozyme data in evolutionary inference and character evolution. Syst. Zool. 25:260-70.

Milliman, J. D., and K. O. Emery. 1968. Sea levels during the past 35,000 years. Science 162:1121-23.

Milne, A. 1962. On a theory of natural control of insect population. J. Theor. Biol. 3:19-50.

Mitton, J. B. 1977. Genetic differentiation of races of man as judged by single-locus and multilocus analyses. Am. Nat. 111:203-12.

Mitton, J. B. 1978. Relationship between heterozygosity for enzyme loci and variation of morphological characters in natural populations. Nature 273:661-2.

Mukai, T., L. E. Mettler, and S. I. Chigusa. 1971. Linkage disequilibrium in a local population of Drosophila melanogaster. Proc. Natl. Acad. Sci. USA 68:1065-69.

Muller, H. J. 1956. On the relation between chromosome changes and gene mutations. Brookhaven Symp. Biol. 8:126-47.

Myers, J. H. 1974. Genetic and social structure of feral house mouse populations on Grizzly Island, California. Ecology 55:747-59.

LITERATURE CITED

Myers, J. H., and C. J. Krebs. 1971a. Genetic, behavioral, and reproductive attributes of dispersing field voles Microtus pennsylvanicus and Microtus ochrogaster. Ecol. Monogr. 41:53-78.

Myers, J. H., and C. J. Krebs. 1971b. Sex ratios in open and enclosed vole populations: demographic implications. Am. Nat. 105:325-44.

Naevdal, G. 1966a. Protein polymorphism used for identification of harp seal populations. Arbok f. Univ. i. Bergen, Mat.-Naturv. Seria 9:3-20, 1965.

Naevdal, G. 1966b. Hemoglobins and serum proteins in four North Atlantic seals, studied by electrophoresis. Fiskeridir. Skrifter Ser. Havundersokelser 14:37-50.

Naevdal, G. 1969. Blood protein polymorphism in harp seals off eastern Canada. J. Fish. Res. Bd. Can. 26:1397-99.

Naylor, A. F. 1963. A theorem on possible kinds of mating systems which tend to increase heterozygosity. Evolution 17:369-70.

Neel, J. V., M. Layrisse, and F. M. Salzano. 1977. Man in the tropics: the Yanomama Indians. pp. 109-42. In Population Structure and Human Variation. G. A. Harrison (ed.). Cambridge University Press, Cambridge.

Nei, M. 1971. Interspecific gene differences and evolutionary time estimated from electrophoretic data on protein identity. Am. Nat. 105:385-98.

Nei, M. 1972. Genetic distance between populations. Am. Nat. 106:283-92.

Nei, M. 1975. Molecular Population Genetics and Evolution. North-Holland Press, Amsterdam. 288 p.

Nei, M. 1978. Estimation of average heterozygosity and genetic distance from a small number of individuals. Genetics 89:583-90.

LITERATURE CITED

Nei, M., and W. Li. 1973. Linkage disequilibrium in subdivided populations. Genetics 75:213-19.

Nei, M., and T. Maruyama. 1975. Lewontin-Krakauer test for neutral genes. Genetics 80:395.

Nei, M., and A. K. Roychoudhury. 1972. Gene differences between Caucasian, Negro, and Japanese populations. Science 177:434-36.

Nei, M., and A. K. Roychoudhury. 1974. Genic variation within and between the three major races of man, caucasoids, negroids, and mongoloids. Am. J. Hum. Genet. 26:421-43.

Nei, M., T. Maruyama, and R. Chakraborty. 1975. The bottleneck effect and genetic variability in populations. Evolution 29:1-10.

Nelson, G. 1979. Cladistic analysis and synthesis: principles and definitions, with a historical note on Adanson's Familles des Plantes (1763-1764). Syst. Zool. 28:1-21.

Nelson, G. J. 1972. Comments on Hennig's "phylogenetic systematics" and its influence on ichthyology. Syst. Zool. 21:364-74.

Nelson, G. J. 1973. The higher-level phylogeny of vertebrates. Syst. Zool. 22:87-91.

Neter, J., and W. Wasserman. 1974. Applied Linear Statistical Models. Irwin, Homewood, Illinois. 842 p.

Nevo, E. 1978. Genetic variation in natural populations: patterns and theory. Theor. Pop. Biol. 13:121-77.

Nevo, E., and H. Bar-El. 1976. Hybridization and speciation in fossorial mole rats. Evolution 30:831-40.

Nevo, E., and C. R. Shaw. 1972. Genetic variation in a subterranean mammal, Spalax ehrenbergi. Biochem. Genet. 7:235-41.

LITERATURE CITED

Nevo, E., Y. J. Kim, C. R. Shaw, and C. S. Thaeler, Jr. 1974. Genetic variation, selection and speciation in *Thomomys talpoides* pocket gophers. *Evolution* 28:1-23.

Newsome, A. E. 1969a. A population study of house-mice temporarily inhabiting a South Australian wheat-field. *J. Anim. Ecol.* 38:341-59.

Newsome, A. E. 1969b. A population study of house-mice permanently inhabiting a reed-bed in South Australia. *J. Anim. Ecol.* 38:361-77.

Nichols, E. A., and F. H. Ruddle. 1973. A review of enzyme polymorphism, linkage and electrophoretic conditions for mouse and somatic cell hybrids in starch gels. *J. Histochem. Cytochem.* 21:1066-81.

Nicholson, A. J. 1954. An outline of the dynamics of animal populations. *Aust. J. Zool.* 2:9-65.

Nielsen, J. T. 1977. Variation in the number of genes coding for salivary amylase in the bank vole, *Clethrionomys glareola*. *Genetics* 85:155-69.

Nixon, S. E., and R. J. Taylor. 1977. Large genetic distances associated with little morphological variation in *Polycelis coronata* and *Dugesia tigrina* (Planaria). *Syst. Zool.* 26:152-64.

Nozawa, K. 1972. Population genetics of Japanese monkeys. I. Estimation of effective troop size. *Primates* 13:381-93.

Nozawa, K., T. Shotake, and Y. Okura. 1975. Blood protein polymorphisms and population structure of the Japanese macaque, *Macaca fuscata fuscata*. pp. 225-41. *In Isozymes IV: Genetics and Evolution*. C. L. Markert (ed.). Academic Press, New York.

Ohta, T. 1976. Role of very slightly deleterious mutations in molecular evolution and polymorphism. *Theor. Pop. Biol.* 10:254-75.

LITERATURE CITED

Ohta, T., and M. Kimura. 1974. Simulation studies on electrophoretically detectable genetic variability in a finite population. Genetics 76:615-24.

Ohta, T., and M. Kimura. 1975. Theoretical analysis of electrophoretically detectable polymorphisms: models of very slightly deleterious mutations. Am. Nat. 109:137-45.

Orians, G. H. 1962. Natural selection and ecological theory. Am. Nat. 96:257-63.

Oxnard, C. E. 1968. Primate evolution -- a method of investigation. Am. J. Phys. Anthr. 28:289-302.

Oxnard, C. E. 1969. Mathematics, shape and function: a study in primate anatomy. Am. Sci. 57:75-96.

Parker, E. D., Jr. 1978. Genetic and morphological consequences of parthenogenesis in the hybrid lizard Cnemidophorus tesselatus. Ph.D. diss., Univ. Rochester, Rochester, New York.

Pathak, S., T. C. Hsu, and F. E. Arrighi. 1973. Chromosomes of Peromyscus (Rodentia, Cricetidae). IV. The role of heterochromatin in karyotypic evolution. Cytogenet. Cell Genet. 12:315-26.

Patton, J. L. 1970. Karyotypic variation following an elevational gradient in the pocket gopher, Thomomys bottae grahamensis Goldman. Chromosoma 31:41-50.

Patton, J. L. 1972. Patterns of geographic variation in karyotype in the pocket gopher Thomomys bottae (Eydoux and Gervais). Evolution 26:574-86.

Patton, J. L., and J. H. Feder. 1978. Genetic divergence between populations of the pocket gopher, Thomomys umbrinus (Richardson). Zeit. f. Säugetierk. 43:17-30.

Patton, J. L., and T. C. Hsu. 1967. Chromosomes of the golden mouse, Peromyscus (Ochrotomys) nuttalli (Harlan). J. Mammal. 48:637-39.

LITERATURE CITED

Patton, J. L., and P. Myers. 1974. Chromosome identity of black rats (Rattus rattus) from the Galapagos Islands, Ecuador. Experientia 30:1140-41.

Patton, J. L., and S. Y. Yang. 1977. Genetic variation in Thomomys bottae pocket gophers: macrogeographic patterns. Evolution 31:697-720.

Patton, J. L., R. K. Selander, and M. H. Smith. 1972. Genic variation in hybridizing populations of gophers (genus Thomomys). Syst. Zool. 21:263-70.

Patton, J. L., S. Y. Yang, and P. Myers. 1975. Genetic and morphologic divergence among introduced rat populations (Rattus rattus) of the Galapagos Archipelago, Ecuador. Syst. Zool. 24:296-310.

Pearson, K. 1897. Mathematical contributions to the theory of evolution. On a form of spurious correlation which may arise when indices are used in the measurement of organs. Proc. Roy. Soc. Lond. 60: 489-502.

Pearson, O. P., and C. P. Pearson. 1978. The diversity and abundance of vertebrates along an altudinal gradient in Peru. Mem. Mus. Hist. Nat. "Javier Prado" 18:1-97.

Penney, D. F., and E. G. Zimmerman. 1976. Genic divergence and local population differentiation by random drift in the pocket gopher genus Geomys. Evolution 30:473-83.

Peters, J., and H. R. Nash. 1977. Polymorphism of esterase 11 in Mus musculus, a further esterase locus on chromosome 8. Biochem. Genet. 15:217-26.

Peters, J., and H. R. Nash. 1978. Esterases of Mus musculus: substrate and inhibition characteristics, new isozymes and homologies with man. Biochem. Genet. 16:553-69.

Petersen, W. 1969. Population. The MacMillan Company Collier-MacMillan Ltd., London. 735 p.

LITERATURE CITED

Petras, M. L. 1967. Studies of natural populations of Mus. I. Biochemical polymorphisms and their bearing on breeding structure. Evolution 21:259-74.

Petticrew, B. G., and R. M. F. S. Sadleir. 1974. The ecology of the deer mouse Peromyscus maniculatus in a coastal coniferous forest. I. Population dynamics. Can. J. Zool. 52:107-18.

Pickering, J., L. L. Getz, and G. S. Whitt. 1974. An esterase phenotype correlated with dispersal in Microtus. Trans. Ill. State Acad. Sci. 67:471-75.

Pimentel, D. 1964. Population ecology and genetic feedback mechanism. In Genetics Today: Proc. 11th Int. Congr. Genetics. p. 483-88.

Pimentel, D. 1968. Population regulation and genetic feedback. Science 159:1432-37.

Pizzimenti, J. J. 1976. Genetic divergence and morphological convergence in the prairie dogs, Cynomys gunnisoni and Cynomys leucurus. II. Genetic analysis. Evolution 30:367-79.

Platnick, N. I. and G. Nelson. 1979. The purposes of biological classification. In PSA 1978, Vol. 2. P. D. Asquith and I. Hacking (eds.).

Powell, J. R. 1971. Genetic polymorphisms in varied environments. Science 174:1035-36.

Powell, J. R. 1975. Protein variation in natural populations of animals. pp. 79-119. In Evolutionary Biology, Vol. 8. T. Dobzhansky, M. K. Hecht, and W. C. Steere (eds.). Plenum Press, New York.

Prakash, S. 1972. Origin of reproductive isolation in the absence of apparent genic differentiation in a geographic isolate of Drosophila pseudoobscura. Genetics 72:143-55.

Ramsey, P. R. 1973. Spatial and temporal variation in genetic structure of insular and mainland popula-

LITERATURE CITED

tions of Peromyscus polionotus. Ph.D. diss., Univ. Georgia, Athens.

Rasmussen, B., M. Rasmussen, and J. Nygren. 1977. Genetically controlled differences in behavior between cycling and non-cycling populations of field vole (Microtus agrestis). Hereditas 87:33-42.

Rasmussen, D. I. 1964. Blood group polymorphism and inbreeding in natural populations of the deer mouse Peromyscus maniculatus. Evolution 18:219-29.

Rasmussen, D. I. 1970. Biochemical polymorphisms and genetic structure in populations of Peromyscus. pp. 335-49. In Variation in Mammalian Populations. R. J. Berry, and H. N. Southern (eds.). Symp. Zool. Soc. Lond., No. 26. Academic Press, London.

Redfield, J. A. 1976. Distribution, abundance, size, and genetic variation of Peromyscus maniculatus on the Gulf Islands of British Columbia. Can. J. Zool. 54:463-74.

Redfield, J. A., M. J. Taitt, and C. J. Krebs. 1978a. Experimental alterations of sex-ratios in populations of Microtus oregoni, the creeping vole. J. Anim. Ecol. 47:55-69.

Redfield, J. A., M. J. Tiatt, and C. J. Krebs. 1978b. Experimental alteration of sex ratios in populations of Microtus townsendii, a field vole. Can. J. Zool. 56:17-27.

Reeve, E. C. R., and F. W. Robertson. 1953. Studies in quantitative inheritance. II. Analysis of a strain of Drosophila melanogaster selected for long wings. J. Genet. 51:276-316.

Reimer, J. D., and M. L. Petras. 1967. Breeding structure of the house mouse, Mus musculus, in a population cage. J. Mammal. 48:88-99.

Reimer, J. D., and M. L. Petras. 1968. Some aspects of commensal populations of Mus musculus in Southwestern Ontario. Can. Field-Nat. 82:32-42.

LITERATURE CITED

Ridgeway, G. J., S. W. Sherburne, and R. D. Lewis. 1970. Polymorphisms in the esterases of Atlantic herring. Trans. Am. Fish. Soc. 99:147-51.

Rinker, G. C. 1942. An extension of the range of the Texas cotton rat in Kansas. J. Mammal. 23:439.

Rinker, G. C. 1963. A comparative myological study of three subgenera of Peromyscus. Occ. Pap. Mus. Zool. Univ. Mich. 632:1-18.

Robertson, A. 1975. Remarks on the Lewontin-Krakauer test. Genetics 80:396.

Robertson, F. W., and E. C. R. Reeve. 1952. Heterozygosity, environmental variation and heterosis. Nature 170:286.

Rogers, J. S. 1972. Measures of genetic similarity and genetic distance. Univ. Tex. Publ. 7213, Stud. Genet. 7:145-53.

Rohlf, F. J., J. Kispaugh, and R. Bartcher. 1969. Numerical taxonomy system of multivariate statistical programs. Version of September, 1969. St. Univ. New York, Stony Brook.

Rohlf, F. J., J. Kispaugh, and D. Kirk. 1974. Numerical taxonomy system of multivariate statistical programs. St. Univ. New York, Stony Brook.

Romero-Herrera, A. E., H. Lehmann, K. A. Joysey, and A. E. Friday. 1978. On the evolution of myoglobin. Phil. Trans. Royal Soc. (Lond.) 283:61-163.

Rose, R. K. 1973. A small mammal live trap. Trans. Kan. Acad. Sci. 78:14-17.

Rose, R. K. 1979. Levels of wounding in the meadow vole, Microtus pennsylvanicus. J. Mammal. 60:37-45.

Rose, R. K., and M. S. Gaines. 1976. Levels of aggression in fluctuating populations of the prairie vole, Microtus ochrogaster, in eastern Kansas. J. Mammal. 57:43-57.

LITERATURE CITED

Rose, R. K., and M. S. Gaines. 1978. The reproductive cycle of Microtus ochrogaster in eastern Kansas. Ecol. Monogr. 48:21-42.

Rose, R. K., and W. D. Hueston. 1978. Wound healing in meadow voles. J. Mammal. 59:186-88.

Rosen, D. E. 1974. Cladism or gradism?: A reply to Ernst Mayr. Syst. Zool. 23:446-51.

Rosen, D. E. 1976. A vicariance model of Caribbean biogeography. Syst. Zool. 24:431-464.

Rosen, D. E. 1978. Vicariant patterns and historical explanation in biogeography. Syst. Zool. 27:159-88.

Roughgarden, J. 1972. Evolution of niche width. Am. Nat. 106:683-718.

Ruddle, F. H., T. H. Roderick, T. B. Shows, P. G. Weigl, R. K. Chipman, and P. K. Anderson. 1969a. Measurement of genetic heterogeneity by means of enzyme polymorphisms in wild populations of the mouse. J. Hered. 60:321-22.

Ruddle, F. H., T. B. Shows, and T. H. Roderick. 1969b. Esterase genetics in Mus musculus: expression, linkage, and polymorphism of locus Es-2. Genetics 62:393-99.

Sadleir, R. M. F. S. 1965. The relationship between agonistic behavior and population changes in the deer mouse, Peromyscus maniculatus. J. Anim. Ecol. 34:331-52.

Sadleir, R. M. F. S. 1974. The ecology of the deer mouse, Peromyscus maniculatus in a coastal coniferous forest. II. Reproduction. Can. J. Zool. 52:119-31.

Sage, R. D. 1978. Genetic heterogeneity of Spanish house mice (Mus musculus complex). pp. 519-553. In Origins of Inbred Mice. H. C. Morse (ed.). Academic Press, N. Y.

LITERATURE CITED

Salthe, S. N., and M. L. Crump. 1977. A Darwinian interpretation of hind-limb variability in frog populations. Evolution 31:737-49.

Sarich, V. M. 1977. Rates, sample sizes, and the neutrality hypothesis for electrophoresis in evolutionary studies. Nature 265:24-28.

Sarich, V. M., and J. E. Cronin. 1976. Molecular systematics of the primates. pp. 141-170. In Molecular Anthropology. M. Goodman and R. E. Tashian (eds.). Plenum Press, N. Y. 466 p.

Saura, A., O. Halkka, and J. Lokki. 1973. Enzyme gene heterozygosity in small island populations of Philaenus spumarius (L.) (Homoptera). Genetica 44:459-73.

Schendel, R. R. 1940. Life history notes of Sigmodon hispidus texianus with special emphasis on populations and nesting habits. M. S. thesis, Okla. State Univ., Stillwater.

Schmidly, D. J., and G. L. Schroeter. 1974. Karyotypic variation in Peromyscus boylii (Rodentia: Cricetidae) from Mexico and corresponding taxonomic implications. Syst. Zool. 23:333-42.

Schmitt, L. H. 1977. Mitochondrial isocitrate dehydrogenase variation in the Australian bush-rat, Rattus fuscipes greyii. Anim. Blood Grps. Biochem. Genet. 8:73-78.

Schmitt, L. H. 1978. Genetic variation in isolated populations of the Australian bush-rat, Rattus fuscipes. Evolution 32:1-14.

Schnell, G. D., T. L. Best, and M. L. Kennedy. 1978. Interspecific morphologic variation in kangaroo rats (Dipodomys): degree of concordance with genic variation. Syst. Zool. 27:34-48.

Schnell, G. D., J. S. Millar, M. L. Kennedy, and R. K. Selander. 1980. Concordance of non-metric morphological and genic variation in hybridizing house mice. In prep.

LITERATURE CITED

Schoener, T. W. 1974. Resource partitioning in ecological communities. Science 185:27-39.

Schwartz, O. A. 1979. The distribution of genetic variation in a social mammal, the yellow-bellied marmot. Ph.D. diss., Univ. Kansas, Lawrence.

Schwartz, O. A., and K. B. Armitage. 1980. Genetic variation in a social mammal: the marmot model. Science 207:665-66.

Seal, U. S., A. W. Erikson, D. B. Siniff, and D. R. Cline. 1971. Blood chemistry and protein polymorphisms in three species of Antarctic seals (Lobodon carcinophagus, Leptonychotes weddelli, and Mirounga leonina). In Antarctic Pinnipedia. W. H. Burt (ed.). Antarctic Res. Ser. 226 p.

Selander, R. K. 1970a. Behavior and genetic variation in natural populations. Am. Zool. 10:53-66.

Selander, R. K. 1970b. Biochemical polymorphism in populations of the house mouse and old-field mouse. pp. 73-91. In Variation in Mammalian Populations. R. J. Berry, and H. N. Southern (eds.). Academic Press, London.

Selander, R. K. 1976. Genic variation in natural populations. pp. 21-45. In Molecular Evolution. F. J. Ayala (ed.). Sinauer Assoc., Inc., Sunderland, Mass.

Selander, R. K., and R. O. Hudson. 1976. Animal population structure under close inbreeding: the land snail Rumina in southern France. Am. Nat. 110:695-718.

Selander, R. K., and W. E. Johnson. 1973. Genetic variation among vertebrate species. Ann. Rev. Ecol. Syst. 4:75-91.

Selander, R. K., and D. W. Kaufman. 1973a. Genic variability and strategies of adaptation in animals. Proc. Natl. Acad. Sci. USA 70:1875-77.

LITERATURE CITED

Selander, R. K., and D. W. Kaufman. 1973b. Self-fertilization and genetic population structure in a colonizing land snail. Proc. Natl. Acad. Sci. USA 70:1186-90.

Selander, R. K., and D. W. Kaufman. 1975. Genetic structure of populations of the brown snail (Helix aspersa). I. Microgeographic variation. Evolution 29:385-401.

Selander, R. K., and S. Y. Yang. 1969. Protein polymorphism and genic heterozygosity in a wild population of the house mouse (Mus musculus). Genetics 63:653-67.

Selander, R. K., W. G. Hunt, and S. Y. Yang. 1969a. Protein polymorphism and genic heterozygosity in two European subspecies of the house mouse. Evolution 23:379-90.

Selander, R. K., S. Y. Yang, and W. G. Hunt. 1969b. Polymorphism in esterases and hemoglobin in wild populations of the house mouse (Mus musculus). Univ. Texas Publ. 6918, Stud. Genet. 5:271-338.

Selander, R. K., D. W. Kaufman, R. J. Baker, and S. L. Williams. 1974. Genic and chromosomal differentiation in pocket gophers of the Geomys bursarius group. Evolution 28:557-64.

Selander, R. K., M. H. Smith, S. Y. Yang, W. E. Johnson, and J. B. Gentry. 1971. Biochemical polymorphism and systematics in the genus Peromyscus. I. Variation in the old-field mouse (Peromyscus polionotus). Stud. Genet. 6:49-90.

Semeonoff, R., and F. W. Robertson. 1968. A biochemical and ecological study of plasma esterase polymorphism in natural populations of the field vole, Microtus agrestis L. Biochem. Genet. 1:205-27.

Setzer, H. W. 1949. Subspeciation in the kangaroo rat, Dipodomys ordii. Univ. Kansas Publ. Mus. Nat. Hist. 1:473-573.

LITERATURE CITED

Shaughnessy, P. D. 1969. Transferrin polymorphism and population structure of the Weddell seal *Leptonychotes weddelli* (Lesson). *Aust. J. Biol. Sci.* 22:1581-84.

Shaughnessy, P. D. 1970. Serum protein variation in southern fur seals, *Arctocephalus* spp., in relation to their taxonomy. *Aust. J. Zool.* 18:331-43.

Shaughnessy, P. D. 1974. An electrophoretic study of blood and milk proteins of the southern elephant seal, *Mirounga leonina*. *J. Mammal.* 55:796-808.

Shaughnessy, P. D. 1975. Biochemical comparison of the harbour seals *Phoca vitulina richardi* and *P. v. largha*. *In Biology of the Seal*. K. Ronald, and A. W. Mansfield (eds.). Rapp. Proc.-ver. Réun., Cons. Intern. Explor. Mer 169. 557 p.

Shaw, C. R., and R. Prasad. 1970. Starch gel electrophoresis of enzymes: a compilation of recipes. *Biochem. Genet.* 4:297-320.

Sheppe, W. 1965. Island populations and gene flow in the deer mouse, *Peromyscus leucopus*. *Evolution* 19:480-95.

Simberloff, D. S. 1974. Equilibrium theory of island biogeography and ecology. *Ann. Rev. Ecol. Syst.* 5:161-82.

Simberloff, D. S., and E. O. Wilson. 1969. Experimental zoogeography of islands: the colonization of empty islands. *Ecology* 50:278-96.

Simpson, G. G. 1945. The principles of classification and a classification of mammals. *Bull. Am. Mus. Nat. Hist.* 85:1-350.

Simpson, G. G., A. Roe, and R. C. Lewontin. 1960. *Quantitative Zoology*. Harcourt, Brace and Co., Inc., New York. 440 p.

Sinnock, P. 1975. The Wahlund effect for the two-locus model. *Am. Nat.* 109:565-70.

LITERATURE CITED

Sinnock, P., and C. F. Sing. 1972. Analysis of multilocus genetic systems in Tecumseh, Michigan. II. Consideration of the correlation between nonalleles in gametes. *Am. J. Hum. Genet.* 24:393-415.

Slatkin, M. 1970. Selection and polygenic characters. *Proc. Natl. Acad. Sci. USA* 66:87-93.

Smith, M. F. 1978. Relationships between genetic variability and niche dimensions among coexisting species of *Peromyscus*. Ph.D. diss., Univ. Calif., Berkeley.

Smith, M. H., C. T. Garten, Jr., and P. R. Ramsey. 1975. Genic heterozygosity and population dynamics in small mammals. pp. 85-102. *In Isozymes IV: Genetics and Evolution*. C. L. Markert (ed.). Academic Press, New York.

Smith, M. H., H. O. Hillestad, M. N. Manlove, and R. L. Marchinton. 1976. Use of population genetics data for the management of fish and wildlife populations. *Trans. N. Am. Wildl. Nat. Res. Conf.* 41: 119-33.

Smith, M. H., M. N. Manlove, and J. Joule. 1978. Spatial and temporal dynamics of genetic organization of small mammal populations. pp. 99-103. *In Populations of Small Mammals Under Natural Conditions*. D. P. Snyder (ed.). Univ. of Pittsburgh, Pittsburgh.

Smith, M. H., R. K. Selander, and W. E. Johnson. 1973. Biochemical polymorphism and systematics in the genus *Peromyscus*. III. Variation in the Florida deer mouse (*Peromyscus floridanus*), a pleistocene relict. *J. Mammal.* 54:1-13.

Sneath, P. H. A., and R. R. Sokal. 1973. *Numerical Taxonomy*. W. H. Freeman and Co., San Francisco. 573 p.

Snyder, L. R. G. 1978a. Genetics of hemoglobin in the deer mouse, *Peromyscus maniculatus*. I. Multiple α- and β-globin structural loci. *Genetics* 89:511-30.

LITERATURE CITED

Snyder, L. R. G. 1978b. Genetics of hemoglobin in the deer mouse, Peromyscus maniculatus. II. Multiple alleles at regulatory loci. Genetics 89:531-50.

Sokal, R. R., and F. J. Rohlf. 1969. Biometry. W. H. Freeman and Co., San Francisco. 776 p.

Sokal, R. R., and P. H. A. Sneath. 1963. Principles of Numerical Taxonomy. W. H. Freeman and Co., San Francisco. 359 p.

Soule, M. 1971. The variation problem: the gene flow-variation hypothesis. Taxon 20:37-50.

Soule, M. 1972. Phenetics of natural populations. III. Variation in insular populations of a lizard. Am. Nat. 106:429-46.

Soule, M. 1973a. Phenetics of natural populations. III. Variation in insular populations of a lizard. Am. Nat. 106:429-46.

Soule, M. 1973b. The epistasis cycle: a theory of marginal populations. Ann. Rev. Ecol. Syst. 4:165-87.

Soule, M. 1976. Allozyme variation: its determinants in space and time. pp. 60-70. In Molecular Evolution. F. J. Ayala (ed.). Sinauer Assoc., Inc., Sunderland, Mass.

Soule, M. 1979. Heterozygosity and developmental stability: another look. Evolution 33:396-401.

Soule, M., and B. R. Stewart. 1970. The "niche-variation" hypothesis: a test and alternatives. Am. Nat. 104:85-97.

Soule, M., and S. Y. Yang. 1973. Genetic variation in side-blotched lizards on islands in the Gulf of California. Evolution 27:593-600.

Soule, M., S. Y. Yang, M. G. W. Weiler, and G. C. Gorman. 1973. Island lizards: the genetic-phenetic variation correlation. Nature 242:191-93.

LITERATURE CITED

Spassky, B., R. C. Richmond, S. Perez-Salas, O. Pavlovsky, C. A. Mourao, A. S. Hunter, H. Hoenigsberg, T. Dobzhansky, and F. J. Ayala. 1971. Geography of the sibling species related to Drosophila willistoni, and of the semispecies of the Drosophila paulistorum complex. Evolution 25:129-43.

Spiess, E. B. 1977. Genes in Populations. John Wiley and Sons, New York. 780 p.

Stanley, S. M. 1973. An explanation for Cope's Rule. Evolution 27:1-26.

Stenseth, N. C. 1977. Evolutionary aspects of demographic cycles: the relevance of some models of cycles for microtine fluctuations. Oikos 29:525-38.

Stenseth, N. C. 1978a. Demographic strategies in fluctuating populations of small rodents. Oecologia 33:149-72.

Stenseth, N. C. 1978b. Is the female biased sex ratio in wood lemming Myopus schisticolor maintained by cyclic inbreeding? Oikos 30:83-89.

Stickel, L. F. 1968. Home range and travels. pp. 373-411. Biology of Peromyscus (Rodentia). King (ed.). Am. Soc. Mammal. Spec. Publ. No. 2.

Stock, A. D. 1974. Chromosome evolution in the genus Dipodomys and its taxonomic and phylogenetic implications. J. Mammal. 55:505-26.

Storer, T. I., F. C. Evans, and F. G. Palmer. 1944. Some rodent populations in the Sierra Nevada of California. Ecol. Monogr. 14:165-92.

Straney, D. O. 1978. Variance partitioning and nongeographic variation. J. Mammal. 59:1-11.

Straney, D. O. 1980. Chaos or order: simple statistical tests for cladistic synthesis. Syst. Zool. (submitted).

Straney, D. O., M. H. Smith, I. F. Greenbaum, and R. J. Baker. 1979. Biochemical genetics. pp. 157-76.

LITERATURE CITED

In Biology of the Bats of the New World Family Phyllostomatidae. Part III. R. J. Baker, J. K. Jones, Jr., and D. C. Carter (eds.). Spec. Publ. 16. Texas Tech Press, Lubbock.

Stuiver, M., and H. W. Borns, Jr. 1975. Late quaternary marine invasion in Maine: its chronology and associated crustal movement. Geol. Soc. Am. Bull. 86:99-104.

Sullivan, T. P. 1977. Demography and dispersal in island and mainland populations of the deer mouse, Peromyscus maniculatus. Ecology 58:964-78.

Sumner, F. B. 1930. Genetic and distributional studies of three sub-species of Peromyscus. J. Genet. 23:275-76.

Sumner, F. B. 1932. Genetics, distributional, and evolutionary studies of the sub-species of deer-mice (Peromyscus). Bibl. Genet. 9:1-106.

Sutherland, T. M. 1965. The correlation between feed efficiency and rate of gain, a ratio and its denominator. Biometrics 21:739-49.

Svendsen, G. E. 1974. Behavioral and environmental factors in the spatial distribution and population dynamics of a yellow-bellied marmot population. Ecology 55:760-71.

Svendsen, G. E., and K. B. Armitage. 1973. Mirror-image stimulation applied to field behavioral studies. Ecology 54:623-27.

Swank, R. T., K. Paigen, and R. E. Ganschow. 1973. Genetic control of glucuronidase induction in mice. J. Mol. Biol. 81:225-43.

Tamarin, R. H. 1977a. Dispersal in island and mainland voles. Ecology 58:1044-54.

Tamarin, R. H. 1977b. Demography of the beach vole (Microtus breweri) and the meadow vole (Microtus pennsylvanicus) in southeastern Massachusetts. Ecology 58:1310-21.

LITERATURE CITED

Tamarin, R. H. 1978a. A defense of single-factor models of population regulation. pp. 159-162. In Populations of Small Mammals Under Natural Conditions. Pymatuning Symp. in Ecol., Vol. 5. D. P. Snyder (ed). Pymatuning Lab. of Ecol., Univ. of Pittsburgh.

Tamarin, R. H. 1978b. Dispersal, population regulation, and K-selection in field mice. Am. Nat. 112: 545-55.

Tamarin, R. H., and C. J. Krebs. 1969. Microtus population biology. II. Genetic changes at the transferrin locus in fluctuating populations of two vole species. Evolution 23:183-211.

Taylor, C. E., and J. B. Mitton. 1974. Multivariate analysis of genetic variation. Genetics 76:575-85.

Te, G. A., and W. D. Dawson. 1971. Chromosomal polymorphism in Peromyscus polionotus. Cytogenetics 10:225-34.

Tenczar, P., and R. S. Bader. 1966. Maternal effect in dental traits of the house mouse. Science 152: 1398-1400.

Terman, C. R. 1968. Population Dynamics. pp. 412-45. In Biology of Peromyscus (Rodentia). King (ed.). Am. Soc. Mammal., Spec. Publ. No. 2.

Thomas, B. 1973. Evolutionary implications of karyotypic variation in some insular Peromyscus from British Columbia, Canada. Cytologia 38:485-95.

Thorpe, R. S. 1976. Biometric analysis of geographic variation and racial affinities. Biol. Rev. 51: 407-52.

Throckmorton, L. H. 1968. Concordance and discordance of taxonomic characters in Drosophila classification. Syst. Zool. 17:355-87.

Throckmorton, L. H. 1977. Drosophila systematics and biochemical evolution. Ann. Rev. Ecol. Syst. 8:235-54.

LITERATURE CITED

Throckmorton, L. H. 1978. Molecular phylogenetics. pp. 221-39. In Beltsville Symposia in Agricultural Research 2. Biosystematics in Agriculture. J. A. Romberger, R. H. Foote, L. Knutson, and P. L. Lentz (eds.). John Wiley and Sons, New York.

Turner, B. N., and S. L. Iverson. 1973. The annual cycle of aggression in male Microtus pennsylvanicus, and its relation to population parameters. Ecology 54:967-81.

Turner, H. N. 1959. Ratios as criteria for selection in animal or plant breeding, with particular reference to efficiency of food conversion in sheep. Aust. J. Agric. Res. 10:565-80.

Turner, H. N., and S. S. Y. Young. 1969. Quantitative Genetics in Sheep Breeding. Cornell Univ. Press, Ithaca. 322 p.

Underhill, D. K. 1969. Heritability of some linear body measurements and their ratios in the leopard frog Rana pipiens. Evolution 23:268-75.

Ursin, E. 1952. Occurrence of voles, mice, and rats (Muridae) in Denmark, with a special note on a zone of intergradation between two subspecies of the house mouse (Mus musculus L.). Vid. Midd. Dansk Naturhist. Forening 114:217-44.

Valentine, J. W. 1976. Genetic strategies of adaptation. pp. 78-94. In Molecular Evolution. F. J. Ayala (ed.). Sinauer, Sunderland, Mass.

Van Valen, L. 1973. A new evolutionary law. Evol. Theory 1:1-30.

Van Valen, L. 1974. Multivariate structural statistics in natural history. J. Theor. Biol. 45:235-47.

Vavilov, N. I. 1926. Studies on the origin of cultivated plants. Bull. Appl. Bot. Plant. Breed. Lennigrad 16:1-284.

LITERATURE CITED

Vuilleumier, F. 1970. Insular biogeography in continental regions. I. The northern Andes of South America. Am. Nat. 104:373-88.

Waddington, C. H. 1968. The basic ideas of biology. pp. 1-32. In Towards a Theoretical Biology: 1. Prolegomena. C. H. Waddington (ed.). Edinburgh Univ. Press, Edinburgh.

Wahrman, J., R. Goitein, and E. Nevo. 1969. Mole rat Spalax: evolutionary significance of chromosome variation. Science 164:82-4.

Wake, D. B., L. R. Maxson, and G. Z. Wurst. 1978. Genetic differentiation, albumin evolution, and their biogeographic implications in plethodontid salamanders of California and southern Europe. Evolution 32:529-39.

Wallace, A. R. 1869. The Maylay Archipelago. MacMillan Press, London.

Wallace, B. 1968. Polymorphism, population size, and genetic load. pp. 87-107. In Population Biology and Evolution. R. C. Lewontin (ed.). Syracuse Univ. Press, New York.

Wallace, B. 1975. Hard and soft selection revisited. Evolution 29:465-73.

Wallace, M. E. 1953. Affinity: a new genetic phenomenon in the house mouse: evidence from within laboratory stocks. Nature 171:27-8.

Wallace, M. E. 1958. Experimental evidence for a new genetic phenomenon. Phil. Trans. R. Soc. B. 681: 211-54.

Wallace, M. E. 1971. An unprecedented number of mutants in a colony of wild mice. Envir. Poll. 1:175-84.

Wallace, M. E., and R. J. Berry. 1978. Excessive mutational occurrences in wild Peruvian house mice. Mutat. Res. 53:282-83.

LITERATURE CITED

Ward, R. H., and J. V. Neel. 1976. The genetic structure of a tribal population, the Yanomama Indians. XIV. Clines and their interpretation. Genetics 82:103-21.

Wartofsky, M. W. 1968. Conceptual Foundations of Scientific Thought. MacMillan, New York. 560 p.

Webster, T. P. 1973. Adaptive linkage disequilibrium between two esterase loci of a salamander. Proc. Natl. Acad. Sci. USA 70:1156-60.

Webster, T. P., R. K. Selander, and S. Y. Yang. 1972. Genetic variability and similarity in the Anolis lizards of Bimini. Evolution 26:523-35.

Wheeler, L. L., and R. K. Selander. 1972. Genetic variation in populations of the house mouse, Mus musculus, in the Hawaiian Islands. Stud. Genet. 7:269-96.

White, M. J. D. 1968. Models of speciation. Science 159:1065-70.

White, M. J. D. 1973. Animal Cytology and Evolution. Third ed. Cambridge Univ. Press, Cambridge. 961 p.

White, M. J. D. 1978. Modes of Speciation. W. H. Freeman and Co., San Francisco. 455 p.

Williams, S. L., and H. H. Genoways. 1977. Morphometric variation in the tropical pocket gopher (Geomys tropicalis). Ann. Carnegie Mus. 46:245-64.

Wilson, A. C. 1975. Evolutionary importance of gene regulation. Stadler Genet. Symp. 7:117-33.

Wilson, A. C., S. S. Carlson, and T. J. White. 1977a. Biochemical evolution. Ann. Rev. Biochem. 46:573-639.

Wilson, A. C., L. L. Maxson, and V. M. Sarich. 1974a. Two types of molecular evolution. Evidence from studies of interspecific hybridization. Proc. Natl. Acad. Sci. USA 71:2843-47.

LITERATURE CITED

Wilson, A. C., V. M. Sarich, and L. R. Maxson. 1974b. The importance of gene rearrangement in evolution: evidence from studies on rates of chromosomal, protein, and anatomical evolution. Proc. Natl. Acad. Sci. USA 71:3028-30.

Wilson, A. C., G. L. Bush, S. M. Case, and M. C. King. 1975. Social structuring of mammalian populations and rate of chromosomal evolution. Proc. Natl. Acad. Sci. USA 72:5061-65.

Wilson, A. C., T. J. White, S. S. Carlson, and L. M. Cherry. 1977b. Molecular evolution and cytogenetic evolution. pp. 375-393. In Molecular Human Cytogenetics. R. S. Sparkes et al. (eds.). Academic Press, New York.

Wilson, C. M., E. G. Erdos, J. F. Dunn, and J. D. Wilson. 1977. Genetic control of renin activity in the submaxillary gland of the mouse. Proc. Natl. Acad. Sci. USA 74:1185-89.

Wilson, D. E. 1973. The systematic status of Perognathus merriami Allen. Proc. Biol. Soc. Wash. 86:175-92.

Wilson, D. S. 1975. A theory of group selection. Proc. Nat. Acad. Sci. USA 72:143-46.

Wilson, D. S. 1977. Structured demes and the evolution of group-advantageous traits. Am. Nat. 111:157-85.

Wilson, E. O. 1975. Sociobiology: the New Synthesis. Harvard Univ. Press, Cambridge. 697 p.

Womack, J. E., and T. H. Roderick. 1974. T-alleles in the mouse are probably not inversions. J. Hered. 65:308-10.

Workman, P. L., and J. D. Niswander. 1970. Population studies on southwestern Indian tribes. II. Local genetic differentiation in the Papago. Am. J. Hum. Genet. 22:24-29.

Wright, S. 1931. Evolution in mendelian populations. Genetics 16:97-159.

LITERATURE CITED

Wright, S. 1943. Isolation by distance. Genetics 28:114-38.

Wright, S. 1965. The interpretation of population structure by F-statistics with special regard to systems of mating. Evolution 19:395-420.

Wright, S. 1969. Evolution and the Genetics of Populations. Vol. 2, The theory of gene frequencies. Univ. Chicago Press, Chicago.

Wright, S. 1977. Evolution and the Genetics of Populations. Vol. 3, Experimental Results and Evolutionary Deductions. Univ. Chicago Press, Chicago.

Wright, S. 1978. Evolution and the Genetics of Populations. Vol. 4, Variability within and among natural populations. Univ. Chicago Press, Chicago.

Yablokov, A. V. 1974. Variability of Mammals. Amerind Publishing Co., Washington, D. C. 350 p.

Yang, S. Y., M. Soule, and G. C. Gorman. 1974. Anolis lizards of the eastern Caribbean: a case study in evolution. I. Genetic relationships, phylogeny, and colonization sequence of the roquet group. Syst. Zool. 23:38-99.

Yates, T. L., R. J. Baker, and R. K. Barnett. 1979. Phylogenetic analysis of karyological variation in three genera of peromyscine rodents. Syst. Zool. 28:40-8.

Yosida, T. H., K. Tsuchiya, and K. Moriwaki. 1971. Frequency of chromosome polymorphosm in Rattus rattus collected in Japan. Chromosoma 33:30-40.

Youngman, P. M. 1967. Insular populations of the meadow vole, Microtus pennsylvanicus, from northeastern North America, with descriptions of two new subspecies. J. Mammal. 48:579-88.

Yunis, J. J., M. Y. Tsai, and A. M. Willey. 1977. Molecular organization and function of the human

LITERATURE CITED

genome. pp. 1-34. In Molecular Structure of Human Chromosomes. J. J. Yunis (ed.). Academic Press, New York.

Zimmerman, E. G., C. W. Kilpatrick, and B. J. Hart. 1978. The genetics of speciation in the rodent genus Peromyscus. Evolution 32:565-79.

Zimmerman, E. G., R. L. Merritt, and M. C. Wooten. 1980. Genic variation and ecology of stoneroller minnows, genus Campostoma (Cyprinidae). Biochem. Syst. Ecol. in press.

INDEX

INDEX